A CULTURAL HISTORY OF PLANTS

VOLUME 3

A Cultural History of Plants
General Editors: Annette Giesecke and David J. Mabberley

Volume 1
A Cultural History of Plants in Antiquity
Edited by Annette Giesecke

Volume 2
A Cultural History of Plants in the Post-Classical Era
Edited by Alain Touwaide

Volume 3
A Cultural History of Plants in the Early Modern Era
Edited by Andrew Dalby and Annette Giesecke

Volume 4
A Cultural History of Plants in the Seventeenth and Eighteenth Centuries
Edited by Jennifer Milam

Volume 5
A Cultural History of Plants in the Nineteenth Century
Edited by David J. Mabberley

Volume 6
A Cultural History of Plants in the Modern Era
Edited by Stephen Forbes

A CULTURAL HISTORY OF PLANTS

IN THE EARLY MODERN ERA

VOLUME 3

Edited by Andrew Dalby and Annette Giesecke

BLOOMSBURY ACADEMIC
LONDON • NEW YORK • OXFORD • NEW DELHI • SYDNEY

BLOOMSBURY ACADEMIC
Bloomsbury Publishing Plc
50 Bedford Square, London, WC1B 3DP, UK
1385 Broadway, New York, NY 10018, USA
29 Earlsfort Terrace, Dublin 2, Ireland

BLOOMSBURY, BLOOMSBURY ACADEMIC and the Diana logo
are trademarks of Bloomsbury Publishing Plc

First published in Great Britain 2022
Paperback edition published in 2025

Series design: Raven Design
Cover image © British Library Board. All Rights Reserved/Bridgeman Images

A catalogue record for this book is available from the British Library.

Library of Congress Control Number: 2021932844

ISBN: PB: 978-1-3505-5060-5
Set: 978-1-3505-5064-3
ePUB: 978-1-3502-5930-0
ePDF: 978-1-3502-5931-7

Series: The Cultural History Series

Typeset by Integra Software Services Pvt. Ltd.
Printed and bound in Great Britain

CONTENTS

ILLUSTRATIONS

SERIES PREFACE

The connectedness of humans to plants is the most fundamental of human relationships. Plants are, and historically have been, sources of food, shelter, bedding, tools, medicine, and, most importantly, the very air we breathe. Plants have inspired awe, a sense of well-being, religious fervor, and acquisitiveness alike. They have been collected, propagated, and mutated, as well as endangered or driven into extinction by human impacts such as global warming, deforestation, fire suppression, and over-grazing. *A Cultural History of Plants* traces the global dependence of human life and civilization on plants from antiquity to the twenty-first century and comprises contributions by experts and scholars in a wide range of fields, including anthropology, archaeology, art history, botany, classics, garden history, history, literature, and environmental studies more broadly. The series consists of six illustrated volumes, each devoted to an examination of plants as grounded in, and shaping, the cultural experiences of a particular historical period. Each of the six volumes, in turn, is structured in the same way, beginning with an introductory chapter that offers a sweeping view of the cultural history of plants in the period in question, followed by chapters on plants as staple foods, plants as luxury foods, trade and exploration, plant technology and science, plants and medicine, plants in (popular) culture, plants as natural ornaments, and the representation of plants. This cohesive structure offers readers the opportunity both to explore a meaningful cross-section of humans' uses of plants in a given period and to trace a particular use—as in medicine, for example—through time from volume to volume. The six volumes comprising *A Cultural History of Plants* are as follows:

Volume 1: *A Cultural History of Plants in Antiquity* (*c.* 10,000 BCE–500 CE)
Volume 2: *A Cultural History of Plants in the Post-Classical Era* (500–1400)
Volume 3: *A Cultural History of Plants in the Early Modern Era* (1400–1650)
Volume 4: *A Cultural History of Plants in the Seventeenth and Eighteenth Centuries* (1650–1800)
Volume 5: *A Cultural History of Plants in the Nineteenth Century* (1800–1920)
Volume 6: *A Cultural History of Plants in the Modern Era* (1920–present).

By way of guidance to our readers, it should be noted that the plant names used in these volumes accord with those in the fourth edition of *Mabberley's Plant-book* (Cambridge University Press, 2017). When they are discussed, individual plants are identified using their common names and, at their first mention in each chapter, with their scientific names: e.g. bay laurel (*Laurus nobilis*). As is recommended for general works such as this, the authorities to whom the scientific names are attributed (e.g. *Laurus nobilis* L., where L. identifies Linnaeus as the identifying authority) have been omitted.

Annette Giesecke and David J. Mabberley,
General Editors

EDITORS' NOTE

For works originating in the Early Modern Period we have aimed, in compiling the Bibliography, to cite original texts as well as translations. Several chapter authors had already chosen to do the same.

Almost every printed book cited here that is no longer in copyright can be found online, often in more than one copy. These digital copies are multiplying as we write, and are increasingly easy to find. Hence the Bibliography and Notes do not ordinarily cite online copies of printed texts, but only those few that are uniquely useful or unusually hard to trace. Manuscripts (unless available in printed facsimile) are not included in the Bibliography: instead, details are given in the text or, where necessary, in the Notes, and references to online digital copies are added when possible. Readers looking for further guidance to online copies of printed books and manuscripts could do worse than to look up authors or titles on the Latin Vicipaedia (la.wikipedia.org), on which we have systematically inserted links to online copies known to us of Early Modern texts cited in this volume.

In quotations we have usually retained original spellings but we have often modernized punctuation and capitalization. We have regularized through the volume the names of Early Modern authors that are seen in several forms, choosing (for example) Matthias de Lobel, Garcia de Orta, Carolus Clusius. We want to thank the chapter authors for their complaisance in allowing us this editorial prerogative—and, much more important, for the contribution that each of them has made to a wide-ranging survey in Early Modern cultural history.

Andrew Dalby and Annette Giesecke

Introduction

"I am sorry to say that I do not recognise them"

ANDREW DALBY WITH ANNETTE GIESECKE

The Atlantic islands mark a frontier that scarcely a single living plant species had crossed before the Early Modern Era, the age of the Columbian Exchange. Saint Helena, far out in the south Atlantic, was isolated from east and west until its discovery by João da Nova (1460–1509) in May 1502. Its first permanent inhabitant, Fernão Lopes, horribly mutilated in punishment for having opposed the Portuguese reconquest of Goa in India, withdrew to live alone on Saint Helena from 1515 until his death thirty years later. He was a legend among the mariners who regularly called there to take on water, meat, and, increasingly, fruit.

It was not consciously known that fresh plant foods were essential in a sailor's diet, and yet, wherever in this age of discovery ships first landed and seemed likely to continue to land, written reports assiduously cataloged vegetables and fruit. Wherever the supply was susceptible of improvement, it will, sure enough, become clear from either contemporary reports or later evidence that new plants were introduced. In the case of Saint Helena such reports are numerous. Fernão Lopes, though he was very rarely seen, benefited from many discreet gifts of seeds and rooted plants.

A century later, in *Some Yeares Travels into Divers Parts of Asia and Afrique*, Thomas Herbert (1606–82) listed the well-chosen greens that could be harvested on Saint Helena by "a willing hand, directed by an ingenious eye: … wood-sorrell, three-leav'd-grasse, basil, parsly, mints, spinage, fennel, annys, radish, mustard-seed, tobacco." These and others, once introduced, had survived in the wild. Sadly, Herbert continued, it was not so with orchard fruits. The small Portuguese settlement of the late sixteenth century, in whose orchard the trees planted in the time of Fernão Lopes had continued to thrive, had since been "delapidated" by the "churlish" English and Dutch. The "lemmons, orenges, pomgranads, pomcitrons, figgs, dates, etc." were scarcely more than a memory. A single lemon tree remained at the head of what was still known (and is still known today) as Lemon Valley (Herbert 1638: 354).

Herbert, an ingenious eyewitness, coolly lifted his lists of the herbs and lost fruits of Saint Helena from the report of Thomas Cavendish's circumnavigation, though making no mention of the "pompions and melons," nor indeed of the "turkies," that Cavendish

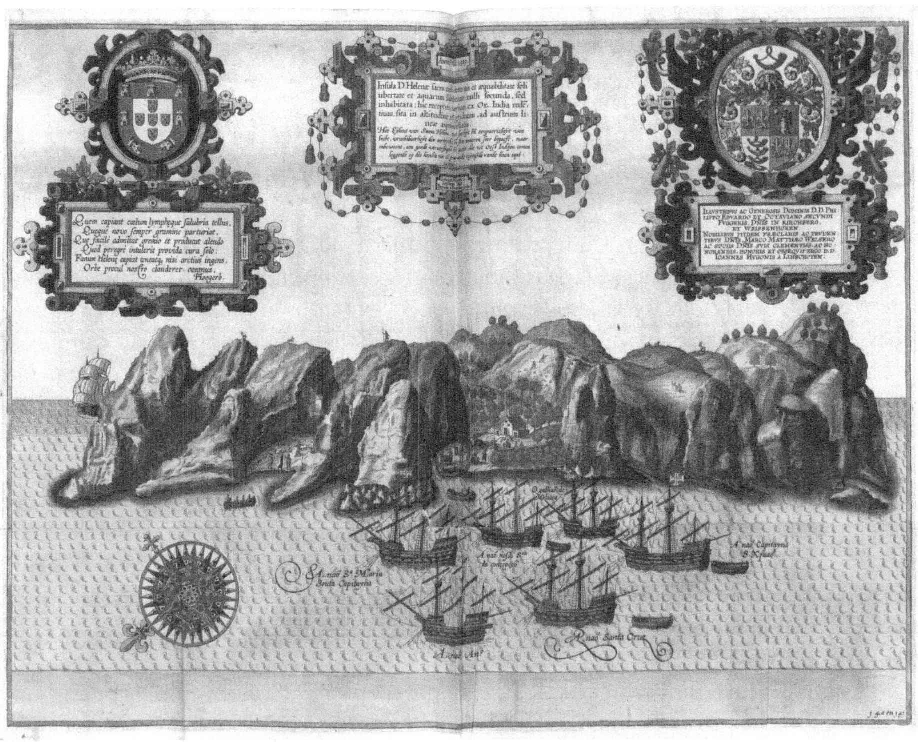

FIGURE 0.1 "View of St Helena." Mid-coast, an orchard flourishes around the little Portuguese church. Hand-colored engraving, copy *c.* 1596 (by Baptista Van Doetichum) from voyages of Jan Huyghen van Linschoten (1563–1611), Koninklijke Bibliotheek. Photo from HuskyBot/Wikimedia Commons.

found in 1588 (Hakluyt 1598–1600: 3:824). Those pumpkins and turkeys had been introduced, we should notice, from the New World.

Instead of the Cavendish narrative Herbert might have plagiarized the report of Duarte Lopez's mission to the kingdom of the Congo. On his way to that African realm Lopez visited Saint Helena in 1578, tasting "*fagiuoli*" (most likely New World haricot beans, *Phaseolus vulgaris*) and admiring the native ebony trees (*Trochetiopsis* spp., now almost extinct), which were used for necessary repairs to visiting vessels (Pigafetta 1591: 3). Again, Herbert might have chosen as his source the narrative of John Lancaster's piratical voyage. Lancaster's crew had rested on Saint Helena for nineteen days in March 1593, making judicious use of the "very holesome and excellent good greene figs, orenges, and lemons very faire" (Hakluyt 1598–1600: 2:108).

Seven months later, marooned for twenty-nine days on Mona, a small and almost deserted island between Puerto Rico and Hispaniola, where there were not enough pumpkins and not much else at all in the garden of the sole indigenous inhabitant, Lancaster's surviving companions could find no better refreshment than "the stalkes of purselaine boyled in water." This reliable and health-giving species, often uprooted as a weed, is one of the very few useful plants that was found in both the Old and New Worlds. Columbus himself had noticed "*verdolagas muchas*," a lot of purslane (*Portulaca oleracea*), on first reaching the coast of Cuba on October 28, 1492

(Navarrete 1858: 192–3); in 1606 Samuel Champlain would find it good to eat *en salade* during a French exploration of the Massachusetts coast (1613: 77). By what agency purslane had been introduced to the North American continent is unknown (Byrne and McAndrews 1985).

THEMES

This is the era in which, for the first time, all other continental cultural zones were in contact with Europe. If Australia was largely an exception, it still finds its place in the index to this volume. The focus, however, in every chapter, is on Europe and to a greater or lesser extent on the contacts of Europeans with other continents and other cultures.

The general editors of this cultural history surely foresaw that a first chapter on staple foods would, by its nature, introduce general issues of climate and population, farming and marketing, food and famine. Without such a foundation cultural history would not, after all, exist.[1] Malcolm Thick's Chapter 1 has a certain statistical emphasis, from which it stands out—the fact might otherwise be overlooked—that too little is known about what plants, and how many of them, people grew in their gardens for their own use and for local exchange. At my nearest weekly market, in the depths of a highly bureaucratic country that forms part of a tightly regulated pluri-national economy, there is a hall where people sell their own produce to their neighbors freely, with no record and no statistics. The importance of food production in town and country gardens in sixteenth-century Europe, likewise largely unrecorded, is rightly emphasized in this chapter. So is the growth of market gardens within easy reach of expanding cities and their busy markets. The first Europeans who observed the markets of Tenochtitlán, though impressed by their size, took their existence as perfectly natural.

Human interaction with plants is a story of cultural continuity mixed with inventiveness and innovation. These, certainly, are the running themes of Chapter 1. It is, after all, not so very safe for a community to abandon a traditional staple foodstuff in favor of one that is locally untried. In what terrain will it flourish? What methods, what timetable of propagation, tending. and harvesting will yield reliable production? Will it continue to thrive every year? Will it fill its intended nutritional place in the long term? These are hard questions.

Continuity would be the theme, it might be thought, in Jill Francis' Chapter 7 on plants as natural ornament. Yet there was change. John Gerard (*c.* 1545–1612), the Elizabethan author of *The Herball*, observed of plants for the garden, "the delight is great, but the use greater, and joined often with necessitie" (1597: Epistle Dedicatorie). "Delight"—the use of plants as ornament—increased in importance in this period. John Parkinson, thirty years after Gerard, was perhaps affirming for the first time that "beautifull flower plants" might be included in the garden, to "set forth a Garden of all the chiefest for choyce, and the fairest for shew" (1629: Epistle). In this context edible fruits were an element of continuity. They were by definition useful and were agreed on all sides to be decorative. In the orchard "profit and pleasure meet and imbrace each other," wrote Ralph Austen (1653: 35–6). John Oglander, cited in Chapter 7, found this double benefit in growing raspberries: they, too, make a pleasant show. It was potentially the same with new, exotic fruits, and this was why Costanzo Felici (1525–85) grew tomatoes (*Solanum lycopersicum*), which were not enthusiastically eaten in sixteenth-century Europe, but were eye-catching. John Parkinson (1567–1650) appreciated them for the same reason: "Although the beautie of this plant consisteth not in the flower, but fruit, yet give me leave to insert it" among garden flowers, he begged the reader, "lest otherwise it have no place" (1629: 379; further quoted in Chapter 2).

Having listed landscape features, structures, and statuary as three potential decorative elements in gardens, Francis notes a paradox. Of all of them it is the plants that are the most changeable and for the historian, therefore, the most elusive. Francis finds sources, nonetheless, little known or unknown until now, in the form of plant lists and layout plans by enthusiastic English gardeners. Rapid change as implied in Sir Robert Sidney's letter to his estate manager at Penshurst ("The little garden ... may goe for this year: if I do not like it, I can alter it the next," more fully quoted in Chapter 7) was all the more necessary with the introduction of new exotics. Gardeners did not know how to deal with them, unless, following the example of Parkinson and others, with each new plant they were prepared to experiment, observe, and record.

Continuities, again, are identified and discussed by Florike Egmond in Chapter 5, on plants and medicine. In the Early Modern period, as in ancient and medieval times, no real boundary was recognized between food and medicine. Plants, more or less edible, were still the chief ingredients of medicinal preparations. Graeco-Roman humoral theory, which positioned every plant on scales of hot/cold and dry/wet, was unquestioned. Newly discovered plants were naturally evaluated on these scales: chilies, for example, were "hot and dry almost to the fourth degree" (Monardes 1574: f. 25r).

These continuities serve as reminders not to accept as universal the themes of rediscovery and new discovery that variously emerge in Chapter 5 and throughout the volume. But it is true that the European Renaissance, a revolution inspired by a fresh reading of ancient Greek and Roman civilization, brought wholesale renewal in many fields, and not least in medicinal botany. Major classical sources, in this as in other scientific subjects, had remained prominent in medieval scholarship, supplemented as they were by Arabic writings, the product of an earlier renaissance. Greek and Arabic had been translated into Latin, the universal language of academic study in medieval Europe. These materials, Pliny's Latin *Natural History*, the Greek *Materia medica* of Dioscorides, and the prolific writings of Galen, the Arabic *Canon* or medical encyclopedia of Avicenna, and others, were available in printed Latin editions by the early sixteenth century. They remained standard textbooks, and were now cheap enough for students and other enquirers to acquire copies for themselves. It became possible as never before to compare and question. The rediscovery of forgotten authors such as Theophrastos on botany and Columella on farming, the publishing of the original Greek texts of Dioscorides and Galen (which had often been traduced in medieval Latin), alongside rapidly multiplying reports of wholly new material from exploration, encouraged the process of confronting old certainties and finding new systems.

Chapter 5 notes a vast increase in the number of documented plants, requiring and encouraging the gradual establishment of patterns in their description, depiction, naming, and classification. As is evident from Andrew Dalby's Chapter 2 and David Marsh's Chapter 3, classical geography, sometimes rightly and sometimes wrongly interpreted, encouraged the great explorations, which were eventually to demonstrate how narrowly limited was the geographical region that the Greeks and Romans could really claim to know.

Classical perspectives are central to the developments in art and culture discussed by Ingrid D. Rowland in Chapter 4, Luke Morgan and Elizabeth Hyde in Chapter 6, Gillian Riley in Chapter 8, as seen not least in the villa architecture set out by Vitruvius, the associated well-planned horticulture and agriculture admired by Pliny the Younger and many another Roman author, and the respectful and imaginative Renaissance recreations of such settings designed by Palladio and his contemporaries.

In these chapters, in the diversions of what may be called natural philosophy, we are taken some way beyond any scientific or technical discipline recognized today. Xerxes, with whom Chapter 4 opens, makes an apposite prelude: the Persian emperor, forever famed for his failure to conquer Greece, was, like the Indian monarch Aśoka and others both ancient and Early Modern, a lover of plane trees. Protagonist of Aeschylus' *Persians* (first performed 472 BCE, the only classical tragedy not taken from Greek mythology, inspiration for the whole sub-genre of neoclassical plays and operas that drew on Oriental themes) Xerxes was also the eponymous hero of the opera *Xersé*, whose libretto, featuring his song to the plane tree "Ombra mai fu," is best known in Händel's later reworking.

Early Modern approaches to the meaning of plants were multiple. Symbolism inspired the typically Renaissance genre of emblem books, by Andreas Alciatus (1531), Joachim Camerarius (1590), and others. Resemblances were basic to the theory of signatures, employed by Paracelsus and Giambattista della Porta, whose *Phytognomonica* (Porta 1588) explained how the appearance of plants signified their utility; and Foucault in *Les mots et les choses* (1966; translated 1970) showed that ramifications of the doctrine of signatures, far from constituting a minor diversion, pervaded Renaissance thought. Theories of sympathy and antipathy, deriving from classical philosophy, had influenced medieval farming and remained current in the sixteenth century, as in Gerolamo Cardano's *De subtilitate* (1550): "Plants have hatreds between themselves ... the olive and vine hate the cabbage." Renewed attention was given to the myths of metamorphosis exemplified in favorite classical literature. A new stage of metamorphosis may be found in the artificial plants imagined in Francesco Colonna's *Hypnerotomachia Poliphili* (1499; Godwin 1999) and in the fictional *pantagruelion* of Rabelais. These examples from Chapter 6 are capped in the last section of Chapter 4, where boundaries are crossed again in the fictional science of Athanasius Kircher (1602–80): his botanical alchemy allowed him to regenerate a plant from its own ashes. Others who tried the experiment somehow failed to reproduce the result.

It must be emphasized that while the chapters of this volume focus on medicinal, nutritional, symbolic, ritual, spiritual, and ornamental uses of plants by the cultures of Europe, it is the case that plants played a critically important role both within cultural groups with which Europeans established contact in the Early Modern period and those with which they did not. For example, the cultural centrality of plants was manifest in the Silk Road kingdoms of Safavid Persia (modern Iran) and Ming China, which experienced an extraordinary flowering of the arts and sciences in this period and which had significant contact with each other. While flowers and plant imagery abounded in Persian literature and the visual arts, the signature plant of Persia was, and still is, the rose, its primacy reflected in the fact that the Persian word for "rose" (*gul*) is also the generic word for "flower." Persian agricultural and horticultural manuals attest to the cultivation of numerous rose varieties. The 1515/16 *Irshād al zirā ʿa* (Guidance on Agriculture) of Qāsem Abūnaṣrī Heravī deals with seventy-nine plant species, among them cereals, vegetables, fruit trees, ornamental trees, flowers, and medicinal plants, as well as the cultivars of these species known to the author. He recorded "over 100 varieties of grapes, approximately 58 varieties of *k̲arboza* (a kind of melon), some 33 varieties of apricot, 19 varieties of wheat and apple, 14 varieties of *kadū* (pumpkin, squash) and barley, 11 varieties of millet, and 6 varieties of watermelon" (Aʿlam 1989); the rose cultivars known to him included many red varieties as well as "musk," "dappled," "five-petaled," and "Baghdad" roses

(Subtelny 2007). The fragrant, pink Damask or "Persian" rose (*Rosa* x *damascena*)—having sprung, according to legend, from the sweat of the prophet Muhammad's brow as he made his ascent to God's throne—was, and still is, used to make Persia's famed rose water and rose oil, which was exported not only to Europe but also to China, India, and Egypt. A symbol of purity and beauty, the rose, together with its extracts, found myriad medicinal applications and, of course, it was extensively cultivated for "commercial" use and as a feature of private gardens.

Among the many varieties of rose cultivated in Persia, it was the fragrant "hundred-petalled" rose (*Rosa* x *centifolia*), which held pride of place both economically and culturally (Subtelny 2007). It is no coincidence that this valuable rose appears to have been introduced into Europe via Holland by Shah 'Abbas I (r. 1587–1629), the most powerful king of Persia's resplendent Safavid Dynasty (Mabberley 2017). In an effort to elevate Persia to its former glory and to stimulate its economy, Shah 'Abbas aggressively courted foreign traders, actively sought commercial ties with European nations, India, and other Silk Road kingdoms, and he made Isfahan his seat of government, transforming it into a cosmopolitan garden city at the heart of which lies the Naqsh-e Jahan Maidan (Image of the World Square), itself surrounded by mosques, markets, palaces, a variety of

FIGURE 0.2 Garden scene and scrolling vegetal ornament, decorative ceramic tile, reputedly from a palace along the Chahar Bagh Avenue, Isfahan, Iran. First quarter of the seventeenth century, Isfahan. Rogers Fund, 1903, The Metropolitan Museum of Art, New York. Image courtesy of The Metropolitan Museum.

public buildings, and a profusion of gardens that, as terrestrial paradises, at once reflected the gardens of the Quranic afterlife and the Shah's position of power in the region (Gharipour 2013; Hobhouse 2004; Kuykendall 2019; Ruggles 2008; Walcher 1997). Buildings that Shah 'Abbas constructed were adorned with foliate and floral decoration in tile and painting, including the lotus and split-palmette arabesque, patterns of great antiquity in Iran but transformed and reinvigorated at the hands of contemporary artisans (Canby 2009; see also Chapter 7 of Volume 2 in this series). These patterns, together with increasingly realistic depictions of flowers and other plants, can be found also on carpets, brocades, and other textiles as well as on tiles and in painted "miniatures" produced in Isfahan's workshops. Trade with Europe, India, and China ensured that Persian arts and artifacts were inevitably subject to diverse outside influences—and a taste for these—continuing a trend already centuries old; thus it came to pass that the Shah brought three hundred Chinese potters, together with their families, settling them near Isfahan so as to capitalize on and satisfy the desire for Chinese ceramics, which played a key role in the transmission and assimilation of floral motifs along the Silk Roads and beyond (Finlay 2010: 6, 245).

If the rose held pride of place in Persia, it was the peony that was queen among flowers in China both in the Ming Dynasty (1368–1644) and earlier (Goody 1993: 356–7; Rawson 1984). Indeed, the prose master Ouang Xiu (1007–72) reported in his *Luoyang mudan ji* (An Account of the Tree Peonies of Luoyang) being told of some ninety cultivars and proceeded to list, rank, and describe (in terms of origin, color, and distinguishing characteristics) some thirty varieties that he considered most famous: Yao Yellows, Wei Flowers, Houan Fine Petals, Waistband Reds, Niu-Family Yellows, Wading Creek Deep Reds, and so forth (Hargett 2020: 136–8). In Ming Dynasty China, there were, of course, still more. Other flowers that garnered especial praise in the literature of the period were numerous and included the apricot, crabapple, pomegranate, plum, chrysanthemum, lotus, cassia, orchid, gardenia, magnolia, and rose. What flowers and plants were planted in Ming gardens, and to what effect, must be gleaned from texts like poems, descriptions of specific gardens and their notable features, and agronomic works, as well as from images, such as woodblock prints and paintings, since no Ming garden survives in its original state, all of them having been reconstructed (Hardie 2004). Agronomic literature, incidentally, flourished in the sixteenth and seventeenth centuries, with some texts offering practical information specifically for the small landowner and other works aimed at the elite (Clunas 1996: 78). Texts and images also shed light on the appreciation of plants in contexts apart from that of the garden (Hardie and Campbell 2020: 93–177). In the symbolic realm, for example, bamboo suggested integrity, the peony wealth, the orchid modesty (or its opposite), and the pine constancy. Individual plants might be displayed in pots either outside or indoors; indeed, the sixteenth century witnessed a growing fashion for potted plants and for dwarfing trees to create portable, miniaturized *pan zai* (bonsai) landscapes (Clunas 1996: 100). Cut flowers, artfully displayed, constituted a particularly important aspect of the ornamental use of plants. In his *Pinghua pu* (A Treatise of Vase Flowers), Zhang Qiande (1577–1643) asserted that "among the things of refined living, flower arrangement is the most difficult" and proceeded to offer detailed instruction on the selection of the most apposite specimens, the best type of vase, and the most suitable water:

> In picking flowers, it is necessary to select first the stem. The stem may be luxuriant above and slender below. It may be taller on the left and shorter on the right, or vice

FIGURE 0.3 "Garden Flowers." Part of a series of sixteen realistic paintings of spring flowers, serving as botanical specimens as well as reminders of the brevity of life, beauty, and material existence. Chinese, *c.* 1540, after Chen Chun (1483–1544). Gift of Douglas Dillon, 1986, The Metropolitan Museum, New York. Image courtesy of The Metropolitan Museum.

> versa. It may have two branches cross-crossing each other, gnarled and crooked in shape. It may have a stout vigorous stalk in the center, sparse atop and fence-like below, covering the mouth of the vase. Whether ascending or hanging, tall or low, sparse or dense, and oblique or upright, the branches that have a natural beauty must show the appealing features of cut flowers as depicted by the painter. Straight branches and windblown flowers are not suitable for refined arrangement ....
>
> (Li 2020: 154–5)

As in Safavid Persia, plants were greatly appreciated for their synaesthetic appeal: their visual beauty; their heady, evocative scents—wintersweet (*Chimonanthus praecox*) and apricot (*Prunus mume*) of late winter and early spring, lotus of summer, and osmanthus of fall; their taste, if edible; and their "sounds," such as the rustle of windblown bamboo and the patter of raindrops on the banana's broad leaves (Hardie and Campbell 2020: 95). In the case of the gardens of the aristocracy, the productive capacity of plants therein was also of significance, as productive gardens were a means of self-representation, bearing proof of the owners' morality and humble self-sufficiency, though as in Renaissance Italy, orchard fruit (which included mulberries for silk production) and an emphasis on gardens' economic yield or value became less prominent over time, giving way to rare and exotic imported specimens that activated the gardens' potential as arenas of social competition in the realms of wealth, display, and taste. Such elite gardens became destinations both for local visitors and, like the gardens of Isfahan, of European travelers who remarked upon their splendor: upon his visit, in the year 1598, to the Duke of Weiguo's garden in the city of Nanjing, the

Italian Jesuit priest Matteo Ricci, who steeped himself in Chinese culture, declared it to be "il più bello de questa città" (the most beautiful in the city) (Clunas 1996: 95 and 49ff. passim).

THE SPIRIT OF ENQUIRY AND ITS TEXTUAL REPRESENTATION

Returning now to a more detailed discussion of this volume's themes, if the intellectual effect of the European Renaissance can be summed up briefly, the summary will highlight the asking of new questions and the finding of new answers on any and all subjects. There is no reason for surprise that Poggio Bracciolini (1380–1459), secretary to six fifteenth-century popes, appears in this volume for two entirely different activities: in Chapter 4 for his rediscovery in 1517 of the long-lost classical Latin agricultural writers, the *Scriptores de re rustica*, in a manuscript at St. Gall, one of several critically important finds made during his frequent absences from the papal court, and in Chapter 2 for his narrative of the adventurous Asian travels of Nicolò de' Conti (*c.* 1395–1469) as they were reported to pope Eugenius IV. Long after the death of all three, this narrative was published under the title *India recognita*, "India explored afresh," on February 15, 1492, early in the same year in which Vasco da Gama and Christopher Columbus set out in their quests for an ocean route to India.

There is no reason to be surprised that it was the playwright Christopher Marlowe, Master of Arts, in his *Tragicall History of the Life and Death of Dr Faustus* (1604: see Chapter 3) who, in or about 1592, expressed the obsession of the great botanical scientists of the sixteenth century as simply and cogently as they had ever expressed it themselves. Doctor Faustus made the apparently reasonable demand for a book "wherein I might see all plants, herbs, and trees, that grow upon the earth." All he required, it would seem, was the never-quite-published omnibus: Dodoens' herbal, Mattioli's Dioscorides, Clusius' *Exoticorum libri decem*, and the Mexican plants of Hernández together with illustrations from Cassiano dal Pozzo's paper museum and the synonyms of Bauhin's *Phytopinax*. In the very next year Walter Raleigh (1552/4–1618), exploring the delta of the Orinoco in his search for the gold of Guiana, showed that it would never be quite so easy: "On the banks of these rivers were divers sorts of fruits good to eat, flowers and trees of such variety, as were sufficient to make tenne volumes of herbals" (in Hakluyt 1598–1600: 3:645).

It is unremarkable that François Rabelais, with his Montpellier medical training and unbounded imagination, should place alongside his fantastic invented herb *pantagruelion* a survey of plant nomenclature that is, in sixteenth-century terms, both serious and up-to-date (see Chapter 6). We should find it natural, too, that parallel descriptions of the wonderfully inventive gardens at the Villa Medicea di Castello are owed to four extremely different authors: Giorgio Vasari, practicing architect and biographer of artists and architects; Michel de Montaigne, moral philosopher in his own uncompromising style; Pierre Belon, gardener, apothecary, and rediscoverer of the ancient pharmacopoeia; and Carolus Clusius, translator of Belon and many others, gatekeeper of sixteenth-century botanical science.[2] Vasari and Montaigne are both quoted in Chapter 6 on the remarkable holm oak (*Quercus ilex*) in these gardens; Vasari and Belon will be quoted later in this introduction on its citrus groves.

"Changes in the media," as Florike Egmond rightly calls them (Chapter 5), had a vast effect on the distribution of knowledge. General literature forms part of this information revolution of the late fifteenth and sixteenth century, but a more striking result of the proliferation of published texts was the emergence of new genres.

EXPLORATION

The Early Modern book-trade found an eager market for reports of exploration, difficult though it often is to fit them into the straitjackets of literary genre and bibliographical description. Their protagonists were sailors or traders, not authors. Informally written in various languages, rapidly translated into several others, destined to be studied by future travelers and their patrons, these reports emphasize topics critical to that audience: geography, including practical details of land and sea travel; natural products, animal, mineral, and vegetable. Plants and their habitat are prominent under both rubrics.

The demand for narratives of exploration, and the supply of them, were nourished by rivalry among Spanish and Portuguese, French, Dutch, and English (not to mention Italians and Catalans, Basques and Bretons, Chinese and Arabs). It was the competition among the first five that directly or indirectly brought new information into the public domain. An anonymous participant in the warlike English expedition to the Azores in 1592, led by Walter Raleigh and others, says this plainly in his report. The English capture of the great ship *Madre de Dios* and its rich cargo, thanks to "Gods great favor towards our nation," had "discovered those secret trades and Indian riches, which hitherto lay strangely hidden and cunningly concealed from us," and these riches are listed at length: "… The spices were pepper, cloves, maces, nutmegs, cinamom, greene ginger: the drugs were benjamim [benzoin], frankincense, galingale, mirabolans, aloes zocotrina, camphire [camphor]" and miscellaneous wares included "… coco-nuts, hides, eben-wood [ebony] as blacke as jet, bedsteds of the same, cloth of the rindes of trees very strange for the matter and artificiall in workemanship." This report was published, a few years after the event, in the expanded edition of Richard Hakluyt's great collection *The Principal Navigations, Voiages, Traffiques and Discoveries of the English Nation, made by sea or over-land, to the remote and farthest distant quarters of the earth, at any time within the compasse of these 1600 yeres* (1598–1600: 2:ii:198).

Hakluyt (1553–1616), a Londoner with roots in Herefordshire, spent much of his life finding, translating, and editing writings on exploration. He frequently lectured on the opportunities they offered, though he himself scarcely traveled further than Paris. In this, as in some other biographical details, he closely resembled the Venetian Giovanni Battista Ramusio (1485–1557), editor of the much-reprinted *Navigationi et viaggi* (1550–9), which, like Hakluyt's work, drew on previously unpublished narratives as well as earlier publications.

Among preceding and less ambitious collections there is the little book on recent discoveries compiled by Fracanzano Montalboddo in 1507, *Paesi novamente retrovati et novo mondo da Alberico Vesputio Florentino* (Newly Discovered Lands and the New World of Amerigo Vespucci of Florence). Early separately published narratives, alongside Poggio Bracciolini's *India recognita* mentioned above, include Ludovico di Varthema's *Itinerario* (1510) and Maximilianus Transylvanus' *De Moluccis insulis* (On the Molucca Islands, 1523), the latter being a partial report of the Magellan circumnavigation. Antonio Pigafetta's fuller report of the same voyage circulated in manuscript (see Robertson 1906) until a version of it appeared in Ramusio's collection. These three sources are all quoted in Chapter 2 for their new information about the Oriental spices that Europeans had been so anxious to find.

The sources for the explorations of Christopher Columbus are much more complex than those for Magellan. The printed report by Columbus to the Catholic monarchs was a brief summary, but contemporary readers had only this, along with a few chapters of Montalboddo's *Paesi novamente retrovati*, and, most important, the opening section of Peter Martyr ab Angleria's *De orbe novo decades* (Histories of the New World), written in Spain and based on first-hand accounts, including those of "that great discoverer of vast regions, Christopher Columbus" (MacNutt 1912: 1:207).[3]

The manuscript logbook of the explorer's first voyage, now lost, was available to Fernando Colón (Ferdinando Colombo), who drew on it for his biography of his father, published only in Italian translation and not until after Fernando's death (Ulloa 1571). The logbook was also available to Bartolomé de Las Casas, but his extensive transcription of it was not published until the nineteenth century (Navarrete 1858). Even when read through these intermediaries the logbook fully demonstrates how closely Columbus attended to plants, their discovery, identification, cultivation, and transplanting. It shows how complex was the relation between his aims, observations, private records, and public reports. It reveals his four imperatives: spice plants, gold, land for colonization, and—selected from any or all of these three—something profitable to report to the Catholic monarchs.

His visit to Chios (Ulloa 1571: f. 9), source of the mastic (*Pistacia lentiscus*) from which his native Genoa drew a steady profit, had supplied him with a fine example of how the sources of exotic plant products were guarded and insights as to how the monopolies might be broken. Hence, in his published report to his patrons after the first Atlantic voyage, he promised

> spices and cotton at once, as much as their Highnesses will order to be shipped, and as much as they will order to be shipped of mastic, which till now has never been found except in Greece, in the island of Chios, and the Seignory [of Genoa] sells it for what it likes; and aloeswood [*Aquilaria malaccensis*] as much as they will order to be shipped. (Navarrete 1858: 320)

The promises were made with fingers crossed, because although Columbus thought he had recognized the aromas of aloeswood and mastic in the West Indies, no one had tried to extract these spices from the allegedly aromatic trees. No one had even made a serious identification. Using a logbook entry of late October 1492 Fernando Colón tells the story of the felling of firewood which was "found to be of the mastic tree, which abounds throughout that country. The leaves and fruit of this tree resemble those of the lentiscus, but it is a much bigger tree" (Ulloa 1571: f. 58). The Chios mastic tree, *Pistacia lentiscus* 'Chia,' is indeed bigger than the lentisk—but no more is ever heard of mastic trees in the New World. Columbus was equally keen to find cinnamon, one of the most expensive spices in medieval Europe, originating as it did in Sri Lanka and Southeast Asia. In this case the tree that he thought to be cinnamon, *Canella winterana* (canella bark), was convincing enough to be taken into cultivation.[4]

Merchants and explorers of this period had in fact a simple method for identifying potential sources of the aromatics that interested them so much. The method had already been used in a different context forty years earlier, in 1454, when the Venetian adventurer Alvise Cadamosto (1432–88) visited Prince Henry the Navigator: "They came to our galleys, by order of the Prince, with some samples of sugar of the Isle of Madeira, and dragon's blood [resin of *Dracaena draco*], and other products of the new islands." Cadamosto, thus encouraged by the Prince to make the voyage to Madeira, had continued from there to explore the West African coast (Montalboddo 1507: sig. a iii). Columbus himself knew the method perfectly well. According to his logbook for November 2, 1492 the party that he sent out to explore the interior of Cuba was given *muestras de especería* (samples of spices) to show to the inhabitants. The latter were understood to have confirmed that these spices were to be found in Cuba (Navarrete 1858: 198; see Chapter 5). In such exchanges, misinterpretations were natural.

Columbus did better when not led astray by the sample method or by his own wish for a positive answer. His identifications could be accurate. At Bariay Bay in Cuba on October 28 he correctly recognized in the same glance *verdolagas*, purslane, one of the rare truly cosmopolitan species, and *bledos* (Navarrete 1858: 192–3): the latter was not precisely the blite (*Blitum bonus-henricus*) that was familiar to him as a European potherb but one of its relatives that were found in Cuba, perhaps *Amaranthus crassipes*, or *A. minimus* (known only from there). Boldly he chose the term *panizo* (millet) for the maize that differed so strongly from any cereal he had ever seen: a justifiable and natural comparison, for millet as a cereal grain, a "disappearing staple" in western Europe by the sixteenth century, was still locally familiar in Columbus' native Italy (see Chapter 1). He immediately saw the resemblance between the *Vigna* beans of the Old World and the *Phaseolus* of the New, which were to be enthusiastically described for good reason by a series of sixteenth-century authors and were eventually to drive their rivals out of cultivation in most of Europe (Clusius 1605: 335–7; Piergiovanni and Lioi 2010).

Beyond such useful approximations Columbus was able to admit ignorance and knew what to do next: "I am sorry to say that I do not recognise them," he wrote on October 19, 1492 of the plants he saw at Crooked Island, and "I am bringing back specimens" of fruit trees and herbs (Navarrete 1858: 186, 188; fuller quotations in Chapter 3). His frequent mistake, which any reader conscious of his overall achievement and aware of his fourth and most pressing imperative can easily forgive, was to overestimate the value of nearly everything he saw. Each successive island was "very fertile," every kind of fruit was "certainly precious." Not quite convinced, by contrast, of the value of the tobacco (*Nicotiana rustica*, not *N. tabacum*) that he saw in use on October 15, 1492, on this subject he contented himself with neutral observation: "He was carrying ... dried leaves which they must appreciate highly since they have already given me some as a present" (Navarrete 1858: 180). His son, Fernando Colón (1488–1539), alluding to the same incident and reading the same logbook, describes "dried leaves which they value highly as being aromatic and healthy" (Ulloa 1571: f. 54v). During the sixteenth century these puzzling leaves, naturalized in many parts of Europe, found a significant place in Old World culture, for better or worse. "Indian Henbane, or Tabacco," according to Parkinson, was "an excellent helpe and remedy for divers diseases, if it were rightly ordered and applied, but the continuall abuse thereof in so many, doth almost abolish all good use in any" (1629: 363–4).

Since Columbus was looking for spices above all, he was immediately interested by the chili, a West Indian condiment quite different from any previously known in Europe. In Hispaniola, he noted on January 15, 1493, there was plenty of "*aji*, which is their pepper and worth more to them than pepper. They never eat without it, and they find it very healthy. Fifty caravels could be loaded with it every year in this island of Hispaniola" (Navarrete 1858: 286). Chilies were no doubt the very "spices" that he promised to the Catholic monarchs to ship home in quantity (above). There is no evidence that he eventually did this, but he certainly took some chilies home with him. In a report of 1494, he mentions "axí, which we call pepper, of which I brought some to your Highnesses from my previous voyage. There is as much of it here as your Highnesses will require: they sow it and grow it in gardens" (Rumeu de Armas 1989: no. 2).

DISCOVERY AND REDISCOVERY

Thomas Cavendish's circumnavigation of 1586–8 has already been mentioned. Cavendish himself appears again at the beginning of Chapter 5, for he was present at the meeting in London on July 28, 1589 at which star anise (*Illicium verum*) made its first appearance in Europe. He was still only twenty-eight years old, and died at thirty-one in the course of a second attempted circumnavigation.

The star anise first seen in London in 1589, soon thereafter brought to the attention of European scholars in general, was nothing new in its native region. A medieval medicinal and culinary spice of southernmost China and the northern edge of Vietnam, it is probably native to the province of Guangxi, though whether a wild population exists there or anywhere is uncertain. It had been known to the Chinese who made the long journey to the southern provinces as early as the twelfth century (Hargett 2010: 108), but only in the late sixteenth century did it become known generally: it was listed in 1596 in the standard (northern) Chinese pharmacopoeia, Li Shizhen's *Bencao gangmu*, five years before its description and illustration by Carolus Clusius in his *Rariorum plantarum historia* of 1601 (Chapter 5).

It can be compared with the plants first reported to Europe by Columbus and those who came after him. The five cultivated *Capsicum* species (chilies) of the West Indies, Mexico, and northern and central South America had all been locally grown and prized. Some hybrids had already developed (this is evident from the first printed illustrations), a clear sign of more-than-local horticulture. Tomatoes had been transmitted from Peru to Mexico, where cultivars were beginning to be selected. Fruit of the central American stands of *Theobroma cacao*, the precious cacao beans, had already for centuries been traded northwards to Mexico (where Europeans first saw them) and as far as Utah. But it was after 1492, and largely in the course of the sixteenth century, that these and many other valuable plants of the New World made their way to Africa, Europe, southern Asia, and the East Indies. During the same period a great number of Old World plants made their way westwards, a major constituent of the Columbian Exchange (Crosby 1972).

Whether they are to be counted as discoveries or rediscoveries, plants that became newly known to Europeans in the Early Modern period jostled for inclusion among collections of images and descriptions. They wait to be noticed, rare intruders among the crowd of habitués, if manuscripts and paintings embellished with botanical miniatures as a form of the late medieval grotesque are studied: not yet in the *Hours of Katherina van Kleef*, in the 1440s (Chapter 8), but definitely in the *Grandes heures d'Anne de Bretagne* painted by Jean Bourdichon in 1508 in which a bitter orange, at that date rare in northern Europe, finds a place alongside what are surely haricot beans newly transplanted from the New World (ff. 168r, 194r; Bilimoff 2015: 148–53). A decade later three pods of haricot beans and three unmistakable cobs of maize (*Zea mays*) are joined by three different kinds of New World pumpkins (*Cucurbita* spp.) and no fewer than four sorts of citrus among the swathes of more familiar plants in the frescoes of Agostino Chigi's villa, painted by Giovanni da Udine in 1518 (nos 69, 75–8, 92–5, 160 in the list at hort.purdue.edu/newcrop/udine/). A tomato almost bursting with juice reclines among the caterpillars in the *Mira calligraphiae monumenta* embellished around 1590 by Joris Hoefnagel (f. 42; Hendrix and Vignau-Wilberg 1992: 134).

A study of manuscript herbals of the sixteenth century shows that artists were beginning to reject schematic and stylized representation and moving toward direct observation of plants and thus toward a new naturalism of which the *Libri picturati* (Chapter 5) are a striking example. It was hard for printed images to reach the level of accuracy achieved by skilled botanical painters, but the scale of improvement, even in printed work, is seen when herbal incunabula, such as the *Herbarium Apulei Platonici* of about 1481, are compared with Giovanni Battista Ferrari's lavish *Hesperides*, devoted entirely to citrus fruits (1646), or with the princely *Hortus Eystettensis*, the work of Basilius Besler financed by his patron Prince-Bishop Johann Konrad von Gemmingen of Eichstätt (1613; Chapter 8). A true landmark, less than halfway through this long period, is the work of Leonhart Fuchs, *De historia stirpium commentarii insignes* (1542): a landmark not only for its images (cleanly drawn and fastidiously presented, without fuss or frame, as if, for the first time, the only imperative is visual accuracy) but also for the generous acknowledgment, in the form of fine portraits on the last page, of the three men who made the images, Heinrich Füllmaurer (1497–*c.* 1547), Albrecht Meyer, and (the engraver) Veit Rudolf Speckle (d. 1550).

The range of printed books about plants, developing out of the herbals and manuals of medical botany of the Middle Ages, grew extensively in the mid- and late sixteenth century (Chapter 5; Arber 1986). Competing for an apparently inexhaustible market, there were general herbals in Latin and various European vernaculars, increasingly well illustrated, appearing in ever longer editions; among the many who fought over this field were Hieronymus Bock, Leonhart Fuchs, William Turner, Matthias de Lobel, and John Gerard, and these names will recur in several chapters of this volume. Quoting in Chapter 3 Faustus' demand for the ultimate plant book, David Marsh reasonably asks, "What was it about plants that commanded a desire for complete knowledge of every single one in the world?" For it is true that in the real milieu of sixteenth-century publishing authors were continually adding to their herbals information on new discoveries, ambitious for completeness and unaware of the unattainability of this aim. Among the most successful of these general works, two big books, similar in their inspiration, may be selected for mention here.

Pietro Andrea Mattioli (1501–77), a practicing physician trained in medical botany at Padua, set out to translate into his native Italian, with commentary, under the title *Di Pedacio Dioscoride Anazarbeo libri cinque della historia & materia medicinale, tradotti in lingua volgare italiana* (1544), the *materia medica* of the ancient Greek pharmacologist Dioscorides, which after fifteen centuries still served as a standard handbook. New editions were wanted every few years, evidence of heavy demand. In each successive edition the commentary was expanded and more woodcuts were added. Mattioli's original commentary meanwhile demanded to be translated into Latin, the language of science. He responded under a Latin title that highlighted his original commentary and illustrations, *Commentarii in libros sex Pedacii Dioscoridis Anazarbei de Materia Medica adjectis quam plurimis plantarum et animalium imaginibus* (1554). Many of Mattioli's annotations took the form of full descriptions of useful plants unknown to Dioscorides, such as the extra Solanaceae—aubergine (*Solanum melongena*), tomato, and chili—inserted immediately after Dioscorides' mandrake (*Mandragora officinarum*).

Dioscorides had arranged his *materia medica* according to medicinal effects. Rembert Dodoens, a physician who had studied at Leiden, chose in his *Cruydeboeck*, literally "herb book," (1554) to arrange by plant habit, making the work more approachable to horticulturalists, though an emphasis on medical uses remained. The Dutch (Flemish) original was translated three years later into French and thirty years later into Latin, introducing it to a wide new readership. For both of these translations Dodoens' friend Carolus Clusius was responsible (Clusius 1557, 1583b). There were eventually two English versions, by Henry Lyte in 1578 (*A Niewe Herball*) and by John Gerard in 1597 (Gerard's *Herball*). Clusius and Gerard both made extensive additions to Dodoens' original, and some of these improvements fed back into later Dutch editions.

Descriptions of plants previously unknown to Europeans are found not only in Early Modern herbals such as these but also, in a completely different context, in histories and geographical surveys of newly explored regions. Among numerous works in a popular sixteenth- and seventeenth-century genre, the *Commentarios reales de los Yncas* are notable as the work of an author of Inca and Spanish parentage, educated in both cultures, who had long lived in Spain. He was describing Peruvian plants and their uses from nostalgic childhood memory (Garcilaso de la Vega 1609; Chapter 2). Several authors on the New World gave extensive information on Mexican and Caribbean plants, not only Petrus Martyr de Angleria but also Gonzalo Fernandez de Oviedo in his *Historia general y natural de las Indias, islas y tierra-firme del mar Océano* (General and Natural History of the Indies, Islands and Continent of the Ocean Sea), incompletely published in the 1550s (Amador de los Rios 1851–5; Chapter 3) and José de Acosta's *Historia natural y moral de las Indias* (1590).

There are fascinating reports of exploratory botanizing by those who traveled specifically to rediscover ancient knowledge. The obvious example here is *Les observations de plusieurs singularitez et choses mémorables trouvées en Grèce, Asie, Judée, Égypte, Arabie et autres pays estranges*, published in 1553 by Pierre Belon (1517–64), scientific attaché, so to speak, in a French embassy to the Ottoman Empire in 1548. In 1589 Clusius issued a Latin translation of Belon's rich narrative, whose focus is the rediscovery of ancient medical botany. Belon was followed by Leonhart Rauwolf (1535–96), whose lengthy report of Middle Eastern travels, rarely read (and deserving better) was entitled *Aigentliche Beschreibung der Raiß, so er vor diser Zeit gegen Auffgang inn die Morgenländer ... selbs volbracht* (1582–3).[5]

Such works have an advantage that they share with general narratives by the more plant-aware of travelers (beginning with Columbus, above): there is room in them for observations of local traditional knowledge of plants, their medical uses, and their cultural significance. Here begins the study of ethnobotany or the "knowledge of herbs and plants ... practiced by wild and barbaric people" as it was defined by Balthasar II Moretus, editor and publisher in 1644 of the last edition of Dodoens' herbal, supplemented with many of Clusius' descriptions of non-European plants (fuller quotation in Chapter 5).

Books of botanical exploration led in turn to another new genre, that of enumerations and catalogs of useful plants with distant origins. One of the greatest such works, bearing the imprint Goa, April 10, 1563, is the Portuguese dialogue by García de Orta (*c.* 1501–86) on the spices and drugs of Asia, *Colóquios dos simples e drogas he cousas medicinaes da India* (quoted several times in Chapter 2). Within two years Nicolás Monardes (1493–1588) of Seville had begun his series of works on medicinal plants and drugs of America. The title may be given as *Historia medicinal de las cosas que se traen de nuestras Indias Occidentales*: the three-part complete edition appeared in 1574. Orta and Monardes served as principal sources for two contemporary authors who also added material of their own: Juan Fragoso of Madrid in *Discursos de las cosas aromáticas, árboles y frutales, y de otras muchas medicinas simples que se traen de la India Oriental* (1572) and Cristóbal Acosta in *Tractado de las drogas y medicinas de las Indias Orientales* (1578).

Just as Monardes' work reached completion, Francisco Hernández (1514–87) of Toledo made his visit to Mexico, financed by the Spanish royal court, to find and describe medicinal plants. Somewhat ahead of his time, he systematically described all the useful plants he could, whether or not they might become available in Europe, and named them in Nahuatl. The resulting compilation may well have been thought unpublishable, or the royal court's enthusiasm for new medicinal plants may have waned during his absence. At any rate his work was not published and his original manuscript was lost, though a transcript was rescued and prepared for publication by Federico Cesi's Accademia dei Lincei (Hernández et al. 1651; see Chapters 3 and 5). The recording of Nahuatl names—a prefiguration of modern ethnobotany—was not unique to Hernández. Many of the names and traditional uses are also given in the illustrated manuscript herbal known as *Libellus de medicinalibus Indorum herbis*, prepared *c.* 1552 (Chapter 5), and in the encyclopedic manuscript on Aztec culture by Bernardino de Sahagún, *Historia general de las cosas de Nueva España*, known as the *Florentine Codex*, compiled during the same period. These works, too, remained unpublished for many years.[6]

The books by Orta, Monardes, Fragoso, and Cristóbal Acosta listed above have several common features: written in the vernacular by practicing physicians, discursive and easy to read, cheaply printed and produced, widely translated, ... and they all caught Clusius' eye. He bought his copy of the *Colóquios* on December 28, 1564 during a stay in Lisbon (Boxer 1963: pl. 2).[7] Clusius has been mentioned often above, and crops up in several chapters below (see also Egmond et al. 2007), all for good reason. If a single scientist may be said to embody the Early Modern botanical revolution it is Clusius (1526–1609), the greatest of the sixteenth-century natural scientists. Trained at Montpellier, for twenty years he directed the imperial botanical garden at Vienna before moving to Leiden in 1593 to create a university botanical garden there. He is notable not only for his own careful and innovative studies of plants and their human uses, but also for the work by others that he never tired of translating and publishing. His *Rariorum plantarum historia* of 1601 was expanded with notes, monographs, and illustrations entrusted to him by botanists and horticulturalists, and there is much more in the notes published after his death, *Curae posteriores*, not least the Capuchin herbalist Gregorio da Reggio's little work on chilies (Clusius 1611: 95–108). Clusius' second great compilation, *Exoticorum libri decem* (1605), consists partly of his own work but embodies revised, well-illustrated Latin editions of Orta, Monardes, and Cristóbal Acosta with expanded commentaries, which include select quotations of the few original passages in Fragoso's book.

Clusius' pivotal role is asserted in the 1587 version of William Harrison's "Description of England," perhaps surprisingly considering the insular focus of this text, which, it should be emphasized, appears not in a botanical or horticultural publication but as general introduction to a work of wider interest, Raphael Holinshed's famous *Chronicles*.

> The chiefe workeman, or as I maie call him the founder of this devise, is Carolus Clusius, the noble herbarist ... For albeit that Matthiolus, Rembert, Lobell, and other have travelled verie farre in this behalfe, yet none hath come neere to Clusius, much lesse gone further in the finding and true descriptions of such herbes as of late are brought to light.
>
> (Edelen 1994; Harrison 1587: 210)

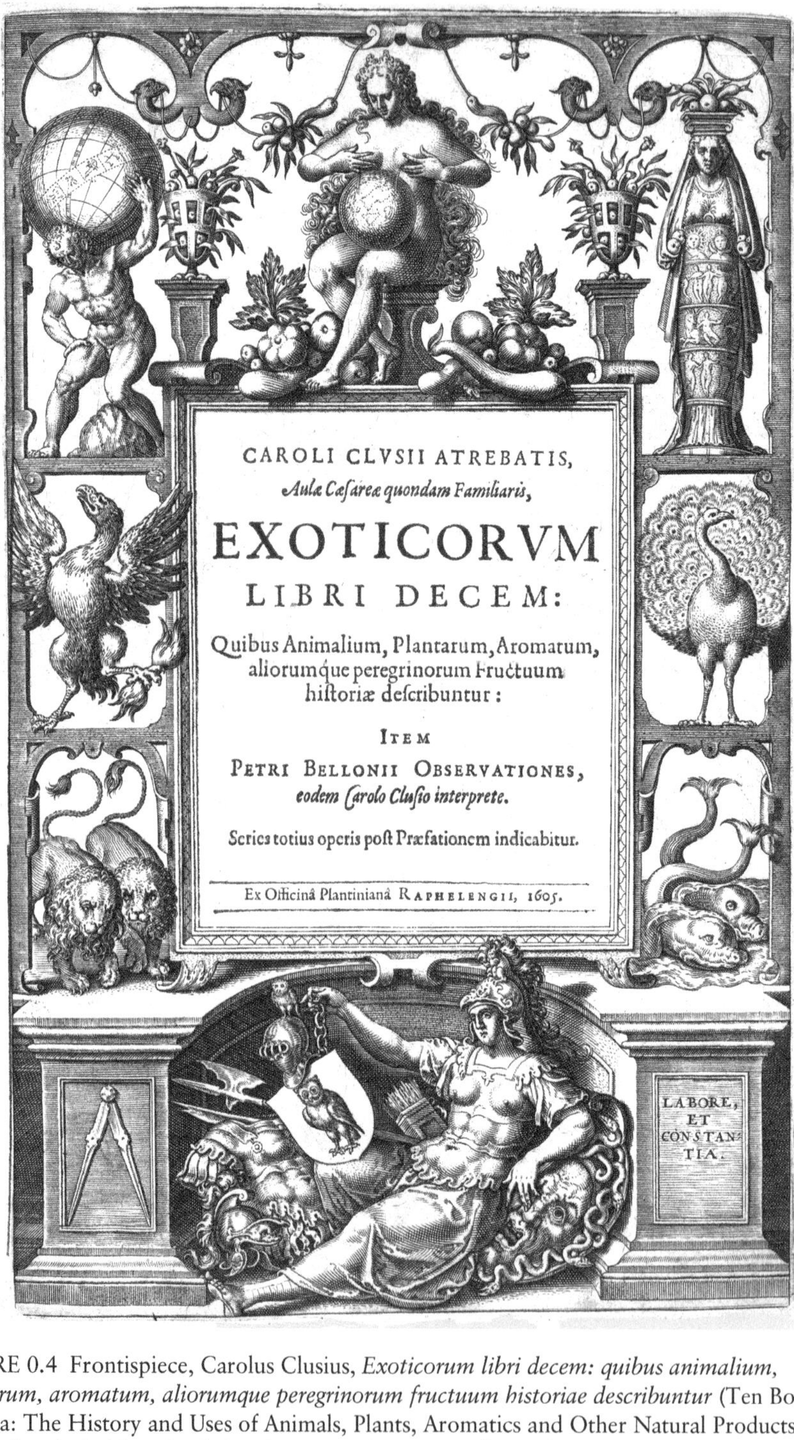

FIGURE 0.4 Frontispiece, Carolus Clusius, *Exoticorum libri decem: quibus animalium, plantarum, aromatum, aliorumque peregrinorum fructuum historiae describuntur* (Ten Books of Exotica: The History and Uses of Animals, Plants, Aromatics and Other Natural Products from Distant Lands, Leiden: Raphelengius, 1605). Tentoonstelling Clusius Universiteitsbibliotheek, Leiden. Photo by Shyamal/Wikimedia Commons.

FARMS AND GARDENS

As this quotation shows, farmers and gardeners appreciated the work of medical botanists, but the sixteenth century also saw the development of a genre of farming and gardening books. It was not wholly new. In fact, it included printed texts of the greatest such works of earlier times, the Greek *Geoponica*, the Latin *rei rusticae scriptores* largely rediscovered by Poggio Bracciolini, and the fourteenth-century *Liber commodorum ruralium* of Petrus de Crescentiis. These editions and translations apparently sold well but were fated to be superseded by up-to-date works in several languages of Europe, as discussed in various chapters below: in Latin by Charles Estienne (Stephanus 1554), then more extensively in French by Estienne (1564) and expanded after Estienne's death by Jean Liebault (1570), translated into English by Richard Surflet (1600) and expanded by Gervase Markham (1616); in French also by Olivier de Serres (1600); in Latin by Conrad Heresbach (Heresbachius 1570), translated and enlarged in English by Barnaby Googe (1577); in English by Thomas Tusser in doggerel verse (1557, 1573) and Gervase Markham in prose (1613), along with books specifically on gardens, more modestly by Thomas Hyll (1563, 1577), more expansively John Parkinson's *Paradisi in sole Paradisus terrestris* (1629). The title means "Park-in-Sun's earthly paradise," playing both on the author's surname and the undertones of the biblical and classical word paradise: a paradise is, in simpler words, a park or garden.

LINES OF TRANSMISSION

The sixteenth-century ferment of discovery and innovation in the world of plants is visible on paper—in the texts and illustrations of printed books, in paintings and drawings, in dried plant collections—but not only on paper. The world of real plants around us today looks very different as a result of the Columbian exchange and other plant introductions. Few people who worked with plants were writers, and not all of them were readers. Let us list those involved in plant medicine—herb women, midwives, folk healers, alchemists, apothecaries, aristocratic ladies, members of religious communities (Chapter 5)—and let us add to the list peasants and farmers and landowners, gardeners and those who employed gardeners and designed gardens, those who carried and sold plants and plant foods, those who traded spices, medicines, and plant products over long distances and those who sailed the wooden ships in which they traveled. Given that any of these people might happen to be involved in the first transmission of any plant from one habitat to another, and that the incident would only become memorable if and when the plant succeeded in its new habitat, the creation of a written record of the incident was rather unlikely. Thus, in the great majority of cases neither the people involved, nor the crucial date at which plants or seeds first traveled, can now be stated with full confidence. Yet some such transfers are indeed recorded in Early Modern manuscripts or printed books, and in a few cases the writers themselves claim responsibility.

To take England as an example, sixteenth-century developments in agriculture and horticulture were far from being the responsibility of any small group of scientists or innovators. The fact is emphasized in Harrison's "Description of England" already quoted above. Harrison outlines recent improvements in English herb gardens and fruit orchards and attributes them to individual gardeners and fruit-growers of all social classes:

> Like thankes be given unto our nobilitie, gentlemen, and others, for their continuall nutriture and cherishing of such homeborne and forren simples in their gardens, for

> hereby they shall not onlie be had at hand and preserved, but also their formes made more familiar to be discerned, and their forces better knowne than hitherto they have béene. And even as it fareth with our gardens, so dooth it with our orchards, which were never furnished with so good fruit, nor with such varietie as at this present. For beside that we have most delicate apples, plummes, peares, walnuts, filberds [*Corylus* spp.], &c., and those of sundrie sorts, planted within fortie yeeres passed, in comparison of which most of the old trees are nothing woorth: so have we no lesse store of strange fruit, as abricotes, almonds, peaches, figges, cornetrees [cornel, *Cornus mas*] in noble mens orchards. I have seene capers, orenges, and lemmons, and heard of wild olives growing here, beside other strange trees, brought from far, whose names I know not. So that England for these commodities was never better furnished.
>
> (Harrison 1587: 210; fuller quotations and commentary in Chapter 3)

In perfect accord with the generous tone of Harrison's survey, Marsh, Egmond, and Francis in this volume show how botanical, medicinal, and horticultural advances in this period depended not on any individual but on networks. There was an ever-growing community of enthusiasts of remarkably varied backgrounds and interests, a few of them individually celebrated, the majority scarcely known beyond their own chain of contacts. Among those involved in trade the work of James and Pieter Garet is highlighted in Chapter 5. Dutch in origin, they were based in London, a center of intellectual exchange. Evidence on the networks of London plantsmen and gardeners is teased out by David Marsh in Chapter 3. Among the enthusiastic proprietors of English gardens beyond London Chapter 7 draws on manuscript lists and plans of Revd Walter Stonehouse, Sir John Reresby, Sir Thomas Temple, Sir John Oglander, and Sir Thomas Hanmer. The most famous of their immediate successors were the garden designers John Tradescant the Elder and the Younger.

Of some innovators in this field no record survives. Of the history of some plant species that crossed the world during the Early Modern period, transferred from country to country and from continent to continent, prominent in humanity's later history, essential in some cases to the health of millions in recent and modern times, the facts are recorded by the merest chance. In the case of coca, *Erythroxylum coca*, two syllables that are now on so many lips in the name of a fizzy drink and an addictive drug, we seem to grasp the exact moment at which knowledge of the plant's stimulant use was transmitted from the peoples of the Andes to the world at large:

> Rodrigo Pantoja, travelling from Cuzco to Rímac, met a poor old soldier with his two-year-old daughter on his back. "I cannot pay an Indian to carry her," he explained, "so I have to do it myself." "But why are you chewing coca as the Indians do," Pantoja demanded, "when Spaniards detest the stuff?" "Necessity has forced me to imitate them. It is because of coca that I feel strong enough to carry my burden."

Yet we cannot, after all, date the moment or name the poor Spanish soldier; Garcilaso de la Vega "el Inca," who said that he was told the story by Pantoja, gave no further help (Garcilaso 1609: f. 212v; Livermore 1966: 1:511, abridged).

For chili (above) and maize, the first two American plant species that are known to have crossed the Atlantic eastwards, our information is relatively good. As we have seen, Columbus called maize "millet" at first. In what was to become the first book of *De orbe novo decades*, drawing directly on Columbus' reports, dated November 5, 1493, and addressed to cardinal Ascanio Sforza in Italy, Peter Martyr ab Angleria described this newly observed staple as

> a kind of millet, similar to that which exists plenteously amongst the Milanese and Andalusians. This millet is a little more than a palm in length, ending in a point, and is about the thickness of the upper part of a man's arm. The grains are about the form and size of peas. While they are growing, they are white, but become black when ripe. When ground they are whiter than snow. This kind of grain is called *maiz*.
>
> (1530: f. 4r; Galinat 1992; MacNutt 1912: 1:64)

On April 29, 1494, dispatching his next report to cardinal Sforza, Peter Martyr hinted at how many botanical specimens Columbus had already distributed to contacts at the Spanish court:

> That you may inform your apothecaries, druggists, and perfumers concerning the products of this country and its high temperature, I send you some seeds of all kinds, as well as the bark and the pith of those trees which are believed to be cinnamon trees. If you wish to taste either the seeds or the pith or the bark, be careful, most illustrious Prince, only to do so with caution; not that they are harmful, but they are very peppery, and if you leave them a long time in your mouth, they will sting the tongue. In case you should burn your tongue a little in tasting them, take some water, and the burning sensation will be allayed. My messenger will also deliver to Your Eminence some of those black and white seeds out of which they make bread. If you cut bits of the wood called aloes, which he brings, you will scent the delicate perfumes it exhales.
>
> (Peter Martyr 1530: f. 7r; MacNutt 1912: 1:84)

The "black and white seeds" are maize: see Chapter 3 on this first transfer and on later confusion over the origin of maize. Cassava (*Manihot esculenta*), which is *iucca* in Peter Martyr's Latin, and sweet potatoes (*Ipomoea batatas*), which he called *ages*, were perhaps not in these first voyages brought home to Spain. They were nonetheless described confidently as "roots destined to become the food of Christians and take the place of wheat bread, radishes, and our other vegetables" (Petrus Martyr 1530: f. 45v; MacNutt 1912: 1:342). Several chapters of this volume offer examples of the long-distance spread of plant species in this Early Modern period: cassava, Chapter 3; the Levant(ine) rose (*Rosa phoenicia*), Chapter 3; bananas (*Musa* spp., Chapter 2, which according to Oviedo leapt from Alexandria in Egypt to Seville, from there to the Canaries, and then in 1516 crossed the Atlantic from Gran Canaría to Santo Domingo (Amador de los Rios 1851: 1:293). The tulip, mentioned in Chapters 4, 6, and 7, is well known as a sixteenth-century introduction from the Ottoman Empire to Europe, though the details of its transmission have been far from clear (Goldgar 2007). The first man in Europe to taste a pineapple (*Ananas comosus*) (see also Chapters 2 and 3 of this volume) may perhaps have been king Ferdinand II of Aragon, as he himself boasted to Peter Martyr:

> It is like a pine-cone in form and colour, covered with scales, and firmer than a melon. Its flavour excels all other garden fruits, for it does not grow on a tree but on a plant similar to an artichoke or an acanthus. The King places it above all others. I myself have not tasted it, for it was the only one which had arrived unspoiled … Spaniards who have eaten them fresh plucked where they grow, speak with the highest appreciation of their delicate flavour.
>
> (1530: f. 33v; MacNutt 1912: 1:262–3)

In the same paragraph Peter Martyr made the first written mention of potatoes (*Solanum tuberosum*), but said nothing of their arrival in Europe. They spread there slowly from south to north, reaching James Garet's garden in London in 1588–9 (Chapter 5; Egmond 2010: 202–3).

Citrus fruits are among the most important of the fruits that crossed from the Old to the New World in the Columbian exchange. The major cultivated species, native to southern and eastern Asia, were at this date still somewhat exotic in Europe. The enthusiasm they aroused is exemplified as fully in Giovanni Battista Ferrari's lavish picture book (Ferrarius 1633) as in the lovingly detailed descriptions of the Villa Medicea di Castello. Vasari in 1568 noticed the orange trees in the labyrinth garden, and had heard of another projected garden that "was to be filled with orange-trees (*Citrus* × *aurantium*) since it is protected by the walls and the mountain from the north and other contrary winds" (Hinds 1927: 3:171–2; Vasari 1568: pt 3, 2:404–5). It was of the labyrinth garden that Belon had written a few years earlier: "All its walls are embroidered with laurel trees interwoven. The citron and lemon trees and the *ponciers* are woven together in the same way" (1558: f. 79v).

The category "citrus fruits" was not what it is now. Tangerines (*Citrus reticulata*) had not yet reached Europe. Grapefruits are the most notable among several now-familiar hybrids that did not yet exist. On the other hand, one fruit now forgotten by nearly everyone was a distinct kind or species in sixteenth-century opinion. This is the *poncier* (so Belon names the tree, as quoted above) that bears *poncires* (so Rabelais names the fruit, as quoted in the epigraph to Chapter 6). It has not been easy to identify: Clusius translated Belon's *ponciers* into Latin as *Adami mali* (1589: 86), mistakenly if by this he meant "bitter orange trees," for they actually correspond to the group of varieties named *limon ponzinus* by Ferrari (Ferrarius 1646: 289–97), lemons (*Citrus* × *limon*) with thick pith, the pulp eaten fresh with sugar, of which several kinds were grown on the west coast of Italy. At least one cultivar remains modestly popular today, named *limone ponzino amalfitano* by nurserymen, quite different from its close neighbor the *limone Costa d'Amalfi* which has the specious dignity of a protected geographical designation.[8]

Citrus fruits are now major crops in the southern United States and in several Latin American countries. Californian and Floridian research stations are at the forefront in the development of citrus cultivars, including hybrids. How and when did oranges (some of the leaders in this migration) first cross the Atlantic and first reach the American continent? Those answers happen to be known. For the first stage the adaptable genius of Christopher Columbus was responsible. Perhaps he took no seeds westwards on his first journey: why would he? He expected to find Oriental civilizations thriving and all useful exotica already growing in the lands he was about to explore. But it was not so. Adapting to circumstances, on his second journey he took a selection of seeds with a view to establishing a colony on Hispaniola. The colonists were wiped out, but not before planting crops including bitter oranges, which were soon to spread to other Caribbean islands.

As for the transfer of oranges to the American mainland, that event is separately recorded. The known name in the chain is that of Bernal Díaz (1492–1584), by no means a hero but a minor *conquistador* whose real ambition was to establish an estate for himself in what is now Guatemala. There he died, not before writing an eyewitness history of the Spanish conquest, in the first draft of which he inserted a personal anecdote:

> I sowed the seeds of some oranges near to another idol house ... seeds which I had brought from Cuba, for there was a rumour that we were coming back to settle. They

> germinated [*nacieron*] very well, for it seems that the priests, when they saw that these plants were unlike any they knew, protected them, watered them and weeded them. All the oranges in that province are the descendants of these plants, and I have called this to mind because these were the first oranges planted in New Spain.
>
> (Maudslay 1908: 1:62; Serés n.d.: 54)

The anecdote recurred with an apologetic note in a later draft, and was erased, having been rejected as trivial, in the final manuscript. It is interesting not least for Díaz's assumption, surely correct, that the priests encouraged the unknown saplings to grow. Experimenting with potentially useful plants has been a human activity since the Neolithic revolution and even before.

Failures should not be overlooked. One of the most notable was the growing of Oriental spices in Mexico in the mid-sixteenth century by Francisco de Mendoza y Vargas (1496–1584; Chapter 2), who, with all the advantages of his position as son of the first Spanish viceroy, and all the resources he was able to dedicate to his experimental plantations, succeeded in establishing no new species except ginger. Against the failures must be counted an astonishing range of successes, some of them undeservedly obscure and unfairly forgotten, like the New World strawberries (*Fragaria virginiana*, *F. chiloensis*, and especially their hybrid, *F. ananassa*) that would soon eclipse native Old World strawberries (*F. vesca*). Parkinson discusses the Virginia kind, tantalizingly difficult to grow and ripen, and then the "Bohemia strawberry" which "hath been with us but of late dayes, but is the goodliest and greatest ... for some of the berries have been measured to be neere five inches about. Master Quester the Postmaster first brought them over into our country, as I understand ... Master Vincent Sion who dwelt on the Banck side, neere the old Paris garden staires" was the best at growing them (Parkinson 1629: 528; Wilhelm 1974: 266).

THE SUNFLOWER

Let the early history of the sunflower serve as a case study in the sixteenth-century transmission of useful plants. *Helianthus annuus* in its wild form ranges from the southern edge of Canada to northern Mesoamerica. The species is well known among pre-Columbian archaeologists as one of the principal early domesticates of North America: it was a member of the Eastern agricultural complex established by about 1800 BCE in the Mississippi valley (Heiser 1951; Smith et al. 2009). Although the sunflower grew wild in Mexico, was used in Aztec religious ritual, and was illustrated in sixteenth-century Mexican codices, there has been debate as to whether it was cultivated as far south as Mexico at this period. Its limited familiarity there might help to explain why several early European sources stated firmly that it came from Peru. Authors who had direct contact with Peru did not share this misconception, for in fact the sunflower was not found in South America at all.

The first-known European image of a sunflower appears in the epilogue to the herbalist Rembert Dodoens' monograph on flowers, published in Latin in 1568. The image speaks for itself, fortunately, because there is no accompanying description. "As we were preparing this edition," Dodoens wrote, an image of the very rare "*chrysanthemum Perunianum*" was forwarded to him by Christina Bertolf from her husband, royal counsellor Joachim Hopper, who was in Spain (Dodoens 1568: 294–5; van Riemsdijk 2015). In a revised edition published in the following year, the state of European sunflower cultivation is

FIGURE 0.5 The sunflower depicted among superficially similar flowers on a page of John Parkinson's *Paradisi in sole Paradisus terrestris* (Park-in-Sun's Terrestrial Paradise, 1629: 297). Hand-colored copy. Photo courtesy of Getty Research Institute.

fully described. Native to "Peru and some other American provinces," this giant annual had been grown in the royal garden in Madrid, where it reached about 7 m (24 ft) in height. Dodoens himself had now seen the sunflower growing in Jean de Brancion's botanical garden at Mechelen (now Belgium), but there it only reached about 3 m (10 or 11 ft) and the flower had not fully opened at the approach of winter, so it could not be grown on. In the greenhouse of Giacomo Antonio Cortuso at Padua, Italy, it had grown even taller than in Madrid, and the flower had opened. The same informant had found the young leaf-stems (petioles) and flower (inflorescence)-bases good to eat and aphrodisiac (Dodoens 1569: 305–9; Wheelock 1999: 25).

The next oldest description is owed to the Madrid medical author Juan Fragoso, who called the sunflower a "well known plant of Peru," giving the names *flor del sol* and *sol de las Indias*. Fragoso, like Cortuso, had eaten the flower-base "as one does an artichoke." He was the first to claim that the head of flowers turns to follow the sun "as if offering worship to it" (1572: ff. 25r–26r).

Soon afterwards Francisco Hernández made his botanical journey to Mexico. His work, not published until long after his death, is impressive (as already noted) for its recording of local names and uses of plants: hence he was the first to note the usefulness of sunflower "seeds" (technically, single-seeded dry fruits), "like melon seeds," which were, after all, the real reason for the plant's domestication. "Eaten rather generously they cause headache, soothe the chest, and cool the libido, and certain people grind them and bake them into bread ... It is cultivated by sowing in plateaus and lowlands in several regions, flourishing better in the lowlands" (Hernández et al. 1651: 228).

While Hernández was at work in Mexico, Nicolás Monardes in Seville was meanwhile enlarging his work on the medicinal plants of the New World. Monardes found room for a short entry on the sunflower, of which "they are still sending me seeds, though we have had it here for some years already." It made a fine show in gardens, he added, though it required support while growing to its full height (1574: f. 109v). Unlike Fragoso, Monardes did not connect the sunflower with Peru, and one notes that he had correspondents in Peru.

Carolus Clusius, who had read Fragoso and Monardes, inserted all that he thus far knew of sunflowers in his expanded Latin version of Dodoens' complete herbal in 1583. Clusius noted two common names for the plant, *sol Indianus, flos solis* (the latter being the exact Latin equivalent of "sunflower," which eventually became the English name). Clusius adjusted Dodoens' Latin name to *chrysanthemum Peruvianum* (1583b: 263–4), and this name gained popularity.

The story of the first English expedition to Virginia in 1585 was told in a narrative by the participant Thomas Hariot (1560–1621). He made careful observations of indigenous horticulture, describing a "great hearbe in forme of a Marigolde, about six foote in height; the head with the floure is a spanne in breadth ... of the seedes heereof they make both a kinde of bread and broth." As to its identification, "some take it to bee *Planta Solis*," evidently a reference to earlier reports of the sunflower (Hariot 1588: quaternion C2; Hariot 1590: 14; Heiser 1951: 435).[9] Cultivation was by now spreading in Europe: for all that, the phrasing suggests that the colonists had heard of the sunflower but had not previously seen one.

Nardo Antonio Recchi, custodian of Hernández's researches in this period, made additions of his own in copying the manuscript on which eventual printed editions would be based. Hernández's description of the sunflower is headed *chimala[ca]tl Peruina*, the keyword (as usual in Hernández's work) being a Nahuatl name for the plant, the whole phrase a Mexican equivalent of Dodoens' *chrysanthemum Perunianum*. The description contains a short passage, omitted in the quotation above, that is strongly reminiscent of Dodoens and Clusius, "... although some would say that they are aphrodisiac. The plant grows naturally in Peru and several other provinces of America" (Hernández et al. 1651: 228).[10]

Clusius eventually expanded his old translation of Monardes' work to include the 1574 additions, issuing the result as part of that landmark in European knowledge of exotic plants, *Exoticorum libri decem* (1605: 347–8). Realizing, perhaps, that the plant's Peruvian origin was not supported by his best sources, Monardes and Hariot, Clusius in this definitive publication dropped his previous term *chrysanthemum Peruvianum* and adopted *herba solis*, suggested by Monardes' *yerva del sol* and Hariot's *planta solis*. He drily gave credit to the "rhapsodies" of Fragoso, whose wordy description had been the second to appear in print.

In London John Gerard had meanwhile published his own English reworking of Dodoens' herbal. His name for the plant is "sunne flower ... in Latine *flos solis*, taking that name from those that have reported it to turne with the sunne, which I could never observe." He himself had grown the plant from seed: "in one sommer being sowen of a seede in Aprill, it hath risen up to the height of fourteene foote in my garden, where one flower was in waight three pounde and two ounces and ... sixteene inches broade." He had tasted it, too. "We have founde by triall, that the buddes before they be flowred, boiled and eaten with butter, vineger, and pepper, after the maner of artichoks, are exceeding pleasant meate, surpassing the artichoke farre in procuring bodilie lust" (1597: 612–14). "They are too strong for my taste," John Parkinson retorted (1629: 296).

No printed source will tell us in whose hands, shortly before Dodoens' illustration of 1568, sunflower "seeds" had first crossed the Atlantic to Spain, but by the end of the century the plant had been "commonly known in almost all of Europe for many years" (Clusius 1605: 348). Its widespread fame explains its place in Joachim Camerarius' illustrated book of plant symbols and emblems, in which, in close accord with the claim made by Fragoso in 1572 and questioned by Gerard in 1597, an open sunflower is shown straining toward the sun (Camerarius 1590: 59; see Chapter 6).[11]

THE COLUMBIAN FUSION

Given that the outstanding event in the cultural history of plants in this Early Modern period was the Columbian exchange, the effects of this fusion, as we may call it, though equally fundamental on both sides of the Atlantic, were most immediately noticeable in the daily lives of those who traveled westwards, for they were the active participants. They sought the plants of the New World, but they could not manage without the familiar plants of the Old. On March 29, 1494, returning to his new foundation of Isabela (now Dominican Republic) on Hispaniola after sharing cassava bread and yams in neighboring villages, Columbus found that the Spanish gardeners' melons, sown less than two months earlier, were already good to eat, and that the cucumbers (*Cucumis sativus*) had come up in three weeks:

> Next day a planter harvested ears of wheat which had been sown at the end of January; they also picked chickpeas larger than the ones they had planted. All the seeds they had sown sprouted in three days and were ready to eat by the twenty-fifth day. Fruit stones

> ... sprouted in seven days; vine shoots sent out leaves at the end of the same period, and by the twenty-fifth day green grapes were being picked.[12]
>
> (Ulloa 1571: f. 106)

A generation later, the Spanish plantations in the Caribbean, well established and flourishing, were available as immediate source when the conquistadors of Mexico needed to stock their new estates. Already in 1521, at a drunken and overcrowded banquet at Coyoacán to celebrate the initial conquests and the capture of Cuauhtémoc, Cortés had Spanish wine and Caribbean pork, "plenty of wine out of a ship from Castille that had berthed at Villa Rica [de la Vera Cruz], also pigs that had been brought from Cuba" (Bernal Díaz in Serés n.d.: 625–7; Sokolov 1991: 29). Soon afterwards, with farming in view, Cortés was sending for cattle, pigs, sheep, goats, asses, and mares from Hispaniola and Jamaica and Cuba, sugar cane, mulberries for silkworms, vines, olives, and wheat from Spain (Thomas 1993: 578, 599). Experiments during that first decade of European settlement in Mexico are reported by Peter Martyr ab Angleria:

> In the country of Temistitana [Mexico], because it is cold owing to the distance from the sea and the neighbourhood of high mountains (though indeed it lies in the torrid zone at eighteen degrees), if our wheat is sown, it flourishes, with ears and grains of larger size than in Europe. But since they have three kinds of maize, white, yellow and red, their own various kinds of flour are better liked and healthier. Wild vines [*Vitis tiliifolia*] grow spontaneously in the forests, producing large and tasty grapes, but wine has not yet been made from them. It is said that Cortés has planted vineyards, with what result time will tell.[13]
>
> (Petrus Martyr 1530: f. 107v; MacNutt 1912: 2:357)

Vitis tiliifolia, though its fruits are indeed good to eat, was prized in Mexican medicine for its leaves. The wild grapes of the New World were deservedly given much attention by early travelers. Those of Florida (*V. rotundifolia*) are illustrated by Jacques Le Moyne de Morgues in the second volume of Theodore de Bry's *America* (Le Moyne de Morgues 1591: pl. 5, see the front cover of this volume). They were to give rise to the cultivars known as Scuppernong and Catawba grapes. Those of Quebec (*Vitis riparia*) impressed Jacques Cartier so highly in 1535 that he named "Ysle de Bacchus" what is now the Île d'Orléans (Cartier 1545: ff. 14r–15r). Little did he know that *V. riparia* stocks would help save the wine industry of Europe in the late nineteenth century.

The Columbian exchange is already seen in full vigor in the Paradise murals of the Augustinian monastery of Malinalco southwest of Mexico City. Painted in the 1540s by native artists in European style, they mingle the flora of Mexico and Europe, pomegranate and rose beside sapote (*Pouteria sapota*), and heart flower (*Magnolia mexicana*) (Peterson 1993: 182–4), shown with generic vines that belong equally to both continents—and this is natural enough, since all may be supposed to flourish in Paradise. Among the rest, notable is a cacao tree, soon to be spread around the world, its fruit here attracting the attention of a spider monkey (Peterson 1993: 86–9, fig. 34). Neither the plant nor the animal was to be seen locally, but the pictorial juxtaposition is a commonplace of Mesoamerican art, a reminder that the two species had moved across tropical America in a kind of symbiosis, the monkeys eating the cacao fruits and excreting the seeds; the trees thus spreading and providing new habitat (Dreiss and Greenhill 2008: 20).

The Columbian fusion is in evidence in another context in 1538 when the new peace treaty between Spain and France was celebrated in banquets given by Cortés and by the

FIGURE 0.6 Paradise murals of the Augustinian monastery of Malinalco, Mexico. Sixteenth century. Image from Album/Alamy Stock Photo.

viceroy, Antonio de Mendoza (father of Francisco de Mendoza whose spice plantations are mentioned above). A narrative by Bernal Díaz of the Mendoza banquet, attended by "more than three hundred knights and more than two hundred ladies," makes a fitting climax to this introductory survey:

> I was one of those who dined at those great celebrations. First, there were some salads made in various ways, and afterwards, kids and hams roasted in the Genoese style; after that quail and pigeon pies, and then wattled cocks and stuffed hens; then blancmange; after that a stew [*pepitoria*]; then a royal cake [*torta real*]; then local fowl and partridges and quail *en escabeche*, and then they replaced the tablecloths twice and left clean ones with napkins; then they brought pasties [*empanadas*] filled with every kind of bird and game; these they did not eat, nor many of the things served previously; then came fish empanadas; neither did they eat any of them; then they brought in roast mutton, and beef and pork, and turnips and cabbage and chickpeas, of which they also ate nothing; and in the midst of these dishes they put down various fruits to be tasted ... Especially when the stewards brought drinks, they served to the ladies who were dining there, many more than had attended Cortés's dinner, many gold goblets, some with *aloja*, others with wine, others with water, and others with chocolate and with *clarea* ... And I haven't mentioned the service of olives and radishes and cheese and cardoons and marzipans and almonds and comfits and lemon drops [*diacitrón*] and other sugar confections and local fruit; but I can only say that the whole table was covered with that service ... And I still haven't talked about the fountains of white wine and red wine that flowed, both especially made ...[14]
>
> (Seres n.d.: 913–14; Sokolov 1991: 29–31)

CHAPTER ONE

Plants as Staple Foods

Europe in the Post-Classical Era

MALCOLM THICK

INTRODUCTION

By the beginning of our period the word "staple" suggested stability, order, consistency, something which could be stored and so be available when needed. It also signified a regular market (Botsford 1989: 1–2). A plant-based food staple is something an identifiable section of a population depends upon to a significant extent for its normal diet (excluding the rich, who could afford a varied diet). Of course, over a two-hundred-year period there will be changes. New staples will emerge, others will decline, and "temporary staples" will be discovered—foods which in times of famine or dearth relieved hunger but were otherwise shunned. This period saw the import of new plants from America into Europe, notably potatoes, peppers, tomatoes, and maize but only the last became a staple food in this time. Garden vegetables will be considered together as one staple (see below). In terms of geographic emphasis, this chapter focuses on staple foods in Europe.

In practical terms, staples were largely bulky foods, rich in carbohydrates, to fill the stomach, providing a source of energy. They were also fairly pleasant to eat day after day. Staple food-plants were usually suited to the growing conditions of the areas where they were eaten and occupied a considerable amount of the available cultivated area. Land might, however, over time, be increasingly devoted to more lucrative crops—industrial raw materials or luxury foods for the rich—leading to imports of staple foods to feed the bulk of the population. Dependence on a narrow range of staple crops, especially when they were imported, multiplied the problems of dearth or famine when harvests failed—the narrower the range of staples, the more danger of catastrophe (Cunningham and Grell 2000: 213–14). People often continued to consume a given staple *food* although the *ingredients* of that staple food changed: polenta for instance was made with a succession of different grains over time. Thus, there were fewer staple *foods over time* than staple *food-plants*. The focus here is on the plants, namely: grains of all types; olive trees; vines; garden vegetables and fruits; nuts. Staple animal fodder includes grains fed to animals, grasses and other plants grazed or harvested for animals, and peas and beans (which were also consumed to some extent by humans). The main staple drink ingredients were apples and pears, barley, and hops.

RES RUSTICAE: DEVELOPMENTS IN AGRICULTURE

After the devastating depopulation of the Black Death, European agriculture was, as a whole, faced with the task of responding to a general and sustained increase in population in the period: there were an estimated 84 million Europeans in 1500 and by 1600 there were 111 million. Agriculture coped in various ways and with varying degrees of success but European food production as a whole did not keep pace with the population rise. So, at the end of these two centuries, the poor, on average, ate less meat than before, and many were forced to accept inferior grains and even pulses in their daily bread (Montanari 1994: 100).

European growing conditions are diverse. Soils and topography vary considerably, as does the climate, from the hot arid plains of Spain to the frozen wastes of northern Scandinavia and Russia. Weather, especially in coastal areas, is unpredictable. Famine was a constant danger during the period. Farmers coped as best they could. (It should be pointed out that most Europeans at this time were farmers, even in southern England where, by 1650, many whose ancestors were subsistence farmers had been reduced to landless laborers, working for farmers with quite large holdings who produced for the market.) Agricultural advance was patchy—some areas made much more progress than others. Human activity helped or hindered food production: wars were the most lethal, but social organization and changes in it facilitated agricultural advance in some countries and hindered it in others.

In the most backward areas of Europe, the agriculture in 1650 was much the same as that two centuries earlier—peasants, working small farms with few livestock and little in the way of equipment, produced crops predominantly for family consumption. They might sell some produce at market to pay their rent: even this obligation might be met by labor services or in kind. Such was the agriculture of Russia. Soils were poor and implements crude, but land was abundant, so agriculture was extensive. Low yields (1:3 in grain growing—three grains harvested for each grain seeded—was common) meant an ever-present threat of dearth. Serfdom reduced most peasants to subsistence farming on their family plots. The peasants reacted to risk by taking decisions communally and this, plus serfdom, held back economic and agricultural advance (Hosking 2001: 8).

Poland also witnessed serfdom during the period. In the fertile areas accessible to the Vistula, large demesne farms were worked by serfs (who also were given small personal plots). From the late fifteenth century until the mid-seventeenth, increasing amounts of grain were exported down the Vistula through the port of Danzig (Gdansk) to other parts of Europe—surpluses produced by the ever more onerous labor services that large landowners placed on the peasants. So, serfdom here increased agricultural productivity, but ultimately, as in Russia, it stifled agricultural advance. The peasants had no surplus to trade from their poorly exploited plots. The landlords exported primary produce and imported foreign luxury goods, and so the area developed little in the way of secondary industries in either luxury goods, food processing, or any other sphere of economic activity (Davies 2005: 197–224). It was similar further down the Vistula and along its tributaries in western Ukraine. While male serfs worked on demesnes, women and younger members of servile families produced food for the family on small plots (Mayhew 1973: 144–8).

Other parts of Europe showed varying degrees of agricultural improvement. In western Germany, peasants developed crops and farming systems to maximize yields and minimize risk of losses: "In general, this meant that peasants cultivated hardy and reliable cereal strains of spelt, rye, oats and barley, instead of riskier and higher yield cereal varieties." But although generally conservative, they did respond to market opportunities: "In the Rhineland, fruit orchards, market gardening, and viticulture developed close to urban centres. The Upper Rhine became well known for its apples, cherries, and vegetables" (Robinsheaux 1998: 144). In the sixteenth century, agricultural prices in Germany rose faster than wages did. The poor could afford less meat and, consequently, demand for grain increased. Arable farming expanded at the expense of pastoralism, and reclaimed coastal lands were largely devoted to arable production. Farmers did well at this time, especially those with large holdings in the north and east. There followed a period of agricultural depression in the early seventeenth century which was exacerbated by the Thirty Years War, 1618–48. The war resulted in famines and the loss of a third of the population. Only in areas unaffected by the war (for example East Friesland and Schleswig-Holstein) did farmers prosper. Overall, German agriculture in this period saw no great changes in techniques but some intensification and the start of agricultural science and literature (Mayhew 1973: 118–69). A notable example of the latter is *Rei rusticae libri quatuor*, written by Konrad Heresbach (1496–1576) of Wesel on the Rhine and first published in Köln in 1570. Much influenced by classical agriculture, the book nevertheless reflects the contemporary state of German farming. Heresbach came from a wealthy family and the book is much concerned with running a family estate. It was quickly translated into English by the early pastoralist poet, Barnabe Googe (1540–94), appearing in 1577 as *Foure Bookes of Husbandry*.

Growing demand from towns and cities in northern Italy gave market stimulation to agriculture, as it did in parts of Spain where larger holdings produced wheat, olive oil, and wine for domestic and international markets, although most peasant holdings in sixteenth-century Castile yielded little saleable surplus. The two-field rotation widely practiced kept productivity low and, although extending cultivation into marginal land increased output in the short term, erosion of light hillside soils became a problem in the sixteenth century. Barley and oats, which are fairly drought-resistant, might have been more suitable for many parts of Castile, but they were looked upon as inferior, and wheat remained the favored grain. These areas were also some of the first to take to new crops introduced from America, maize being the best example (Vassberg 1984: 184–5, 200–4).

Some of the most sophisticated agriculture, by 1650, was to be found in the coastal areas of the Low Countries and northern France. Urbanization and growing industry fed demand for agricultural produce, including industrial crops: hops, flax and hemp, and dyestuffs, such as madder (*Rubia tinctorum*). The area ceased to be self-sufficient in grain, but this was supplied by imports from Poland via Gdansk. Urbanization stimulated the new trade of market gardening (a feature of other agricultural parts of Europe near urban centers). The social structure was favorable and landlords increasingly asked for rents rather than labor services, which gave farmers the incentive to improve efficiency. Already, by the peak of population in the thirteenth century, intensive agricultural techniques such as marling (liming), weeding, and fallow cultivation with legumes or rape (*Brassica napus*) were employed. In the demographic recession following the Black Death, northern France saw poorer quality land abandoned, leading to better overall yields. In the Low Countries the new advanced techniques were not abandoned and by the seventeenth century more intensive agriculture raised productivity to higher levels than before, almost constituting an agricultural revolution (Campbell and Overton 1991: 144; Scott 2015: 400–5).The agricultural geography of the British Isles was, like that of the Continent, diverse. Large areas of Scotland, Wales, and high ground in the north of England are cold with relatively short summers and heavy annual rainfall. Such areas were suited to pasture—cattle and sheep grazed widely. The southern half of England, where the climate was warmer and rainfall less, was largely a mixed farming area although here too there were marginal areas of moorland where little cultivation was possible.

The commonest arable crops were wheat, rye, barley, oats, beans, and peas. Buckwheat (*Fagopyrum esculentum*), vetches and tares (*Vicia* spp.), and lentils were also grown. Fruit orchards were expanding over the period, especially in Kent and the cider-making parts of the West Country, while some areas concentrated on specialized market garden crops such as hops, liquorice (*Glycyrrhiza glabra*), and tobacco (likely *Nicotiana rustica*, not *N. tabacum*). Of these, wheat, rye, barley, oats, and garden produce could be considered staple crops for humans, along with apples and pears, hops, and barley for alcoholic drinks. Barley and oats were commonly fed to animals, as were the pulses. Animals foraged for much of their food, although some attempts at increasing the supply of animal feed occurred in the period (Thirsk 1967a: 168).

The historiography of southern English agriculture has recently been dominated by a debate over the rise of capitalist agriculture and the consequent demise of poor peasants' holdings. In essence, the argument is that greater availability of land after the Black Death hastened commutation of labor services and the advent of relatively large leaseholds held by free tenants with no feudal obligations. They employed wage laborers and paid cash rents to landlords who in turn leased out their demesnes and lowered rents to attract tenants. The farmers, attuned to market conditions, were now the main decision-makers and capitalism replaced feudalism (Spishak 1998). Opponents of this theory claim that significant agricultural advances were seen under feudalism before the Black Death. It has been noted that there has not been much in this debate about agricultural practice: the spread of leases and accompanying enclosure has been assumed to lead to increased output by making conditions for agricultural change possible (Scott 2015: 406).

Historians are agreed that there was a secular increase in agricultural output during the period and that this was stimulated by population increase, especially from the early sixteenth century onward and by urbanization which created regular markets. In the case of the largest urban areas, such as London, specialized forms of agriculture—market gardening, fruit orchards, hops, and industrial crop-growing—came into being. The main reasons for increased output may be summarized as

> the use of more intensive rotations, accompanied by better manuring; the use of improved varieties of grain; and probably the most important of all, the impressive increase in the total acreage of land under the plough as a result of reclamation of waste and conversion of pasture. This increased acreage, it should be noted, was regarded by contemporaries as the main explanation for the improved supply of corn. (Thirsk 1967a: 199)

Shrewd observers, such as John Norden (*c.* 1547–1625), noticed, "the Fennes and low grounds in Lincolnshire, Cambridgeshire, Norfolke, and other places which I did thinke impossible ever to be made drie … And yet I heare, much of it is made lately firme ground" (1618: 195). Those historians, however, who have attempted to assess yields *per acre* in our period have discovered no significant increase in grain yields before 1700 (Scott 2015: 410).

HUMAN FOOD CROPS

Bread Grains

Bread made from grain was the predominant food throughout the period in Europe. It correctly has been observed that "[b]read was quite simply *the* staple food, to an extent no longer seen anywhere in Western Europe today" (Cunningham and Grell 2000: 211, emphasis added). For example, in the case of a certain relatively well-to-do town craftsman, living in Antwerp (now Belgium) around 1600, half of his household expenses were on grain to make bread, or on bread itself, while in the north Italian countryside bread was so ubiquitous that farmlands were referred to as "*terre da pane*" (breadlands) in agricultural contracts and produce was called "*recolta panis*" (bread harvest). Other foods were called "accompaniments to bread" (Cunningham and Grell 2000: 211; Montanari 1984: 211, 307).

As a foodstuff, grain had many advantages: it kept well; provided energy; contained vitamins; was generally low in toxins; and could be easily grown (Prance and Nesbitt 2005: 45). Bread was made from a variety of grains and other plants and, although some people favored certain types of bread because they were accustomed to it, there was a general hierarchy in bread ingredients. At the apex was white wheat-flour, bolted (sieved) to remove all bran, followed by wheat with increasing amounts of bran, then "maslin," which is wheat mixed with the flour of other grains, usually rye and barley. Barley bread and rye bread were lower on the scale, then bread made with the addition of chestnut (*Castanea sativa*) flour, acorn (*Quercus* spp.) flour, flour of dried roots, and so on. While the rich could afford wheat grown locally or imported, the poor ate whatever grain was cheapest. The situation in England was succinctly summed up by William Harrison (1534–93), an English clergyman (1577: f. 95): "... the gentilitie commonly provide themselves sufficiently of wheate for their owne tables, whylest their houshold and poore neighbours are inforced to content themselues with rye, or barley, yea and in tyme of dearth with breade made eyther of beanes, peason [peas], or otes, or of al togither [and some acorns among]."[1] The disdain that the rich had for grains lesser than wheat is unambiguous in the agricultural textbook by Charles Estienne (1504–64) and his son-in-law, Jean Liébault (1535–96), published in France in 1570. In the English edition of 1606, barley is blithely dismissed as "better of the poore people then for the rich" and oatmeal porridge is not derided but millet bread is "unpleasant enough," "of small nourishment," and causes "obstructions," "Turkie wheat," i.e. maize, makes bread which is "grosse, thicke ... of a slimie substance" and rye bread is "unpleasant, fattie, slimie, heavie, like paste, blacke ... such as be wealthie and men living at ease make no reckoning of it" (Surflet 1606: 683–9).

Wheat

Wheat may not have been any more nutritious than other grains, but it was preferred for its taste, texture, and color (see Figure 1.1). It was much used for bread in Mediterranean countries, while those who were rich enough chose to eat wheaten bread in northern and western Europe. It is probable however, that wheaten bread was not eaten by the majority of the urban population in England until the eighteenth century, even though at the time the country was the most agriculturally advanced in Europe (Cunningham and Grell 2000: 211). Londoners may have taken to wheat earlier; the evidence of relative price movements in London during bad harvests of the 1430s has

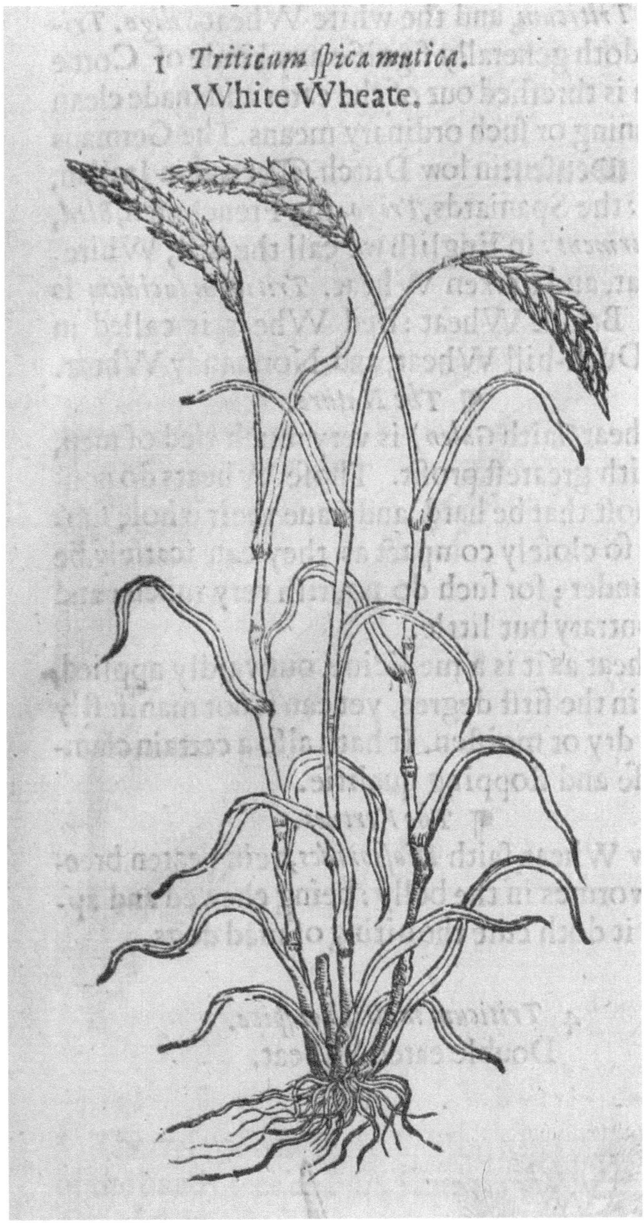

FIGURE 1.1 Wheat: a major staple in Europe. Commonly eaten in towns and cities, it was preferred above other grains because bread made from wheat flour was white and tasty. From John Gerard's *Herball* (1597). Image courtesy of Fisher Library, University of Toronto.

been interpreted as showing that the citizens as a whole were consuming wheaten bread and were reluctant to give it up. Historian and antiquarian John Stow (1524/5–1605) recorded that Londoners at this time eventually "had to eat beans, peas and barley more than in a hundred yeeres before" (Stow, *Annales*, quoted in Miller 1971: 627). There may also have been many in the countryside of southern England who regularly ate wheaten bread at the start of the period, to judge by the high percentage of late medieval corrodies (allowances of provisions for maintenance dispensed as a charity), which specify wheat or wheaten bread.[2] Fynes Moryson (1566–1630), writing in the late sixteenth century, thought that "citizens [i.e. Londoners] and gentlemen eat most pure white bread" (Moryson 1617, quoted in Ashley 1928: 59). Of course, the wheaten bread consumed by many in towns or countryside may not have been "most pure white bread": many grades of wheaten bread were commonly baked, with varying quantities of bran in them, not to mention various maslin mixtures (mixtures of different grains). There is compelling evidence that rye bread was more important than wheaten as a whole in England in the sixteenth century (Ashley 1928: 61–6). In 1696, just after the end of the period, the statistician Gregory King (1648–1712) gave his estimate of grain production in England in millions of bushels: wheat, 14; rye, 10; barley, 27; oats, 16. These are rounded figures, taking no account of imports, and it must be borne in mind that most of the barley was used to make beer and much of the oats was used as fodder, but the figures imply that, while many were eating wheaten bread, the majority were not (Ashley 1921: 290).

Two species of wheat were commonly grown in England at this time: "rivet," *Triticum turgidum*, which produces a soft "biscuit" flour, and *Triticum aestivum*, a hard wheat which yields strong white bread flour and was the most common. There were many locally grown cultivars of these two. Almost all wheat was autumn-sown, although in the north some was spring-sown. Those farmers who sowed maslin often chose early-ripening wheat so that the rye and wheat would mature at the same time (Thirsk 1967a: 168).

In Spain as a whole, more wheat was grown yearly in the period than was any other grain; here, where conditions were favorable to wheat growing, the majority of peasants ate wheaten bread. In many parts of Europe, the consumption of wheat was much higher in the towns and cities than in the countryside where it was actually grown. This was so in England. In late medieval and Early Modern Italy wheat was a major crop in northern Italy, Romagna, Umbria, Sicily, and Sardinia. But it was consumed mostly by the rich, generally, and the more prosperous town and city dwellers: it is likely that the urban poor and peasants made do with rye in central and southern Italy and millet in the north, despite the claim by one historian that "since Roman times the first preoccupation of politicians has been to ensure the availability of large quantities of wheat in order to keep the populations peaceful" (Borgean and Angus 2001: 27).

In western Germany peasants generally played it safe and cultivated grains hardier than wheat. In several parts of France combinations of these minor crops were sown. In sixteenth-century France "the Beauce, the Brie, the north of France and the highlands of Auvergne then specialised in wheat cropping or 'frumentals' with yields of 12 to 15q/ha, significantly higher than the rest of France, where rye at the time was cultivated more than wheat," and between 1549 and 1674, just over a third of available arable land in the agricultural region north of Paris was cropped with wheat at any one time, slightly more than that cropped with all other grains (Borgean and Angus 2001: 138–9).

Romania, like Poland, was a long-standing exporter of wheat to other European countries. The Ukrainian and Polish grain trade through Gdansk has already been noted: the volumes exported varied from year to year, but the trade was an important contribution to the food supply of many parts of Europe accessible by coastal shipping. Estimates of annual exports from Gdanks for sample years in tonnes (in metric tons) are: 1491/2, 12,000 tonnes; 1563, 138,000 tonnes; 1618, 248,000 tonnes (the greatest annual total). In the second half of the sixteenth century, Polish exports from Gdansk reached 150,000 tonnes in several years (Borgean and Angus 2001: 226, 337; Mayhew 1973: 149).

Agricultural improvements affected the yield from wheat cultivation and therefore the availability of this grain. At the beginning of the period, the seed-to-yield ratio for wheat in England could be as low as 1:3 with yields of 6 bushels per acre. George Fussell thought the average ratio had increased in England to 1:8 in the sixteenth century. In this and the subsequent century yields are estimated to have been between 11 and 16 bushels per acre (Fussell 1966, cited in Bell 1987).

Rye

Rye (*Secale cereale*) will grow on much poorer soils than other bread grains will, producing a crop on peaty or sandy soils (see Figure 1.2). In sixteenth-century France, rye "craveth not so industrious and carefull ordering nor yet so fat a ground, and so well inriched as doth the wheate" (Estienne 1606: 683). It will also withstand much colder conditions than wheat will after sowing, tolerating snow cover and frost because it contains proteins which protect it from low temperatures. Rye, noted above, was often mixed with wheat—either in the sowing, where the hardier rye might yield a crop in years when the wheat succumbed, or in a mixture of flours with which to make bread. Two printed books on English agriculture, *The Boke of Husbandry* (1523/34), by John Fitzherbert (1470–1538), and *A Hundreth Pointes of Good Husbandrie* (1557) by Thomas Tusser (*c.* 1524–80), both favored sowing a mixture of rye and wheat (Ashley 1928: 61–6). Together with barley, much rye was eaten by late-sixteenth-century English country people. "Travel-writer" Fynes Moryson (1566–1630) opined in his *Itinerary* that "the English husbandmen, eat barley and rye brown bread" (1617: Pt III, ch. iii "Of England") and William Harrison (1577: f. 95v) concurred: "In champeigne countryes much rye and barly breade i[s] eaten, but especiallye where wheate is scant." Gregory King's estimate shows the importance of this crop at the end of the period in England; it was a major bread grain of the poor in most parts of Europe, northern and eastern Europe especially. It was, throughout this period, the most common bread grain of Russia, eaten by rich and poor alike, albeit that the rich ate bread made from rye bolted to remove all bran. When not eaten as bread, rye could be used for a variety of porridge-like meals and to thicken soups. Russian rye was usually winter-sown (Smith and Christian 1984: 6). Much rye was also eaten in Poland, although there bread was often made of a mixture of grains (Dembinska 1999: 106–7). Rye was often cultivated in western Germany and northern France (Robinsheaux 1998: 114); it was consumed by poor people especially in Normandy and Brittany, where it was also malted and used to make ale (Astill and Langdon 1999: 11). In the 1570s it was sown particularly in "Beauce Solongnois" [Sologne], "notwithstanding that the ground therof for the most part be leane, gravelly and very slenderly husbanded and tilled by the inhabitants" who spent most of their time sheep farming. It was also grown in "Auvergne, Limosin, Perigord, and Farest [Forez]" (Surflet 1606: 683). In Spain and Portugal, rye was limited to those areas cool and wet enough for it to grow, but it may have been more common than was once thought (Ruiz 1998: 66). The urban poor and peasants made do with rye in central Italy and it was grown in alpine regions there (Montanari 1984: 308). In England, two kinds of rye were sown, winter and spring, and it was often sown in small patches (as well, as noted above, as being sown with early wheat). For many it was their major bread grain or was part of a "maslin" (Thirsk 1967a: 169).

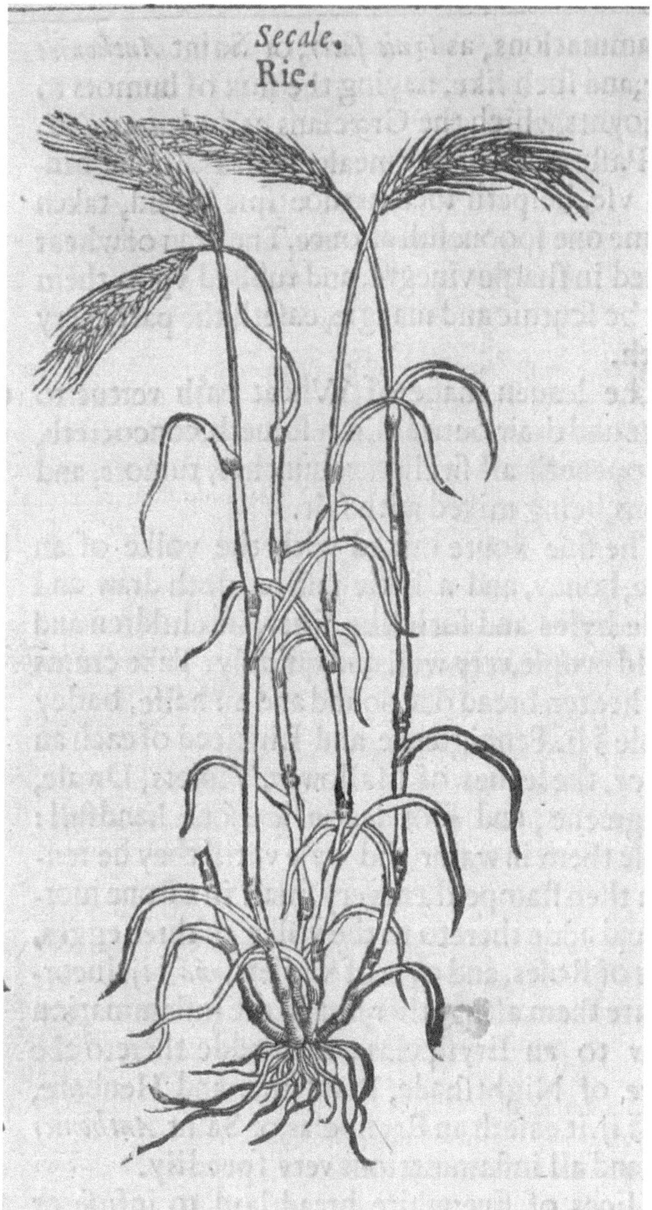

FIGURE 1.2 Rye: a hardy grain crop much eaten in eastern Europe and in other parts of the Continent often mixed with other grains to make "maslin" bread. From John Gerard's *Herball* (1597). Image courtesy of Fisher Library, University of Toronto.

Barley

Barley (*Hordeum vulgare*) has long been a staple crop in Europe, cultivated for human consumption, animal feed, and to process into malt from which ale and beer are made (see Figure 1.3). Noted above are comments made in the late sixteenth century that barley was the bread grain "where wheat is scant" in England. In the century before this

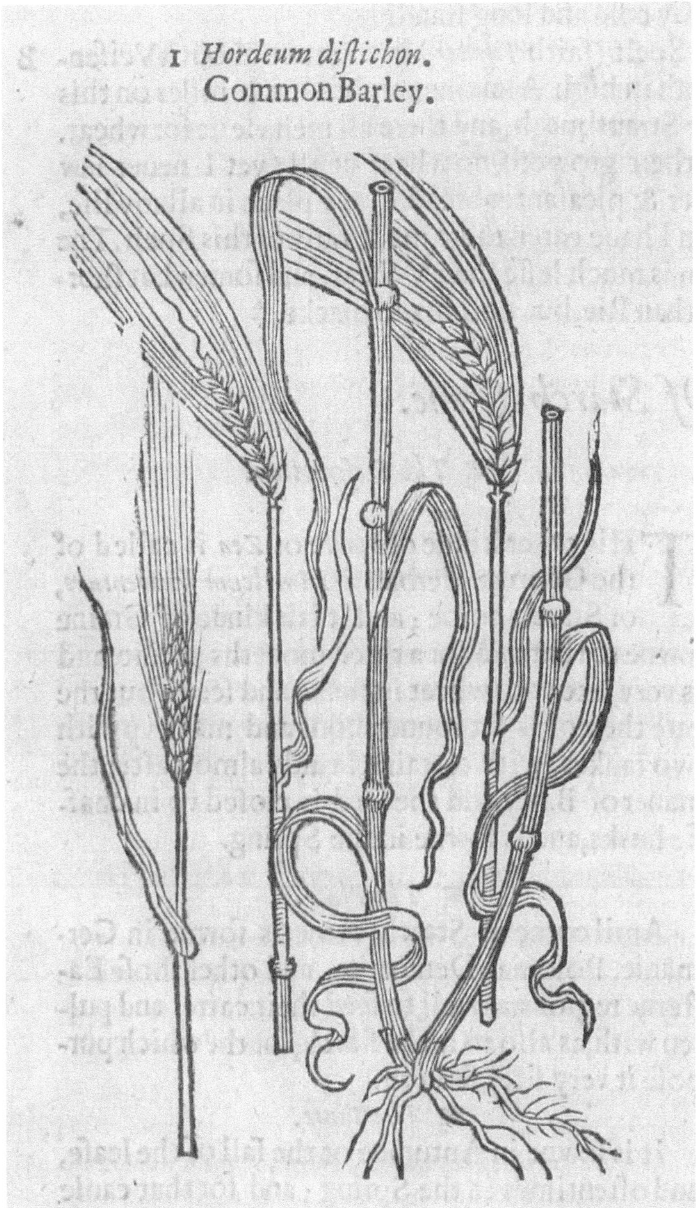

FIGURE 1.3 Barley: a grain widely used by poorer people to make bread (often mixed with other grains) and also, as malt, a major constituent of ale and beer. From John Gerard's *Herball* (1597). Image courtesy of Fisher Library, University of Toronto.

period begins, barley was widely eaten in many parts of southern England although, at harvest time, when labor was at a premium, workers were given wheaten bread. It seems that there was a see-saw between these two grains as the income of ordinary English people waxed and waned. Much more barley than wheat was grown in England; it was more suited to the less fertile common fields. With good reason, then, it was dubbed "the countryman's tillage." The care Thomas Tusser in the 1570s devotes to instructions about sowing barley are an indication of its importance to the English countryman:

> Sow barlie in March, in April and Maie,
> The latter in sand, and the sooner in claie.
> What worser for barlie than wetnes and cold?
> What better to skilfull than time to be bold?
> Who soweth his barlie too soone or in raine,
> Of otes and of thistles shall after complaine:
> I speake not of Maie weed, cockle and such,
> That noieth the Barlie so often and much.
> Let barlie be harrowed, finelie as dust,
> Then workmanly trench it, and fence it ye must.
> This season well plyed, set sowing an end,
> And praise and praie God a good harvest to send.
> Some rowleth their barlie straight after a raine,
> When first it appeareth to leavell it plaine.
> The barley so used the better doth growe,
> And handsome ye make it at harvest to mowe
> (Payne and Herrtage 1878: 99; Tusser 1573)

In sixteenth-century England, two-rowed cultivars were common; of these there were two main sorts, the *long eared* and the *spratt*. The less popular *naked* barley was common in parts of Staffordshire and the worst kind was the six-rowed *bere* or *big* which could yield a harvest, albeit a poor one, on poor dry soils. In the seventeenth century a raith- (early-) ripe type was developed; it could be sown in late May (two months later than usual) and harvested just two months afterwards. This allowed barley to be grown in cooler areas of northern England. Barley made a hard, greyish bread but could be mixed with superior grains such as wheat. It was also widely fed, usually in the form of a boiled mash, to livestock—pigs, capons, and sheep (where grass was scarce) and cattle being fattened for market. In areas suited to barley which lacked good river transport to take it to market, such as Leicestershire, the grain was much used as livestock feed. A version of boiled mash, a barley pottage, was eaten by those people without access to an oven. The use of malted barley to make beer accounts for its high output in the beer-drinking areas of England (Thirsk 1967b: 170–1).

In many areas elsewhere in Europe, barley was of less importance as a bread grain than as a major beer ingredient. Production of it in Poland was dwarfed by the output of rye, and it was significantly less cultivated there than wheat. In Scandinavia barley was the main crop in medieval times but rye dominated in this period. There barley was eaten as a flatbread. It was fairly common in western Germany, where much was used in beer production (Robinsheaux 1998: 114; Teutenberg 1992: 217). In the Castile region of Spain in the sixteenth century barley was grown and, because of its drought resistance, could probably have been more widely sown but it was disparaged as a human food and so was only fed to animals (Vassberg 1984: 200).

Oats

Of the major bread grains, oats (*Avena sativa*) usually yielded the heaviest crop (see Figure 1.4). Best suited to colder and wetter conditions than the other grains, they were the grain of choice in highland areas and those with poor soils, especially in the north

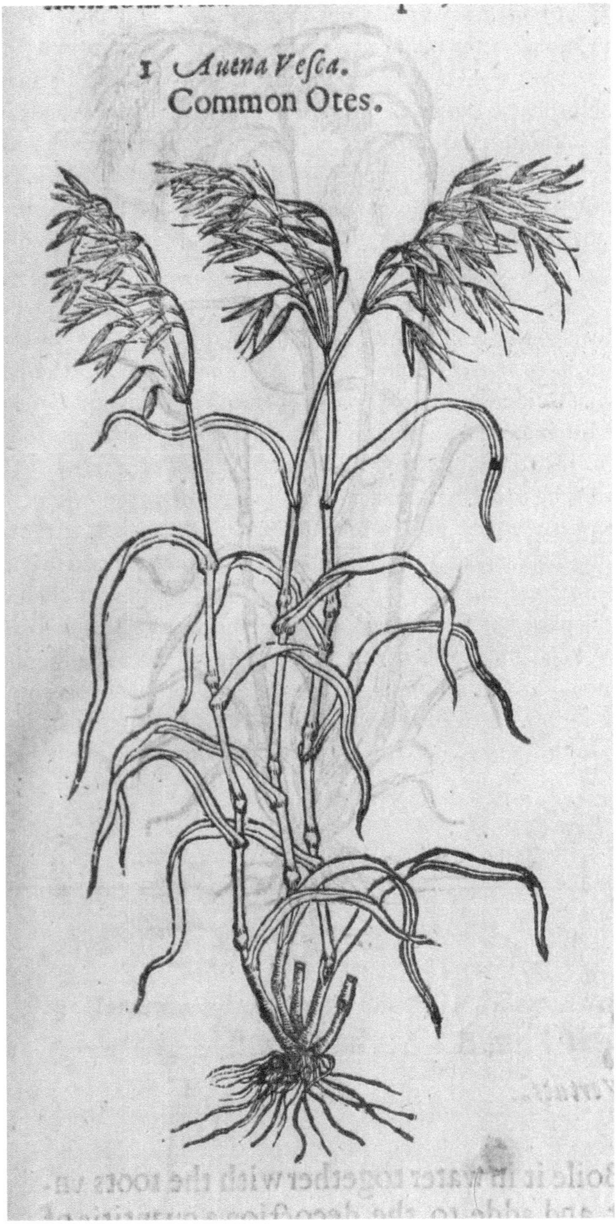

FIGURE 1.4 Oats: a grain widely grown in cooler and wetter parts of Europe, it was important both as a human staple and as animal fodder. From John Gerard's *Herball* (1597). Image courtesy of Fisher Library, University of Toronto.

of Europe. Konrad Heresbach commented: "The Oate ... groweth lyke a good fellowe in every place, where no seed els wyll growe" (Googe 1577: 31). In England, such areas as the Peak District, Yorkshire Moors, Wales, and Dartmoor were cropped with oats, which were eaten by the populace as porridge, gruel, or oatcakes (Thirsk 1967a: 168). The oat was dominant in Scottish cuisine, not only eaten in these forms, but also as a major constituent of boiled meat puddings such as haggis and white pudding (Lockhart 1997: 1–3, 45). Oats were the major grain grown and consumed in Ireland throughout the period (Lucas 1960: 8–43). They were also grown in western Germany and some in Scandinavia (Robinsheaux 1998: 114). Their high energy content made oats a satisfying food for humans (although in southern England in the late sixteenth century they were regarded as famine food) and also a useful grain for livestock, especially horses. In the seventeenth century oats were grown in Hertfordshire in southern England, because they found a ready sale as horse-fodder in nearby London (Thirsk 1967a: 168).

Maize

Although American food-plants eventually had a significant effect on worldwide eating habits, only maize (*Zea mays*) became a staple in this period, spreading to China and Africa as well as becoming a staple in parts of Europe (Kiple and Ornelas 2000: 1:73) (see Figure 1.5). Introduced by Columbus to Spain in 1493/4 (see Introductory Chapter and Chapter 3 of this volume), it was first grown there soon afterwards; by 1520 it was being grown in Portugal and had reached the Veneto in northern Italy by 1530. By the end of the century it was also grown in southern France. In the seventeenth century it had penetrated the Balkans, and in the eighteenth century it was an important crop in Hungary. Maize does not tolerate cold, and is usually sown in Europe in springtime. It is also drought-intolerant, but it is a versatile plant and would eventually become a major staple in these regions, both for humans and animals. Its spread in the Iberian Peninsula was gradual—initially it was grown largely in gardens, probably because many of these were tithe and/or rent free as well as being easier to irrigate (Montanari 1994: 102; Palma and Reis 2014). In this period only in northern Italy did it become a staple, where "the

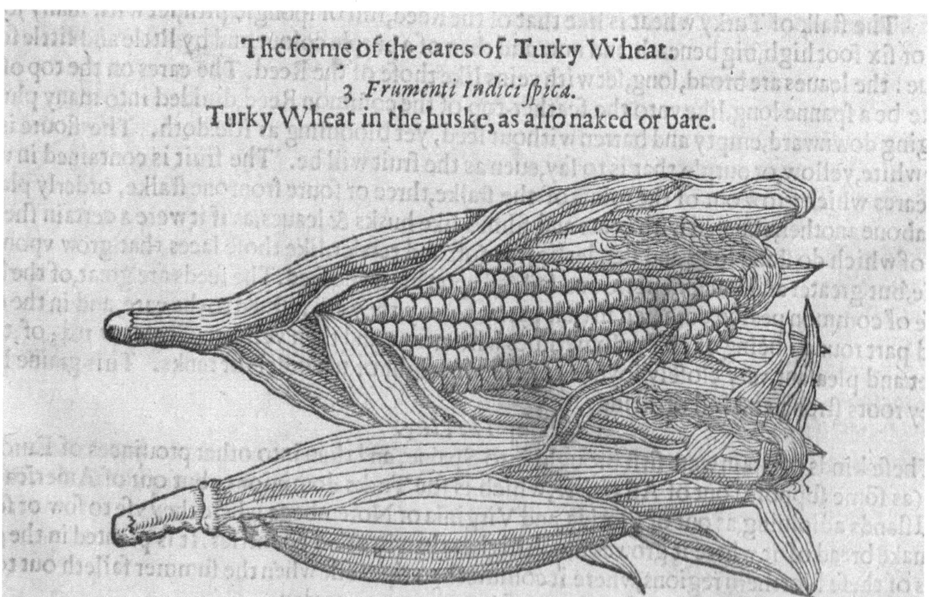

FIGURE 1.5 "Turkey corne" or maize: the only plant introduced from America that became a staple in parts of Europe before 1650. From John Gerard's *Herball* (1597). Image courtesy of Fisher Library, University of Toronto.

impact of maize on rural productivity and consumption patterns ... is comparable to that of the potato in nineteenth-century Northern Europe ... by the 1650s it had replaced lesser cereals in Venetian peasants' cropping patterns and was becoming a staple of the urban and rural poor" (Epstein 1998: 107). It replaced other grains as the raw material for polenta, long the staple food of much of the region (Epstein 1998: 107)

Chestnuts

A minor staple, but an important one in the areas in which they were produced, were chestnuts. These were grown "in mountainous areas that stretched in a belt from Portugal to Turkey" (Kiple and Ornelas 2000: 361). They grew in the cold uplands of northern Italy where, in plantations they were called "bread trees," and also grew in the mountains of Sicily (Montanari 1984: 312). They were a staple crop in Cantal and Périgord and the hills of the Cevennes in France. The trees thrived in very poor soil and a rocky terrain, with scorching heat in August and a wet September. The woods were carefully managed and, in France, the nuts were dried in special oasthouses (hop-drying kiln) over a slow fire, although in Périgord they were dried in sand instead (Stacy 1991: 268–72).

In the 1616 English edition of Estienne and Liébault's work, first published in Paris in 1570, it was noted: "Especially in Perigord, the greatest parts of the forests are of chestnut trees, and an infinite number of people live not of any other thing but of this fruit, eating it sometimes boyled, sometimes roasted, sometimes made into bread, sometimes into broth with milke, sometimes in meale baked after another sort" (Surflet 1616: 390–1). In Italy chestnuts were ground and used to make polenta or breadcakes (Montanari 1984: 312). Outside the areas where they were a regular staple they were sometimes offered as famine food, with mixed success—the poor of Naples refused flour made from chestnuts in 1585 (Montanari 1994: 106).

Millet and Others

Millet (*Setaria italica*, and, in some places, *Panicum miliaceum*) as a bread grain was a "disappearing staple" in this period. In medieval northern Italy it was quite common but by the sixteenth century was out of favor (Epstein 1998: 73). It was similar in the Low Countries, although in Poland "porridge made with finely ground millet or millet grits was the staff of life for ... princes and commoners alike. Millet was also one of the distinctive features of medieval Polish cookery that differentiated it from cookery in Western Europe"; until potatoes replaced millet in the Polish diet in the late seventeenth century, "millet kasha was one of Poland's defining regional foods" (Dembinska 1999: 103). In the 1570s it was reported that "the Russians and Moscovians are chiefly nourished with this kind of pottage" (Googe 1577: f. 31v) and at about the same time it was common in "Bearne, Bigorre, and Armignac" in France, having the advantage that it "growth not onely in gravellie ground, but also in sand, when the countrie is wet and moist" (Surflet 1606: 691). Millet was also common in Romania until, in the seventeenth century, it was supplanted by maize as a staple crop (Kiple and Ornelas 2000: 103).There were other staple grain crops in Europe which were locally important, such as spelt (*Triticum aestivum* Spelta Group) and buckwheat—the latter made into a form of porridge much eaten in Russia and Poland (see Figures 1.6 and 1.7). Its main advantage was that "it may be sowed in any ground how bade so ever it be." Heresbach in the 1570s wrote that "this grayne hath not long since ben brought from Russia and the northerne partes into Germanie: now is it become common, and used for fatting of hogges, and serveth the common people in deare seasons to make bread and drinke withal" (Googe 1577: 31).

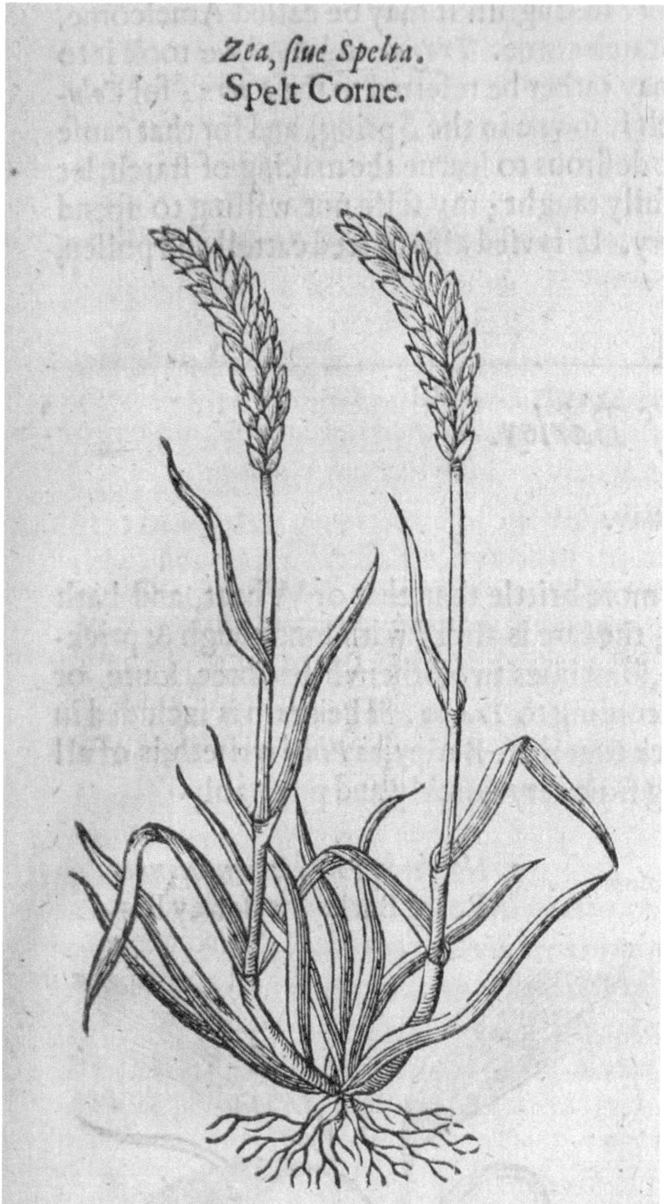

FIGURE 1.6 Spelt: a grain commonly grown in parts of medieval Germany, it was hardier than many grains and therefore a safer crop to grow. From John Gerard's *Herball* (1597). Image courtesy of Fisher Library, University of Toronto.

Garden Produce

Garden vegetables, taken as a whole, are considered here to constitute a staple crop in Early Modern Europe. The extent of their production is very difficult to establish, resulting in an underestimate of their importance in the past. This has been emphasized by several writers. Where there was commercial production, as in the Rhineland in

FIGURE 1.7 Buckwheat: a staple in parts of eastern Europe and also fed to animals. It will grow on very poor soil. From John Gerard's *Herball* (1597). Image courtesy of Fisher Library, University of Toronto.

Germany, in parts of the Low Countries, in the suburbs of London and in specialist areas of southern England such as Sandwich in Kent or near Colchester in Essex, there are records of such activity (Robinsheaux 1998: 114; Thick 1985: 503–32). But, time and time again, historians write of their frustration at not knowing the extent of production from *private* gardens. Of Polish estate records there is the comment: "The common feature of these accounts was, however, their limited content, that they generally did not list minor garden products such as vegetables, fruit and the like, these products were not entered since they normally were not sold" (Teutenberg 1992: 215–16). One historian of Polish food maintains that garden vegetables were an established part of the diet in the sixteenth century (Dembinska 1999: 120–1). Another writer commented: "The gardens of medieval Northern France are still not well understood, but clearly grew in number from the thirteenth century onwards, especially in urban settings. They were seemingly a place where intensive agriculture was performed, often with irrigation. They were also generally given first priority for manure and other fertilizers" (Astill and Langdon 1999: 19). Further south, in northern Castile, peasants grew vegetables and legumes in their gardens to supplement their diet (Ruiz 1998: 66).

It is known that, especially in southern Europe, vegetables were *consumed* in this period because it was commented upon. The Italian naturalist, Costanzo Felici (1525–85), wrote that for "this food salad ... Italians are truly greedy (or so say those from beyond the Alps), and they have snatched away sustenance from the animals who eat raw greens" (Arbizzoni 1986; Montanari 1994: 113). Early in the following century the Italian protestant exile Giacomo Castelvetro (1546–1616) wrote a treatise, *A Brief Account of the Fruit, Hearbs and Vegetables of Italy,* to encourage his English hosts to consume more garden produce (Grigson 1989: 28–44). That the English were utilizing their gardens was commented upon later in the century by English agriculturist, John Worlidge (1640–1700): "There is scarce a Cottage in most of the southern parts of England but hath its proportionable Garden, so great a delight do most of men take in it." Not only did they grow flowers "but furnish themselves and their neighbours upon extraordinary occasions; as nuptials, feasts, and funerals, with the proper products of their gardens. It furnishes our kitchen and tables with various esculents, as well satisfying nature, as pleasing our appetites." Some vegetables were common: of cabbages (*Brassica oleracea*) he said that in England there is "not a village without them"; "the constant and vulgar use of them is manifest" (Worlidge 1700: 5–6, 158, 168, 175, 183).

The English botanist, John Parkinson (1567–1650), noted that "our ordinary Beanes [*Vicia faba*] ... [serve] for foode for the poorer sort for the most part" whereas kidney beans (*Phaseolus vulgaris*), an introduction from America, "are esteemed more savory meat to many mens pallates" (1629: 521). From this and other references Thirsk (2007: 171–7) detected an increased consumption of peas and beans in England in this period and there is evidence of a similar shift in diet in Germany. The beans consumed included newly imported South American cultivars, which were rapidly disseminated to Europe, Africa, and Asia.

Olives

By 1450 olives (*Olea europaea*) had long been cultivated in coastal areas around the Mediterranean and here they can be considered a staple crop. Most were crushed to produce oil. The technology of pressing had been developed long before and there were no major developments during this period. Olive oil was the staple cooking medium for frying, roasting, and baking in these areas, and was also used to dress salads of all kinds. The Venetians used olive oil to make soap, and it was also used as lamp oil to provide light. Stimulated by trade in lamp oil and soap, plus demand for the oil in clothmaking (see below), Italian olive-oil production rose steadily after the Black Death. All classes used the oil in their food but, as with bread and wine, much was of low quality. Good-quality oil was beyond the pocket of the poor (Epstein 1998: 76). Roman techniques of production had been lost and they were not rediscovered until the sixteenth century (Lanza 2011: 12–28, 62). In Languedoc, too, increasing numbers of olive trees were planted throughout the sixteenth century, often with wheat sown between the trees (Le Roy Ladurie 1976: 57). Southern Italy exported olive oil to northern Europe, not only as a condiment for the rich but also as an important raw material in cloth production. After washing and scouring cloth, it was oiled to make it supple and fine quality cloth was oiled with olive oil (Sakellariou 2012: 239).

ALCOHOLIC BEVERAGES

Here this chapter is concerned mainly with drinks processed from grain (principally barley), hops, grapes, and apples. Alcoholic beverages were consumed all over Europe (apart from Ottoman-controlled areas) in this period—for pleasure, medicinally, and in religious ritual. They were safer to drink than much of the water Europeans had access to, having been boiled in the course of production, while the alcohol in them is antimicrobial. It is difficult to estimate exactly how much beer or wine ordinary people consumed at this time—most statistics come from institutions—hospital inmates' daily rations, workers for government-controlled bodies like the Arsenale in Venice, construction workers, and so forth. The actual figures are less important than the fact, confirmed by these statistics, that workers were consuming beer, ale, or wine every working day—alcohol was a part of their staple diet (Phillips 2014: 85–95). For those who did not receive a daily ration of alcohol there were ample opportunities to buy it; in 1577 there was an alehouse in England for every 142 inhabitants (Phillips 2014: 106) and many households brewed their own beer. A study of the diet of peasants in the Bureba district in northern Spain found that wine consumption was between one and eight cups (two liters) a day (Ruiz 1998: 67). Moralists complained of drunkenness (Surflet 1606: 771–2) and Unger (2004: 126) noted that "in 1995 Belgians, among the most avid beer drinkers in the world, consumed on average nearly 30 gallons (102 liters) per person per year, less than half the amount of urban drinkers in the late Middle Ages or Renaissance."

In considering these drinks, Europe may be split into two regions—northern Europe, where grain-based drinks were consumed by the general population and southern Europe, where grape-based drinks predominated, the dividing line being set by the northern limit of grape-vine growing. This picture was modified by a growing trade in quality wines from the producing areas to northern ports. As a result, the richer sections of society in the north largely drank wine by the end of our period, whereas their servants and the general population drank beer or ale. In some parts of Europe, the grain-based fermented liquor was further refined by distillation into a spirit; *usquabach* or whisky in Ireland and Scotland, vodka in Russia and Poland. Some small areas, such as Normandy in France, or the West Country in England, used apples or pears as the raw material for their drink, producing cider and perry.

The division of Europe between grain- and grape-fermented drinks was noted by the French authors Estienne and Liébault, writing in the mid-sixteenth century. They explained that man, having rejected water as "tastlesse and unsavorie," has,

> ... where the heat of the sunne might bring forth and leade along the grape unto his full ripeness, chosen wine for the most excellent and delightsome drinke of all others' whereas "in other colde countries ... they have either continued drinking of water, or fetched and procured wine from other places, or else have prepared some other kinde

> of drinke .... of beere and ale; some of cyder and perrie; and others of all sorts" .... All France, Italy, Sicilie, Spaine, and all other countries which are farre from the north, do content themselves with wine ... England, Scotland Dalmatia, Polonia, Samatia, and all other northern countries do use partly wine, as procuring the same from other places, and partly beere.
>
> (Surflet 1606: 771–2)

Wine

In France, plagues and wars from 1300 to 1600 led to a decrease in the demand for wine. Lords leased their demesne vineyards to peasants, so that by 1600 there were many small independent producers supplying largely poor-quality, short-lived, local wine for the general population. However, in the sixteenth century higher-quality wines emerged, such as those produced in the Côte d'Or in Burgundy. Concluding their section on wine, Estienne and Liébault included ten pages of detailed description and evaluation of the regional wines of France (Surflet 1606: 795–805). Clearly, by the mid-sixteenth century, wine was not only produced all over France, but producers made wine with distinctive characteristics. There was obviously an internal market for wine in France as well as an export market through seaports such as Bordeaux.

Wine was the drink of choice of all classes in Mediterranean countries in this period, but as with so many other staples, the quality of that drunk by the rich was far superior to that of the poor. The planting of vineyards in Spain was stimulated in the sixteenth century by a robust domestic and overseas demand. Spanish wine was exported principally to England and The Low Countries while at home there was a rising demand from towns and the court which demand stimulated good-quality, low-yielding vineyards on the high plateau south of Valladolid (Unwin 1991: 227). Wine consumption in Bureba, Spain, was between one and two liters of wine per person per day (Ruiz 1998: 67). German vineyards were to be found as far north as Stettin (today's Szczecin, Poland) and Königsberg (now Kalingrad, Russia) and some close to Berlin were producing wine throughout the period. The "Little Ice Age" of the sixteenth and seventeenth centuries, however, hit production of northern vineyards; many in Flanders went out of production in the face of competition from those in warmer southern regions (Unwin 1991: 221–34).

Estienne and Liébault rightly recognized the flow of wine from southern ports to the north. As a result, English consumers were well acquainted with wines from the south and developed a taste for some wines in particular. English poet and writer, Gervase Markham (1568–1637) included a chapter on the management of wine in later editions of his *The English Huswife*. In it he covered: the heavy sweet wines of the eastern Mediterranean—Malmsey and "Muskadine"; "Bastard" from Madeira; Sack from Jerez (i.e. sherry) in Spain; sweet Malaga; Bordeaux wines, especially claret, from France; Rhenish wines from southern Germany and Alsace; and other Portuguese, Spanish, and Italian wines (Markham 1623: 143–53; 1656: 112–22). English consumers who could afford wine had many to choose from; William Harrison knew of eighty-six different wines available in London in the 1570s (Harrison 1587: 167; Phillips 2014: 96).

In the fifteenth century Bruges (now Belgium) was an entrepot for imports of French wine to northern Europe, as were Köln and Antwerp, but from the sixteenth century onward this market demanded stronger, longer-lasting wines and so increasing amounts were supplied from Spain and other Mediterranean countries. Much came to England directly from the ports near wine-producing areas.

Beer, Hops, and Brewing

The period from about 1450 to the early seventeenth century has been characterized as a "golden age" for brewing, which increased in tandem with increases in population growth; indeed, it has been remarked, that "[i]n some places in northern Europe, it grew faster than the population. It enjoyed unprecedented economic success. Beer invaded new parts of Europe, claiming new territory where wine was the preferred drink" (Unger 2004: 106).

Beer is made using hops (*Humulus lupulus*) to give a bitter flavor and also to keep it in good condition. Before hops were widely used, beer was commonly called "ale" and was flavored with a variety of herbs. Hops were used on the Continent much earlier than in England: cultivation probably originated in Bavaria and Bohemia and by 1300 they were widely grown in northern Europe. Large estates in the Low Countries grew hops in some quantities from the fifteenth century onwards (Unger 2004: 125; Verhulst 1990: 123). Tradition has it that in the year 1400 Dutch merchants at Winchelsea in Sussex, England, tired of sweet English ale, first imported hopped beer into England for their own use. But, given the amount of trade between England and northern Europe, it is highly likely that both beer and hops were imported before this: there is archaeological evidence for hop imports at an earlier date and some historians have detected a small hop-growing industry in England from the mid-fourteenth century. Certainly, Flemish hop-growers were planting hop-gardens in England in the 1520s (Hornsey 2003: 303, 306, 308–9).

Two centuries later, in the 1570s William Harrison explained:

> Of late years we have found and taken up a great trade in planting hops, whereof our moory, hitherto unprofitable grounds do yield such plenty and increase that there are few farmers or occupiers in the country which have not gardens and hops growing of their own, and those far better than do come from Flanders unto us.
>
> (Harrison quoted in Hornsey 2003: 324–5)

In the same decade a well-produced pamphlet on hop-growing was published in England and quickly went through several editions: in his preface the author complained that Englishmen still imported hops from Flanders when they should have been grown at home (Scot 1578: Bii). By 1700 there were an estimated 20,000 acres in Britain under hops (Hornsey 2003: 325–6).

As in Europe, in England, outside London and some other large towns, beer- and ale-brewing was, throughout the period, a domestic activity—in most households it was women's work. Brewing in London was in transition: much domestic activity continued but the ever-growing demand for ale and beer led to the development of commercial brewing (Phillips 2014: 101). The city company which regulated commercial brewing in London dates from at least the thirteenth century. Enterprising men could, and did, make their fortunes from brewing. Richard Plat (1528–1600), father of the Elizabethan scientist Sir Hugh Plat (1552–1608), came to London as a brewer's apprentice and died in 1600 a rich man, bequeathing his son a coat of arms, the lease of a brewery, and substantial property holdings. But at the same time alewives continued to brew for their household and to sell the surplus to thirsty Londoners (Thick 2010: 11–14).

In much the same areas where vodka was produced (see "spirits," below), *kvas* was drunk. Made mostly from fermented stale bread, this drink was only slightly alcoholic and was the equivalent of "small beer" produced by many domestic brewers—refreshing and safer than unpurified water, but not strong enough to interfere with a day's work (Kiple and Ornelas 2000: 2:1240).

Cider

As noted above, in some areas where weather was too cool or unreliable for vines, apple- and pear-growing was common, and cider and perry were the alcoholic drinks of choice. They were the staple drinks of the French provinces of "Normandie and Britaine" and the counties of "Mans, Chartraine, and Touraine." The best French cider was made from specially bred cider apples (Estienne and Liebault listed twelve named cultivars in the 1570s) "growen and bred in orchards carefully and dilligently dressed, kept, husbanded and ordered" (Surflet 1606: 527). In 1655 an English author identified "Biscay in Spain" (a province of the Basque country in Northern Spain) as an area "where wine is scarce, they make Cider of a certain sweet apple … and the Cider is very Good" (Hartlib 1655: 19).

In England, cider was made in the English West Country—Worcestershire, Herefordshire, Gloucestershire, and Devon and also in Kent. The English Civil War (1642–51) saw an increase in the output of West Country cider because the opposing armies ranged over the area and soldiers developed a taste for the drink (Thirsk 2007: 103). As in Normandy, by the start of our period, apple cultivars had been bred specially for cider production.

Spirits

In the coldest parts of northern Europe—now the countries Russia, Ukraine, Belorussia, Poland, Scandinavia, Scotland, and Ireland, the staple alcoholic drink was a strong spirit obtained by distilling the liquor obtained by fermentation. In all cases, the origins of distillation appear to have been the work of local monks in their monasteries. In Russia and neighboring Slavonic countries this spirit was known as vodka. To obtain this drink, fermented liquor was distilled one or more times, increasing its alcoholic content. Redistilling also removed impurities which had an unpleasant taste—sometimes flavorings were added to do this. A variety of vegetables were used as the raw material for vodka, as well as a range of grains, cabbage, fruits, and, when introduced from America, potatoes. As well as an intoxicant, vodka was widely believed to have medicinal properties (Herlihy 2012: 12–19, 20–4). Aquavit, the staple spirit of Scandinavia, was also credited with curative powers. Distilled since the fifteenth century, it was flavored with a variety of herbs and spices.

Scottish and Irish spirits were made from malted barley. Production of whisky in Scotland was well established by the fifteenth century, as was production of Irish whiskey. Both were sometimes flavored with herbs or spices.

FAMINE

While it is difficult to class times in this period as "normal," phases of dearth and famine can clearly be distinguished. Such times could be local or regional—given limitations of transport and trade sufficient imports from unaffected areas might not alleviate a local problem—or they might be widespread, caused by, for example, bad weather over a wide area. Sometimes a variety of factors conspired together to produce a "general crisis" such as that of the 1590s in Europe. In that decade a series of wet winters and summers devastated grain harvests in many parts of Europe: England, Ireland, Scotland, Scandinavia, and Italy were the main affected areas. These problems were compounded by devastating warfare in northern France, after-effects of war in Russia, financial collapse in Spain, and epidemics of disease in France, Italy, and England. All these factors combined to produce conditions which were at best dearth and at worse famine with widespread mortality. The 1590s are just one example—western Europe in particular was afflicted by sporadic warfare throughout the period—causing disruption of food-trading, destruction of crops, and in some cases abandonment of agriculture altogether (Clark 1985: 4–12).

In such circumstances of dearth (and famine in some cases), staple foods came to be even more relied on, at the expense of more desirable ones. In more extreme cases, where the staple crop failed, people turned to new staples for the duration of the famine, elevating a minor staple into a major one, or resorting to very unpalatable alternatives (O Grada 2009: 73). Charles Estienne and Jean Liébault vividly described the bread consumed in France "in such times of hard distress, bread of oats, barly, beanes, rice, millet, and panicke [*Panicum miliaceum* or proso millet] ... yea of bran, fishes dried in the sun, acornes, chestnuts, and ferne rootes" (Surflet 1606: 705). Sometimes the alternative foods became a part of the diet when the famine was over. This seems to have happened in London and maybe other parts of southern England after the dearth of the 1590s. In the middle years of this decade the wet weather caused the grain harvest to fail three years consecutively and all grain prices rose dramatically. One response was the shipping of many tons of root-crops—carrots, parsnips, and turnips—from East Anglian and Kent ports to London: much of this produce came from the commercial gardens of the Low Countries managed by French protestant refugees who had settled in those areas. The poor consumed these roots, probably initially with reluctance. But in the 1620s the London botanist and apothecary John Parkinson (1629: 509) observed that they now had a place on London dining tables:

> Being boyled in salt broth, they all of them eat most kindly, and by reason of their sweeteness are much esteemed, and often seene as a dish at good mens tables: but the greater quantitie of them are spent at poore mens feasts.

Parkinson's comment that turnips "engender moist and loose flesh," a symptom of a diet largely based on root-crops, indicates that some people were still largely dependent on them, and half a century later foreign visitors still remarked on the quantities of boiled roots seen on London dining tables (Misson 1698, quoted in Earle 1989: 273).

ANIMAL FODDER

In many parts of Europe, throughout the period, farm animals were little more than scavengers—they grazed on weeds and grass on roadsides and common lands (see Figure 1.8). In some areas, such as Russia, there was ample common land, while in more heavily populated areas, the Low Countries for example, or southern England, such land was becoming over-grazed. When communal arable fields were fallowed, animals were put on to this land to eat the weeds. Near large towns some closes were set aside for growing hay to feed the many draught animals used there. Animals were also fed on the less-regarded staple grain crops—notably oats and barley but also buckwheat, millet, peas, and beans, although availability depended on competing human demand. The generally limited amount of animal feed restricted livestock numbers. As soil fertility was maintained principally by animal dung, this in turn restricted arable output. Attempts to improve animal feeding over the period were essentially the search for a way out of this "vicious circle" (Lane 1980: 179–95). Some areas of Europe had made considerable strides in animal husbandry by 1650. The Low Countries had by then a long history of soil improvement affecting both pasture and arable yields: labor-intensive techniques

FIGURE 1.8 *The Great Turf*, 1603. Most pasture throughout the Early Modern period was not sown with the seeds of specific fodder plants, and this picture by Albrecht Dürer emphasizes the mixture of wild grasses and other plants on which animals grazed. From the Albertina Museum, Vienna. Photo by GraphicaArtis/Getty Images.

such as marling, and fallow cultivation with legumes and rape. Clovers were sown, often as *leys* (temporary pastures), improving soil fertility for subsequent arable crops as well as providing more cattle feed. Roots—turnips and carrots—although still predominantly for human consumption were sometimes fed to animals (Astill and Langdon 1999: 69–80).

To elaborate on the summary above, in England—like most of Europe—cattle and other farm animals were left to pasture where they could: no provision was made for their sustenance and little effort was made to store any feed for the winter months until the beginning of the seventeenth century. As earlier intimated, various types of land were grazed by livestock: common wasteland on the outskirts of villages and parishes; stubble after the harvest; weeds growing on fallows; woodlands where animals could browse; wet meadows near streams and rivers; and dry pastures, many of them common land. The meadows and pastures commonly were also mown for a hay crop. In the upland areas of northern Britain this persisted but it did not in the southern half of the country. There, new techniques of land management and deliberate sowing of new or better types of herbage greatly improved the provision of animal fodder. Various types of clovers were introduced, mostly from France and the Low Countries (Hartlib 1650; Lane 1980: 179–95).

The change can be charted via husbandry instruction books. In the sixteenth century, John Fitzherbert, Leonard Mascall (d. 1589), and Barnabe Googe made no mention of improving grazing land: their main emphasis was on veterinary skills. John Norden, a professional surveyor, was the first English author to advocate improving pasture land, in 1610. In the same year, Rowland Vaughan (1559–1629) published a book about his experiments with water meadows in Herefordshire, damming chalk streams and drowning pastures in early spring to encourage the sward to grow. Early experiments in pasture improvement were made by other gentlemen who, by the mid-seventeenth century, were recording their results in print. Sir Richard Weston (1591–1652), who had seen the new methods used in Brabant and Flanders in the 1640s, came home to Surrey and reported on his own success with clover. In the 1650s many other reports of the use of such plants were published by Samuel Hartlib (*c.* 1600–62).

The new crops were predominantly legumes which, as well as producing more calories per acre of feed, fixed nitrogen via nodules on their roots and enriched the land for subsequent crops. They were increasingly sown as *leys*. When these were plowed up again, the enriched soil produced much better arable crops. The growing amount of enclosed land made experimenting with new fodder crops and the control of leys much easier. If husbandmen did not go as far as acquiring clover seeds, more did at least encourage their swards by sowing "hay dust," the ripened seeds which fell out of their haystacks. The net result of these changes was an increase in the numbers and quality of livestock and, eventually, an increase in arable yields, both from nitrogen enrichment and the dung shed by the animals (Lane 1980: 179–95).

Establishing precisely what new fodder crops were sown in these improved pastures is difficult—many names were quoted by agricultural writers at the time; various names were given for different kinds of "clover" and "the four important fodder and grazing genera (*Medicago, Melilotus, Onobrychis*, and *Trifolium*) contain over thirty-one cultivated species"; as well as clover, certain types were known as Lucerne (usually *Medicago sativa*) or St. Foine (sainfoin, *Onobrychis viciifolia)* (Lane 1980: 179–95).

On a much more modest scale, some few enterprising husbandmen were sowing turnips and carrots as fodder crops. Pioneers in this enterprise were Protestant refugees from the Low Counties and northern France who had settled in some parts of southern England, bringing their agriculture with them (Hartlib 1655: 8–9). Near London, specialist growers made a living from cultivating grass fields specifically to mow, carefully manuring them and sending the grass and hay to the capital to sell at hay markets. London's transport depended on the large number of horses employed in carrying all sorts of cargo and humans into and about the city. Elsewhere, by the end of our period, some farmers were spreading marl, dung, and other fertilizers on their dry pastures (Thirsk 1967a: 397).

The driving force behind the improvement of pasture was a sustained increase in population from early in the sixteenth century until the mid-seventeenth. Land was increasingly at a premium, rents rose, and land had to be more carefully managed. Wastes were nibbled away by enclosures and stints (limitations) had to be imposed in many areas to control animal numbers on common land. Even so, by the end of this period most farmers had probably not done much to improve their grazing land, so the introduction of these new crops is a staple which reached its full potential in the next period.

CONCLUSION

Underlying agricultural advances and the increased production of staple crops, both for human and animal consumption, was the growth of demand expressed in the development of markets. It was the market which induced farmers to produce a surplus for sale. The exact way in which the market mechanism worked varied across Europe. What might be called the "English model"—a trend towards leases and other secure tenures, held by farmers who paid market rents, had relatively large holdings, employed wage laborers, and produced for local or national markets—contrasts with the "eastern European model." There serfs lived a subsistence existence cultivating small family plots and rendered ever more onerous labor services to their lords, who owned large demesne farms which produced surpluses, most notably grain, for sale to merchants who traded with western and southern Europe.

Markets existed in various concrete and abstract forms. Local markets, usually held weekly, were numerous in western Europe—there were about 760 market towns in England by the sixteenth century and about fifty in Wales (Thirsk 1967a: 467). It was explained above how large towns and cities sucked in staple foods from their hinterlands—there many wealthy people lived; the general population had higher wages and more disposable income than in the countryside, and city authorities were keen to ensure a steady flow of food to their citizens. Sixteenth-century Barcelona drew into its markets agricultural produce from a wide area. Markets in the industrial towns of the coastal Low Countries had affected agriculture on several levels by the seventeenth century. Horticulture on lands nearby fed a demand for garden produce and fruit. As noted above, industrial crops and hops supplied towns with raw materials. These demands caused agriculture to become more efficient and, as parts of Europe were no longer self-sufficient in grain, these same demands created the biggest long-distance trade in Europe in staple foods—the Baltic grain trade. Other international trades in wine and oil from southern to northern Europe, as well as a trade in beer and other foodstuffs across the North Sea, were well-established. The influence of city markets was summed up by a historian of sixteenth-century Spain, "the prosperity of rural areas ... was linked to the economic well-being of the cities that served as markets for their surpluses" (Scott 2015: 403; Vassberg 1984: 188–9). In England, city retail markets spread demand deep into the countryside and induced close to built-up areas novel types of agriculture such as market gardening, animal fodder, fattening cattle and sheep. Some markets became regional centers for particular foodstuffs—there were malt markets, cattle markets, poultry markets, corn markets, hop markets, and more. London, a vast city by European standards, exerted a regional or even national influence. Geese and turkeys, fed on grain, were driven from East Anglia to London, cheese came from Cheshire, grain and malt was sent down the Thames from as far as Oxford and Abingdon, market-garden produce came from Sandwich in Kent and Colchester in Essex by sea, and cattle were driven to London from Wales and Scotland, fattened in the Home Counties before sale and slaughter in London.

Market mechanisms became more sophisticated as trade routes lengthened. Despite official disapproval, wholesale markets developed with middlemen buying from producers and selling to other wholesalers. Crops were bought whilst still in the ground or stored pending expected price rises. But, despite this growth of the market, many peasant farmers still grew staple crops largely for their own families. The "agricultural revolutions," which enabled output to outstrip population increases and the countryside to shed labor to work in factories and services, were still in the future at the end of this period (Thirsk 1967a: 403).

CHAPTER TWO

Plants as Luxury Foods

"And they germinated very well"

ANDREW DALBY

This is a story of discovery, rediscovery, and transplanting, as a result of which the range of plant foods in most parts of the world in 1650 was to be greater, more varied, more versatile than it had been in 1400. It is told, in large part, by the discoverers themselves, who in those exciting decades of the first flourishing of the printing-press were able to record their observations and exchange knowledge more rapidly and to greater effect than ever before. The theme, in this chapter at least, overshadows other contemporary developments, such as the spread of new ideas in gardening and horticulture and the breeding of new plant cultivars. On both of these topics much is said in other chapters.

Discovery and rediscovery were not infrequently an end in themselves. Pierre Belon (1517–64), a French apothecary turned natural historian, traveled to Constantinople and beyond in search of plants about which he had read in classical sources. His rediscovery of the brightly colored, oil-rich terebinth nut, fruit of *Pistacia atlantica*, known to him previously from Greek and Arabic texts, could give him little more than academic pleasure:

> In the next village I came across an Arab peasant leading a camel laden with terebinth seeds. The nearby mountains are covered with these trees, whose resin they take to Damascus to be sold … I have the evidence of reliable authors that over two thousand years ago people used to eat terebinth seeds, and that the Persians lived on them before they knew of bread. The seed is of an exquisite blue colour surpassing any other. Hence all the old Arabic authors call it *granum viride*, because the shade is between green and sky-blue.[1]
>
> (Belon 1553: f. 153r)

On the other hand, Belon's careful identifications of Cretan wild herbs were eventually not irrelevant to food and health. "In May they likewise pick the flowers of the thorny caper and take them to market, simply boiled and lightly salted," he noted incidentally of the buds of *Capparis spinosa* (1553: f. 18r). Another local wild herb, *Petromarula pinnata*, was soon afterwards reported to the scientific world by Onorio Belli (d. 1604), a Venetian physician working in Crete. "Eating the root, and likewise the young and tender leaves, is popularly thought to stimulate libido," he noted after giving a botanical description (Belli in Clusius 1601: ccxcix). As Belon knew, Crete had been the source of the finest dried medicinal herbs under the Roman Empire: in the late second century, the Roman physician Galen had written of this trade in some detail. Cretan wild greens have continued to the present day to be marketed on a small scale as far afield as fresh herbs will travel. Far beyond that, they are now world renowned as an indispensable part of the macrobiotic Cretan diet.

Discoveries and rediscoveries might lead directly to the expansion of trade, as has been the case with two Chinese spices of which Carolus Clusius (1526–1609) made early mention. One is *Illicium verum*, star anise (see Chapter 5 of this volume); the other is *Zanthoxylum* sp. (likely *Z. bungei*), Sichuan pepper or "fagara Avicennae" as Clusius himself called it. He gave what is probably the oldest illustration of Sichuan pepper (Clusius 1605: 185).[2] These spices were both to become part of the medicinal drug trade and eventually the worldwide trade in Chinese food products.

The case is similar with the coconut (*Cocos nucifera*) (perhaps not allowed by all to be a luxury food, but coconut milk, coconut cream, and the dusting of grated coconut on European cakes excuse its mention in this chapter). The coconut had been reported to Europeans as long ago as the sixth century in the form of text and illustration in the *Christian Topography* of Cosmas Indicopleustes, but had been completely forgotten: "The ancient Greeks wrote nothing about it that I have seen," Portuguese physician and herbalist Garcia de Orta (1501?–68) admitted, writing of his investigations in Portuguese India, "and the Arabs very little" (1563: ch. 16).[3] Recently it had been fully if incoherently described to Europeans by Ludovico di Varthema (*c.* 1470–1517), who was in southern India in the first years of the sixteenth century (1510: f. 24v). Medicinal and culinary detail was added by Orta himself: "They pound the flesh to make a sort of milk, and cook rice with it, which is like rice boiled in goat's milk" (Orta 1563: ch. 16). Coconut was soon to be traded as a medicinal drug, later also as a food.

As examples given later will show, the discoveries and rediscoveries of the Early Modern period also led to the spread of major food plants across the world as introduced crops, an aspect of what is now called the "Columbian exchange" (Andrews 1992;

*Cubebarum facultates in memoriam mihi revocârunt Fagaræ historiam apud Avicennam, quæ ferè iisdem cum Cubebis viribus prædita est. Cùm igitur commodiorem in hac Epitome locum non haberem, cui eius iconem & historiam insererem; operæpretium me facturum existimavi, si eam hîc darem.

FAGARA AVICENNÆ.

EST itaque Fagara fructus Ciceris arietini magnitudine, cortice tenui constans ex cinereo nigricante, cui subest tenue putamen nucleum continens satis solidum, & tenui nigráque membranâ obductum. Integer fructus cum magnitudine tum formâ & colore adeò similis est ei quem nostræ officinæ Cuculum Indum, Itali Coccole di levante appellant, ut primo aspectu fallere, & pro eo sumi possit.

MEMINIT huius Avicenna cap. 266. in hunc modum. Fagara quid est? granum simile Ciceri habens granum Mahaleb, & in concavitate eius est granum nigrum sicut Scebedenegi, & asportatur de Sofale. Collocat verò eam in tertio calefacientium & siccantium ordine, atque stomachi & hepatis frigiditati utilem esse, concoctionémque iuvare, & alvum constringere scribit.

FIGURE 2.1 "Fagara Avicennae," Sichuan pepper, perhaps the earliest illustration (Clusius, *Exoticorum libri decem*, 1605: 185). The surrounding text consists of Clusius' comments, beginning with a quotation from Avicenna. Image courtesy of Missouri Botanical Garden.

Birkestad n.d.; Crosby 1972, 1987; Hendry 1934). Some such experiments succeeded so thoroughly that later authors—still within the Early Modern period—no longer knew that the plant was an introduction, still less what its original home was. German physician and botanist Leonhart Fuchs (1501–66), describing and illustrating three kinds of chili in a general flora, applied to the group the name *siliquastrum* found in the *Natural History* of Roman encyclopedist Pliny the Elder (23/24–79 CE), claimed to find a mention in the Persian polymath Avicenna (980–1037), copied a pharmacological analysis from Greek physician Dioscorides (first century CE), and asserted that the fruits may be called "Calicut pepper" as if originating on the west coast of India (Fuchs 1542: 731–5); yet they were quite unknown to those ancient and medieval authors, the first seeds having reached the Old World from the Caribbean within living memory, scarcely fifty years earlier, when Columbus returned from his first voyage.

ANCIENT KNOWLEDGE REDISCOVERED

Long-distance trade across the Old World had been active for many centuries before the age of discovery revealed to Europeans the true origins of the exotic spices that they had learned to prize. Orta was not alone in knowing of the trade that had brought Indian and Eastern products to the markets of the Near East, but, from his vantage point in Goa, he was unusual among Europeans in his full awareness of the Chinese involvement in this spice route:

> The Chinese came here by sea long ago, and since the people were barbarous and ignorant they adopted Chinese customs and practices including going to sea in big ships ... The Chinese had so many ships, so the people of Ormuz say, that their books record one tide that brought four hundred junks ... In Calicut they had a factory in the form of a fortress, still to be seen, called Chinacota, meaning "Chinese fortress". At Cochin they left a stone as mark and memorial of their presence ... They brought from their country gold, silk, porcelain, musk, copper, seed-pearls, alum, and much else. They sold some of it in Malacca, and took on sandalwood, nutmeg and mace, cloves and aloeswood. They sold some of it on the way, that is, in Ceylon and Malabar, and in Ceylon they got plenty of good cinnamon that cost them very little money (while the sailors, for no money at all, got wild and bad cinnamon from the Malabar mountains, and also brought some of it from Java). Landing in Malabar they took on pepper, cardamom and other spices and carried it all to Ormuz or the Arabian coast. The merchants came there to buy their goods, which they then sold in Alexandria, Aleppo and Damascus.[4]
>
> (Orta 1563: ch. 15)

Chinese traders did not open up the spice route, as Orta perhaps believed. No single group can be claimed to have done that. But they had indeed explored it end to end before the Portuguese arrived.

The most famous and mysterious of the ancient spices traded across the Indian Ocean was cinnamon, known to Greek poets as early as the time of Sappho of Lesbos (sixth century BCE), highly valued in medicine and ritual, and surrounded by legends that no ancient or medieval European writer had been able to dispel. Lower grades of cinnamon and cassia from various *Cinnamomum* species were found in southern China (as Marco Polo already knew) and in the Indonesian islands (as reported from the Magellan expedition by Maximilianus Transylvanus [*c.* 1490–*c.* 1538] and Antonio Pigafetta [*c*. 1491–*c*. 1531], confirmed by Orta above). But the first European to report on the harvest of *Cinnamomum verum*, the true cinnamon of Ceylon, was Italian traveler and diarist Ludovico di Varthema: "This canella or cinnamon is the bark of the tree: every three years they cut the branches, and then take off the bark, but not for anything do they cut the trunk ... The cinnamon has not the excellence when they cut it that it will have a month later" (1510: f. 27v; Jones and Badger 1863: 191–2).

Not all spices are luxury foods, but cinnamon certainly was. "If any spice can be tasted with pleasure it is cinnamon. I admit that I see the Germans and Flemings eat pepper, and your black women here eat cloves, but the Spanish eat none of these, only cinnamon," asserted Ruano, Orta's European interlocutor (Orta 1563: ch. 15). It is not surprising that Columbus and his successors would invest so much effort in the search for cinnamon in the New World. But Ruano here forgot his earlier mention of another food spice, the grains of paradise or melegueta pepper of West Africa, *Aframomum melegueta*:

> Why do they consume so much pepper in Europe, and so little melegueta, when melegueta tastes so much better, especially with fish?
> "I asked the German and French merchants this question, and they told me that melegueta did not bear cooking and would not flavour cooked dishes: it could only be used on raw or already-cooked foods. Let us leave the question there, and eat the fish that has been cooked for us. It takes cardamom just as well."
>
> (Orta 1563: ch. 13)

The question here raised would be eventually settled by market forces in favor of cardamom (*Elettaria cardamomum*), more exotic, more expensive, more mysterious. Grains of paradise, though increasingly easy to get as the West African coast opened up to trade, were, for nearly every purpose, not so good.

Ever since Roman times black pepper (*Piper nigrum*) had been the most important Indian spice in European trade. It was not the most costly by weight, a distinction reserved to cinnamon, but it was the spice transported in greatest volume in exchange for the greatest annual money payment. Its production had already been sketched for a European readership by Greek merchant Cosmas Indicopleustes (from Alexandria, Egypt) in the sixth century and by Marco Polo in the thirteenth. In turn Chinese observers reported on the pepper trade of Malabar in the early fifteenth century:

> The land [of Malabar] has no other product, producing only pepper. The people mostly establish gardens to cultivate pepper for a living. Every year when the pepper is ripe, of course, local pepper-traders make their purchases and establish warehouses to store it, waiting for foreign merchants from various places to come and buy it. When fixing the price they calculate by the *bohe* [*bahar*].[5]
>
> (Mills 1970: 135)

In the same period as these Chinese voyages Italian merchant and explorer Nicolò de' Conti (*c*. 1395–1469) visited Malabar in southern India. In his discussion of pepper, as reported by (Gian Francesco) Poggio Bracciolini (1380–1459), he accurately noted a resemblance so European and unexotic that it was ignored by later observers: "Its green grains have the form of juniper berries" (Poggius 1492: 4th page of text). Varthema gave further details from personal observation:

> It is planted near some other tree, because, like the vine, it cannot stand erect. It grows like ivy ... The leaves resemble those of oranges [*melangoli*], but are drier and full of minute veins on the underside. From each branch grow five, six or eight clusters, a little

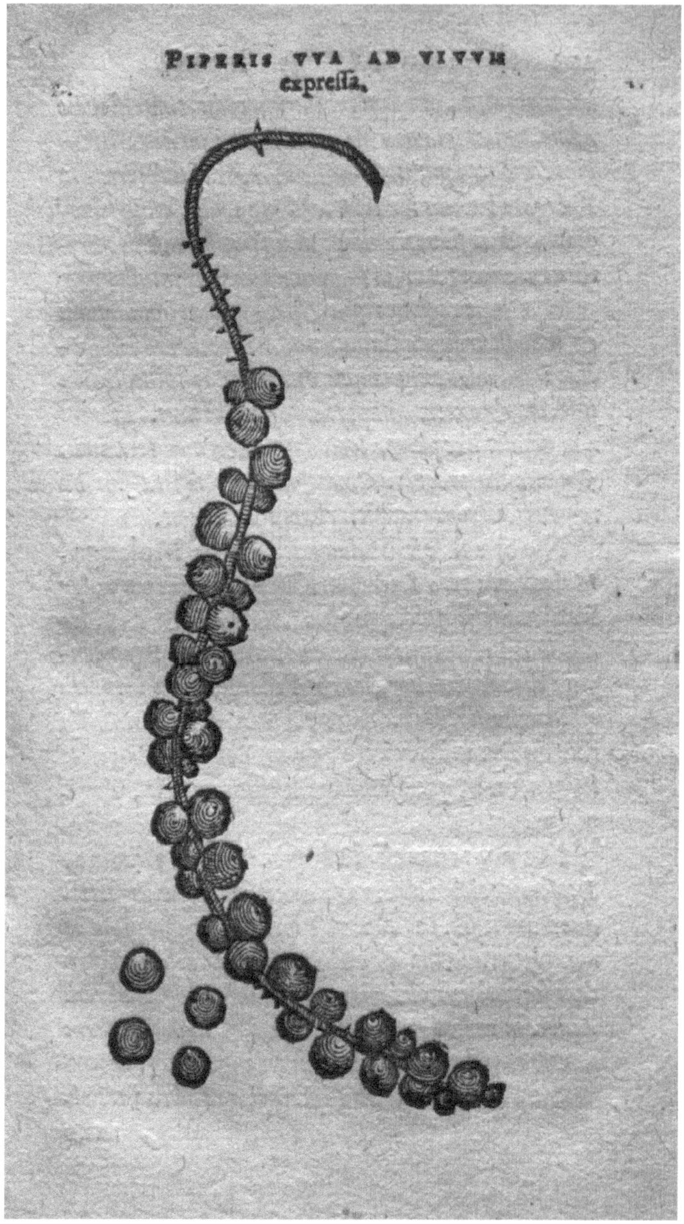

FIGURE 2.2 “The pepper berry, drawn from a real specimen” for Clusius in his translation of Garcia de Orta, *Aromatum et simplicium aliquot medicamentorum apud Indos nascentium historia* (A History of Several Aromatics, Simples, and *Materia Medica* Growing in the Indies, 1567: 107). Image courtesy of Missouri Botanical Garden.

longer than a man’s finger, of fruit like a small raisin, but more regularly arranged, and as green as unripe grapes ... They leave them in the sun for three or four days, when they become as black as they are seen amongst us, without doing anything else to them. These people neither prune nor hoe the tree.

(Jones and Badger 1863: 157; Varthema 1510: f. 23v)

For many centuries *Piper nigrum* had had rivals. Medieval sources, even as they disputed the distinction between black and white pepper (which are variant products from the same plant) correctly made the further distinction with long pepper (*Piper longum*), from Bengal, and cubebs, *P. cubeba*, from Sumatra and Java. In the ancient listings such as Pliny's and Galen's and the medieval handbooks such as Avicenna's *Canon* these alternative and more exotic peppers had been allocated at least as high a price and at least as great a medicinal value as black pepper. On the other hand, early-fifteenth-century Chinese observers gave no separate name to cubebs, treating them as a variant of black pepper:

> People who live near the mountains [of Sumatra] establish gardens for its cultivation. It climbs and creeps as it grows ... it produces flowers which are yellow and white in colour. The grains of pepper are the fruits. They are green when unripe, red when mature. They wait until the berries are half-ripe, then pick them, dry them in the sun, and sell them in the trade. Pepper in the form of hollow and large berries comes from here.[6]
>
> (Mills 1970: 118)

Orta had a few asides on long pepper and a whole chapter on cubebs, which came to India in Chinese ships. Cubebs imported to India "are much used by the Moors, sprinkled in wine, to increase libido, while in their native country (which is Java) they are customarily used for coldness of the stomach" (Orta 1563: ch. 19). Clusius, alongside his Latin translation of Orta's text, added his own independent information on black and long pepper. The end result, in spite of or because of the openness of Renaissance investigators to new information from distant sources, was that cubebs and long pepper, whatever their special qualities, from this point onward steadily fell into obscurity as far as the long-distance spice trade was concerned. *Piper nigrum*, meanwhile, was to be successfully transplanted to the Indian Ocean islands, the West Indies, and beyond, eventually becoming cheaper and more commonplace than any Renaissance explorer could have dreamed.

Another long-standing question was to be settled in a similar way. Ginger (*Zingiber officinale*), as seen by the classical Roman world, had been one of a kind, unique in its flavor and its salutary effect in the diet, but medieval Europe and the Near East had become familiar with other related spices: galanga (in fact two galangas, *Alpinia galanga*, *A. officinarum*) and zedoary and zerumbet (*Curcuma zedoaria*, *Zingiber zerumbet*). Alongside these there was also turmeric (*C. longa*), a rhizome most noticeable for its color, used as cosmetic as well as food, and far from negligible in its dietary and medicinal benefits. Turmeric had been not quite unknown to classical Greeks and Romans, but was never in common use among them.

All these spicy rhizomes were cataloged by medieval Arabic authors, and all were sorted out by Orta and his contemporaries. Orta was the main informant in this period on galanga. "The small kind, very fragrant, is brought from China to India and from here to Portugal. The larger kind from Java ... is occasionally grown here in gardens. It is used locally in pickles and in Indian medicines, especially by the midwives (*daias* as they call them) who act as physicians here" (Orta 1563: ch. 24). As for zerumbet, "they make a conserve of it with sugar, better than ginger" (1563: ch. 57). Orta's names for turmeric were *curcuma* and "earth saffron": "they use it to colour and sauce their food, here and in Arabia and Persia, as we use saffron" (1563: ch. 18). Ginger remained the best known of all, having already in ancient times been transplanted from its original habitat in southern China around the Indian Ocean shores as far as northeastern Africa. Marco Polo therefore naturally found ginger flourishing in several provinces of China, in

Southeast Asia and in India. Maximilianus Transylvanus and Antonio Pigafetta both reported on the ginger encountered on Magellan's circumnavigation. Varthema saw it farmed in India: "When they harvest it, they take an eye of the root, like the eye of a reed, replant it in the hole from which they dug it, and cover it with the same earth" (Jones and Badger 1863: 1582; Varthema 1510: f. 23v). Orta's interlocutor Ruano raised gastronomic questions on ginger, "which gives savour to our dullness on fish days and arouses our appetites when added to pickles (*achar* as your servants call it)," and Orta supplemented his observations: "Green ginger is eaten in pickles mixed with other herbs, oil, vinegar and salt, and in sauces with fried fish. Some eat it with fried meat" (1563: ch. 26).

The long-term result has been to confirm the overriding importance of ginger, which, overshadowing its rivals, has been ever more widely planted and remains universally familiar, its dietary benefits thoroughly confirmed by modern research. The other rhizomes, though they remain important in their own regions, have almost fallen out of knowledge elsewhere (admittedly turmeric, scarcely known beyond southern Asia in ancient and medieval times, rediscovered by Orta, is currently being rediscovered once again).

Cloves (*Syzygium aromaticum*) and mace and nutmeg (different seed-parts of *Myristica fragrans*) came from much further east, from distant and hitherto obscure islands in the Moluccas. Varthema was the first European to reach the region from the west, just a few years before the Magellan expedition arrived from the east. These spices had already played a part in the long-distance spice trade for about a millennium, yet until the early sixteenth century no one in China or Europe had known their real origin:

> The trunk of the nutmeg is formed like a peach tree and it produces its leaves similarly, but the branches grow closer. Before the nutmeg reaches perfection the mace surrounds it like an open rose, and when the nut is ripe the mace clasps it, and at this point they gather it, in September (for in this island the seasons go as with us). Every man gathers as many nutmegs as he can, for they are common property, and no labour is expended on the trees, nature being left to do her own work. The nuts are sold by a measure of twenty-six pounds weight for the price of half a *carlino*: the money of Calicut is current. There is no need for accounting because the people are so barbarous that they would not know how to cheat each other even if they wanted to.
>
> At the end of two days my companion asked the Christians [Hindus] where the cloves grew. They answered that they grew six days away, in an island called Monoch [one of the Maluku islands], and that the people of the island are the same as those of Bandan ... The tree of the cloves is exactly like the box tree, that is, dense, and the leaf is like that of the cinnamon ... When the cloves are ripe the men beat them down with canes, placing mats under the trees to catch them ... Having seen the island and the people we asked the Christians if there was more to see. They replied: "Let us see a little how they sell these cloves." We found that they were sold for twice as much as nutmegs, but by measure, because these people do not understand weights.
>
> (Jones and Badger 1863: 243–6; Varthema 1510: f. 33r–v)

Varthema was evidently not thrilled by his observation of the clove harvest, but Pigafetta had more to say:

> The cloves grow at the end of the twigs, ten or twenty in a cluster. The trees have generally more cloves on one side than on the other, according to the season. When

> the cloves sprout they are white, when ripe, red, and when dried, black. They are gathered twice a year ... No cloves are grown in the world except in the five mountains of those five islands ... Each of the inhabitants possesses clove trees, and each watches over his own trees although they do not cultivate them ...
>
> (De Vos 2006: 412; Robertson 1906: 2:87)

The flower is "just like that of an orange [*malum aureum*]," Maximilianus Transylvanus added, noting, too, that ownership of the clove forests was divided up among the natives "just as with our vineyards." Of nutmeg he observed:

> Close by is another island, called Bandam, longer and wider than the Moluccas, where the nutmeg grows ... Not dissimilar from a walnut, it is protected by a double husk, the outer being like a furry calix. Under it is a thin membrane which embraces the nut as in a net, and this is called *flos muscatae* with us, but *macis* by the Spaniards: it is a noble and wholesome spice. The inner covering is a woody shell, like that of a hazelnut, and inside it is the *nux moscata*, "musky nut."
>
> (Stanley 1874: 207–8; Transylvanus 1523: f. B 6)

The information elicited by Varthema, Pigafetta, and Maximilianus Transylvanus was entirely new to Europeans, and the investigations of Orta (though he perhaps never traveled east of India) were equally important. Clusius translated Orta on mace and nutmeg, and when he came to cloves, he was able to add more still. The expedition of Steven van der Hagen (1563–1621) had seized Ambon from the Portuguese in 1605 for the Dutch East India Company. Clusius' last notes, written in 1609 and published after his death, reported on the beginnings of Dutch management of the clove plantations:

> As soon as the Portuguese had made themselves absolute masters they had arranged the planting of many thousands of seedling clove trees across the region. The inhabitants, following their example, continually started new plantations in great numbers, to good effect, because these trees grow and flourish on Ambon with very little labour ... just as well in the plains and valleys as in the hills and highlands ... In brief, all the work and attention of the inhabitants now goes into these trees, and in the last three years the region has produced 2400 *bazares* of cloves for the Dutch East India Company, each *bazar* equalling 600 Amsterdam pounds.
>
> (Clusius 1611: 130–1)

So it was not true, as Pigafetta had reported, that cloves would only grow in the mountains. With their new lowland plantations Portuguese and Dutch colonists were already laying the groundwork for one of the longest-lived of colonial monopolies. But Clusius, while duly noting any trade secrets reported to him, was a scientist at heart. What really aroused his enthusiasm, when he was in Amsterdam in October 1600, were two clove branchlets still laden with their dried flowers, brought to him on one of the Dutch ships that had left Ternate on August 9, 1599. He described them minutely and illustrated his text with a woodcut, for these were perhaps the first such specimens that had reached Europe (Clusius 1605: 15–16; Ptak 1995). With equal attention to detail he provided cutaway images of nutmeg fruits showing the net-like membrane that is mace and the kernel that is true nutmeg. "Nutmegs also reach us whole as a conserve in sugar," he added in a footnote (Clusius 1574: 86). Conserved nutmeg was recommended in sixteenth-century England as a dietary supplement for students (Dalby and Dalby 2012: 17).

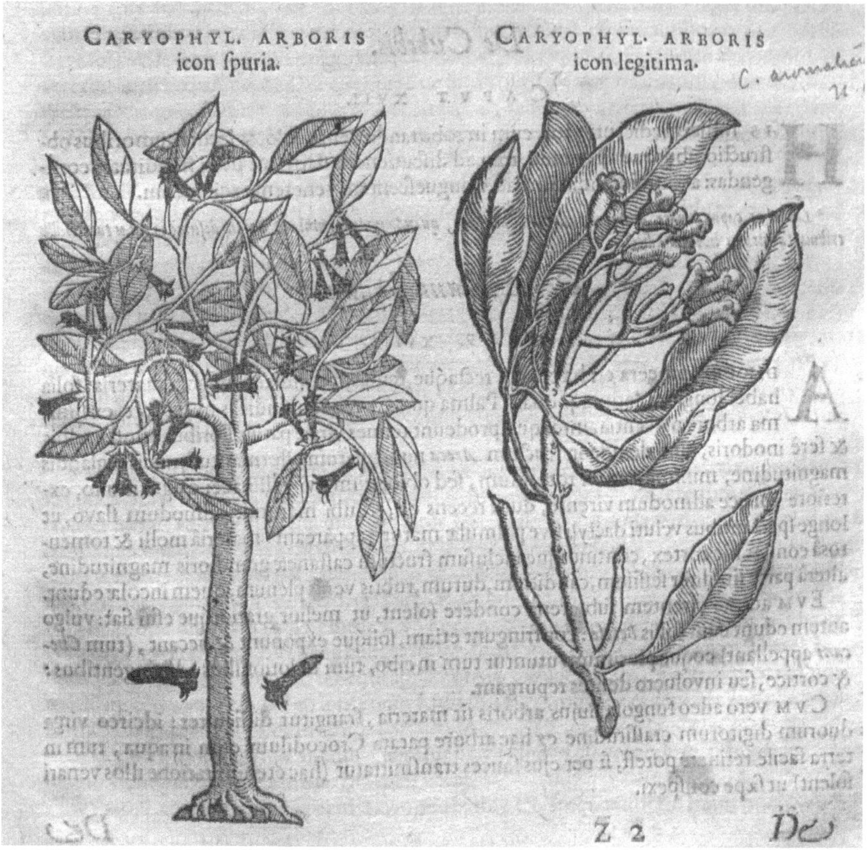

FIGURE 2.3 Cloves. [*Left*] *icon spuria*, the false image of a clove tree that Clusius had used in the first edition of his translation from Acosta (1582: 32), reprinted as an admission of error; [*right*] *icon legitima*, the true image, drawn from the clove branch brought to him in 1600 (Clusius, *Exoticorum libri decem*, 1605: 267). Image courtesy of Missouri Botanical Garden.

DISCOVERIES IN EAST AND WEST

The Chinese and Europeans who were now finding their way to southern Asia discovered a whole range of fruits previously unfamiliar and, in most cases, entirely unknown to them. At last they learned at first-hand what tamarinds (*Tamarindus indicus*) really were, thus far known to them as the source of a tasty and salutary semi-dried pulp, a commodity of long-distance trade (see Chapter 5 of this volume). This fruit, as Clusius rightly wrote, had been evaluated by Avicenna but without any description: "Selection: better are tender fresh fruits that have not aged or dried out, with good acidity. They are cold and dry to the second degree ..." (Avicenna, *Canon* 2.2.699, in Alpagus, ed., 1527: f. 127). Now at last, thanks to Orta and Clusius, a European audience was able to read a real description.

> The rind that covers the fruit is green, but brown when it dries, and easy to strip off. There are pits inside: we discard them and use the pulp, which is sticky. The fruits are like fingers, bowed in shape. When green they are more acid, but not so as to taste unpleasant. Removing the pits, I often eat the pulp with sugar.
>
> (Orta 1563: ch. 53)

To Orta's original Portuguese Clusius, in translating, added the detail that the pits are the same size as edible lupin seeds (Clusius 1605: 196).

During the same period the earliest vague rumours of the durian (*Durio zibethinus*), with its unforgettable but indescribable smell and taste, were confirmed as true. The first known description of this fruit was addressed to a Chinese audience:

> They have a kind of foul-smelling fruit, of which the foreign name is *du'erwu*. It resembles the 'waterfowl head' [*Euryale ferox*] of the Middle Kingdom. It is eight or nine inches long. On the skin grow sharp prickles. When ripe it splits open into five or six sections, whose foul smell resembles that of putrid beef. Inside there are fourteen or fifteen lumps, as big as chestnuts, of milk-white flesh, very sweet and delicious to eat, and these have seeds inside them which, when roasted and eaten, taste like chestnuts.[7]
>
> (Mills 1970: 118–19)

Just a few years later, the explorer, Niccolò de' Conti, described durians to a European listener: "They have a green fruit named *durianum*, as big as a melon, which contains five oblong segments as of oranges [*mala rantia*], of varied flavour, like butter and curdled milk" (Poggius 1492: 4th page of text). Orta in turn described "a thick green rind covered with small prickles like a jackfruit ... It has internal divisions resembling chambers, each containing separate fruits with a flavour like blancmange [*manjar branco*]." He had heard the aroma compared by some to cherries or melons, by others to rotting onions (1563: ch. 20).

The mango (*Mangifera indica*) was noted with interest by Chinese explorers:

> They have a kind of sour fruit, of which the foreign name is *anba*. It is like a large Chinese pear, but rather longer, with a green skin. The flavour is fragrant and strong. When you want to eat it you pare off the skin and shave off slices from the outside of the flesh and eat them. They are sweet-and-sour and very delicious. The stone is as large as a hen's egg.
>
> (Mills 1970: 118)

The earliest mention of mango pickle is owed to Varthema not long afterwards: "From this fruit a conserve is made, as we make from olives, but it is much better" (Varthema 1523: f. 24r). Orta was proud of a mango tree on his estate at Bombay, a tree that gave two harvests every year. He served the flesh freshly sliced: "Try it both ways," he urges his guest, "with wine and without," adding that mangoes were also eaten "in a conserve of sugar; in a pickle of vinegar, oil and salt; stuffed with green ginger; salted; baked" (Orta 1563: ch. 34).

Many other fruits of southern Asia were noticed by travelers for the first time during this period, a few of them already known to Arab medical authors: the langsat (*Lansium domesticum*) and mangosteen (*Garcinia mangostana*) first observed by the Chinese expedition, the jackfruit (*Artocarpus heterophyllus*), jambolan (*Syzygium cumini*), and rose apple (*S. jambos* etc.) first by Varthema; then, in addition to these, the Indian jujube (*Ziziphus mauritiana*), bael (*Aegle marmelos*), carambola (*Averrhoa carambola*), caranda or karamda (*Carissa carandas*), and three kinds of myrobalans (*Terminalia* spp.) all described and discussed by Orta. During the same period Europeans in the Caribbean and tropical America were discovering an entirely different range of fruits, including hog plum (*Spondias* spp. especially *S. mombin*), caimito or star apple (*Chrysophyllum cainito*), soursop (*Annona muricata*), custard apple (*A.* spp. especially *A. squamosa*), prickly pear (*Opuntia* spp.), red and yellow mamey (*Pouteria* spp.). Many of them were first described by the Spanish colonist, Gonzalo Fernández de Oviedo y Valdéz (1478–1557), who wrote in the early sixteenth century, though his *Historia general y natural de las Indias* was not fully published until the 1850s. Among the rest he gave early reports of three species now widely known, papaya (*Carica papaya*; see Chapter 5 of this volume), avocado (*Persea americana*), and guava (*Psidium guajava*). "El Inca" Garcilaso de la Vega (1539–1616), writing in Spain in old age, recalled the guava:

> These are round and about the size of the average apple, which they resemble in having a skin but no rind: inside, the pulp has a great many round seeds or pips rather smaller than those of a grape ... The Spaniards have been making conserves of this and other fruits since the time when I left Peru ... I have seen this conserve in Seville: a traveller, a friend of mine, had brought it from Nombre de Dios [in Panama] and offered me some because it was a fruit of my country.
>
> (Garcilaso 1609: f. 209r; Livermore 1966: 1:502–3)

Peanuts (*Arachis hypogaea*) were described by Oviedo and remembered by Garcilaso as "a fruit which grows underground." Very much like almonds in consistency and taste, peanuts were tasty and wholesome when roasted and could be pressed to produce a nice oil useful in many illnesses: they were a luxury for epicures rather than a staple food. "With honey they make a very good *turrón* [nougat]," he concluded (Garcilaso 1609: f. 208v; Livermore 1966: 1:501).

Molle or Peruvian pepper (*Schinus molle*) was successfully grown in Italy, and even in England, according to friends of Carolus Clusius, but Peru was to remain the only significant source of the fruits. Spanish conquistador Pedro de Cieza de León (1518/20–54), in 1553, was the first to describe the alcoholic drink now known as *chicha de molle* (whose pre-Columbian importance is becoming clear from archaeology). "From a very small fruit that grows on this tree," Clusius wrote, "they make a good wine or beverage, and a vinegar, and a passable honey: all they need do is to rub the quantity of fruit they want in water in a vessel and then warm it" (Cieza de León 1553: f. 125r–v; Clusius 1605: 322). The drink took three or four days to ferment, Garcilaso added: "It is nice to drink, very tasty, and good for illnesses of the urine, liver, kidneys and bladder" (Garcilaso 1609: f. 209v; Livermore 1966: 1:504).[8]

Europeans encountered cacao (*Theobroma cacao*) beans as currency very early in their exploration of the American continent, but learned of the nourishing drink which gave these beans their value only when the expedition of Spaniard Hernán Cortés (1485–1547) approached Tenochtitlán. As Montezuma's after-dinner pleasures were observed by the eyewitness Bernal Díaz, all became clear:

> They brought to Montezuma, in cups of pure gold, a drink made from this cacao. They said it was before he visited his women, and we took no more notice of it than that. I saw them bring in more than fifty big bowls of good cacao, all frothed up, and I saw him drink some of it, the women presenting it to him with great respect. Montezuma enjoyed entertainment and poetry, and he would send the leftovers and the bowls of cacao to the performers.
>
> (Serés n.d.: 285)

Cacao trees did not grow as far north as the Valley of Mexico. As the Aztecs extended their dominion towards tropical Central America, they had demanded as tribute and acquired in trade large quantities of the southern commodities they most desired: feathers, captives for sacrifice, and cacao beans. Oviedo, alone among early sources, described the laborious extraction of cocoa butter, which was used medicinally. He noted that Mexicans appreciated the cacao drink not only for its stimulant effect (at which Bernal Díaz, above, coolly hints) but also its appearance. "Because these people like to drink human blood, they add a little annatto [*Bixa orellana*] to their cacao so that it will turn red: without the annatto it is brown." People in Mexico sometimes flavored their cacao with "spices and chilli; they also prepare it as a paste, which they say is good for the chest and stomach and to treat catarrh" (Amador de los Rios 1851–5: 1:318).

As for the Spaniards, they at first thought cacao disgusting—its color whether red or brown, its texture, its foam—but well before 1590 they had learned to appreciate it. *Los Españoles y mas las Españolas se mueren por el negro chocolate*: "The Spanish men die for this black chocolate, and the Spanish women even more" (Acosta 1590: 251). Here Jesuit missionary and naturalist, José de Acosta (1539/40–1600), adopted the new word "chocolate," of uncertain etymology (perhaps Nahuatl *chocoatl*), coined just early enough to accompany the transatlantic cargoes of cacao beans that began to reach Europe in 1585.[9]

FROM OLD WORLD TO NEW

Experiments in transplanting had been taking place in the north Atlantic as the Portuguese and Spanish settled Madeira, the Canaries, and the Azores. In fact, Islamic Granada had set the example for this agricultural experiment: the resources of Andalusia and neighboring provinces had been enriched with the introduction of several important crops already familiar in Islamic lands to the east (De Vos 2006: 405–6).

It was natural, given these precedents, that already on his first voyage Columbus would be eagerly and gullibly identifying costly spices such as cinnamon and mastic wherever he went, and on later voyages would take familiar food plants from Europe far beyond the already-settled Atlantic islands to the unknown "Indies" that he was "discovering." The Spanish in Hispaniola "have planted much garden stuff [*hortaliza*], which grows more in eight days than in Spain in twenty," wrote Diego Alvarez Chanca, the physician who accompanied the second voyage (Fernandez de Ibarra 1907: 451–2; Navarrete 1858: 368). The biography by the explorer's son, Fernando Colón, gave a second perspective on the same experiments:

> Fruit stones planted in the ground sprouted in seven days; vine shoots sent out leaves at the end of the same period, and by the twenty-fifth day green grapes were ready to be picked. Sugar canes germinated in seven days. This was due to the temperate climate, very like that of Spain.
>
> (Ulloa 1571: ff. 106v–107r)

Thus Hispaniola, site of the first (though ill-fated) Spanish colony, saw many of the earliest experiments in the transplanting of Old World crops to the Americas. Oviedo confirmed and extended the list: fig trees producing more or less throughout the year, pomegranates, quinces, grapes. In a second list he dealt with what Chanca called "*hortaliza*": melons, aubergines, beans (but there were other kinds of beans native to Nicaragua, Oviedo correctly noted), mint and celery, cucumbers, lettuce, and more (Amador de los Rios 1851: 1:291–2, 372–3). Then there were crops that were historically not typical of Europe at all, or only of its southernmost edge. Under this category Oviedo listed dates (but the growers in Hispaniola had not yet learned to dry them), sugar cane (which, as mentioned above, was planted in Hispaniola on Columbus' second voyage), and, at least equally important, the banana (*Musa* spp.), for which Oviedo's name was *plátano*, "plantain":

> As many people have assured me, this plant was brought from Gran Canaría in the year 1516 by the reverend father Fr. Thomas de Berlanga of the Dominican Order to this city of Santo Domingo, from which it has spread to the other settlements of this island and to all the other islands that are inhabited by Christians. They have taken it on to the mainland, and wherever it is planted it does very well ...
>
> I myself saw them in 1520 growing on Gran Canaría at the monastery of St Francis, and they have them elsewhere in the Canaries too. I have heard it said that they are also grown in the city of Almería in the kingdom of Granada. They say that this was the source of the plants that have now reached the Indies, and that they had come to Almería from Alexandria and the Levant, and that they had arrived there from the East Indies.
>
> (Amador de los Rios 1851: 1:293)

The Peruvian native Garcilaso, writing two generations later than Oviedo, appeared no longer to know that "the tree and fruit the Spaniards call *plátano*" were introduced to the New World by Europeans. "The flesh is tender, soft, and sweet; sun-dried it resembles a preserve. They eat it raw and roasted, boiled and stewed in pottages, and it tastes good in all these ways. With a little honey or sugar (a little is enough) they make various banana preserves," he wrote (Garcilaso 1609: f. 211r; Livermore 1966: 1:507–8). The early French settlers in Brazil, on whose vicissitudes French explorer Jean de Léry (1536–1613) reported, found bananas already flourishing there, and called them *figues*, while admitting that they were "sweeter and tastier than the best Marseille figs" (Léry 1578: 206).

Europeans who traveled east were equally impressed by bananas. Varthema saw them in India, noting the fact so convenient to modern global trade that the fruit could be picked unripe and ripened in storage (Jones and Badger 1863: 162; Varthema 1523: f. 24r). Orta knew of an early description by "a Franciscan friar who wrote of the mysteries of the Holy Land" and had asserted that this is the fruit that led Adam to sin. Avoiding theological comment, Orta gave nutritional hints:

> In Bassein [Burma] and elsewhere fruits that are one palm long taste very good baked, swimming in wine, with cinnamon over them: they are like baked quinces, but better ... They also slice them down the middle and fry them in sugar until they well cooked, and with cinnamon over them they taste very good ... They take them on board here as sailors' rations, to be eaten with sugar: it is good food at sea.[10]
>
> (Markham 1913: 196–200; Orta 1563: ch. 22)

Alongside the dates, sugar cane, and bananas that soon began to thrive in Hispaniola we may also with Oviedo list sweet and bitter oranges (ancient hybrids, *Citrus* × *aurantium*, Sour Orange Group and Bitter Orange Group respectively). These citrus fruits, their parental species originally from southern China and Southeast Asia, were now widespread in southern Asia. The fifteenth-century Chinese explorers, finding plentiful bitter oranges in Sumatra, noticed that they fruited "continuously during the four seasons ... The taste is not very sour." Varthema, their European contemporary, found sweet oranges (*melangoli dolci*) in southern India; Orta observed that they were at their best in Sri Lanka (Jones and Badger 1863: 190; Markham 1913: 136, 284; Mills 1970: 119; Orta 1563 ch. 15, 34; Varthema 1523: f. 27r).

FIGURE 2.4 "*Malus Arantia*, The Orenge tree," seen at the top left on a page otherwise devoted to apples. John Parkinson's *Paradisi in sole Paradisus terrestris* (1629: 585). Hand-colored copy. Image courtesy of Getty Research Institute.

Direct evidence for bitter oranges in Europe is found in the botanical illustrations of the *Grandes heures d'Anne de Bretagne* (Anne of Brittany's Book of Hours), completed by French court painter, Jean Bourdichon (1457/9–1521) in 1508, and the botanical frescoes in the Loggia of Psyche in the Villa Farnesina in Rome, painted in 1518 by Giovanni Martini da Udine (1470/5–1535) for Agostino Chigi (see Chapters 5 and 8 of this volume).[11] But by this time such oranges were already flourishing in southern Spain, whence seeds were taken to Hispaniola by Columbus on his second voyage. In 1523, at Rio de Tonalá in the first weeks of Cortes' expedition, the first orange seeds were sown on the American mainland by Bernal Díaz (*c.* 1496–1584) (quoted in the Introductory Chapter of this volume). He had brought seeds with him from Cuba because of a rumor that they would be establishing a colony (Maudslay 1908: 1:62; Serés n.d.: 54).

The idea of transplanting was potentially applicable not only to fruits but also to spices, whose high price had for so long depended on their unique and distant origins. As explained above, sugar cane, originally from the Malay Archipelago, was among the first of Columbus' plantings: it was soon afterwards transmitted to the American mainland, where the French settlers on whom Jean de Léry reported found it growing in Brazil. It was so easy to cultivate that it was briefly an economic staple of Hispaniola, exported in bulk to Spain, until it was realized that ginger would fetch more money and producers turned briefly to ginger, resulting in a shortage of sugar and a glut of ginger in Spain (De Vos 2006: 422–6; Léry 1578: 208). But sugar cane does so well in the Caribbean that it became a monoculture in several islands. Soon after the period dealt with in this volume it was to bring poverty and economic collapse as the price inevitably fell.

Sugar cane, alongside ginger and other higher-value spices, had been planted experimentally in Mexico by Francisco de Mendoza y Vargas (1523–63). In 1558 he was given a royal monopoly to grow black pepper, cloves, cinnamon, ginger, saffron, sandalwood (*Santalum album*), and China root (*Smilax china*) in America. Physician and botanist Nicolás Monardes (1498–1553), back in Seville, first learned of this bold undertaking in conversation with Francisco de Mendoza himself, and was suitably astonished by the green ginger and fresh China root, presented to him directly from New Spain (Monardes 1574: f. 16r). But Mendoza died in 1563, and Monardes reported the sad end of these experiments in a later section of his work:

> Don Francisco de Mendoza, son of the viceroy Antonio de Mendoza, grew in New Spain cloves, pepper, ginger, and other spices brought from the East Indies. The enterprise was lost with his death, except in the case of ginger, which grows very well in those parts.
> (Clusius 1605: 311; Monardes 1574: f. 99v)

The cinnamon plantations at Concepción in Hispaniola (MacNutt 1912: 2:306; Petrus Martyr 1530: f. 100r) belong to a different story. They were of *Canella winterana*, known now as Winter's bark, cinnamon-scented indeed, but quite unable to take the place of the true spice. Disappointment set in quickly and the hunt for New World cinnamon continued. It led eventually to the doomed expedition of Gonzalo Pizarro in search of the rumored "valley of cinnamon" in the upper Amazon basin, but it never resulted in the discovery of a real cinnamon substitute among New World aromatics (Amador de los Rios 1851–5: 1:357; Dalby 2001).

FROM NEW WORLD TO OLD

Naturally, given these examples, the transplanting of useful and luxurious food plants discovered in the New World was immediately undertaken in Europe and southern Asia, wherever the climate seemed suitable and whenever gardeners were able to devise a regime for germination and cultivation. Seeds and saplings began to spread with the very first voyages (on the maize seeds that reached Spain and Italy in 1494 see the Introductory Chapter of this volume).

140 AROMATVM HIST.

ANACARDIVM OFFICINARVM.

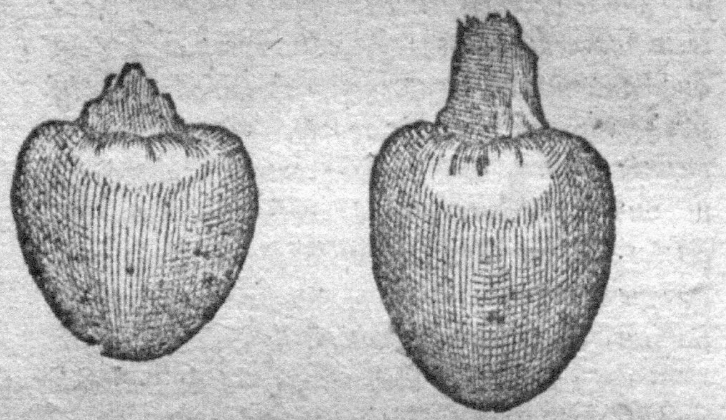

Malaqua, *quod viridis & in arbore pendens ſimilis ſit noſtrati maiori Fabæ, maior tamen.*

MAGNA *eius eſt in Cananor copia, tum in Calicut & reliquis Indiæ prouincijs mihi cognitis, veluti in Cambaya & Decan.*

Serapionis lapſus.

SERAPIO *libr. ſimp. cap. 356. Galenum citat tanquã huius fructus meminerit (cùm tamen illi planè ſit incognitus) & deleteriam facultatẽ habere ait, cui tamen experientia refragatur. Siquidem his in regionibus è lactis ſero maceratum aſthmaticis datur, et aduerſus lumbricos: præterea ijs viridibus & ſale conditus vtimur in cibarijs, oliuarum conditarum modo. Fructu verò ſiccato, cauſtici vice vtuntur in ſtrumis: totáq; India huius cum calce miſti vſus eſt ad pannos obſignandos.*

AVICENNA *lib. 2. cap. 41. id facit fructus Tamarindi oßi ſimile, eiuſq; nucleum Amygdalæ effigie,*

ſine

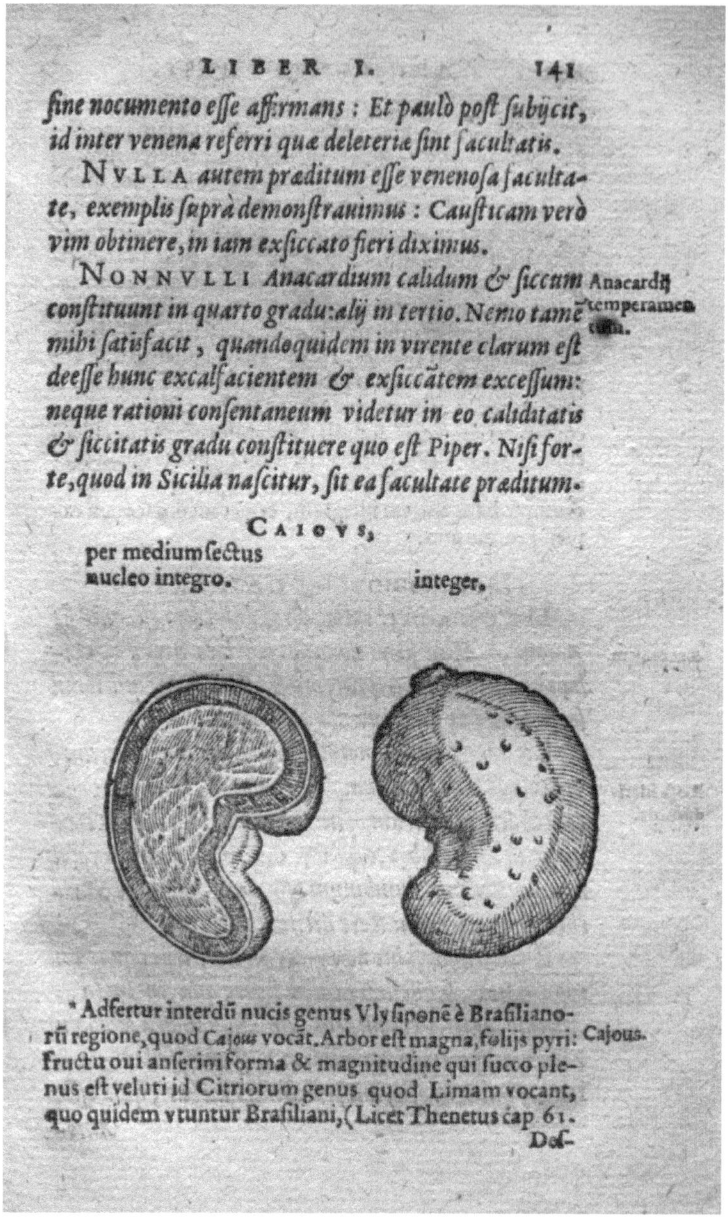
LIBER I. 141

sine nocumento esse affirmans: Et paulò post subijcit, id inter venena referri quæ deleteriæ sint facultatis.

NVLLA autem præditum esse venenosa facultate, exemplis suprà demonstrauimus: Causticam verò vim obtinere, in iam exsiccato fieri diximus.

NONNVLLI Anacardium calidum & siccum constituunt in quarto gradu: alij in tertio. Nemo tamē mihi satisfacit, quandoquidem in virente clarum est deesse hunc excalfacientem & exsiccātem excessum: neque rationi consentaneum videtur in eo caliditatis & siccitatis gradu constituere quo est Piper. Nisi forte, quod in Sicilia nascitur, sit ea facultate præditum. Anacardij temperamentum.

* Adfertur interdū nucis genus Vlysiponē è Brasilianorū regione, quod Caious vocāt. Arbor est magna, folijs pyri: fructu oui anserini forma & magnitudine qui succo plenus est veluti id Citriorum genus quod Limam vocant, quo quidem vtuntur Brasiliani, (Licet Thenetus cap 61. Des- Cajous.

FIGURE 2.5a and b "Anacardium officinarum" and "Caious," the closely related Old World marking nut and New World cashew, as shown on facing pages by Clusius in his translation of Garcia de Orta (*Aromatum et simplicium aliquot medicamentorum apud Indos nascentium historia*, 1567: 140–1). Image courtesy of Missouri Botanical Garden.

Cashew nuts (*Anacardium occidentale*) were much liked by early Europeans in Brazil and were exported to Portugal. They were "as good as pistachios" (Clusius 1605: 198). The cashew tree was flourishing in India by the later sixteenth century: it "is not found everywhere, but is grown in many yards and gardens in Santa Cruz de Cochin" (Acosta

1578: 325). Acosta himself apparently did not realize that it was a newly introduced species, but, since Cochin, like Brazil, was a Portuguese possession, there can be little doubt that Portuguese voyagers were responsible for this bold and successful transplant, though the event is not recorded.[12]

A second tropical New World delicacy that spread early was the pineapple (*Ananas comosus*), a fruit so unlike any other that early authors found it difficult to describe either the shape or the flavor (see also Chapter 3 of this volume). "The Spaniards call [it] *piña* because in appearance it resembles a Spanish pinecone," wrote Garcilaso, remarking "a faintly acid taste which makes it even more palatable" (Garcilaso 1609: f. 211v; Livermore 1966: 1:508). De Léry noted "such an aroma of raspberries that even walking in the forests one is aware of them from a distance" and "their taste melting in the mouth, so sweet that none of the conserves we have in Europe surpasses them. I think it the best fruit in all America." He recalled "extracting more than a glass of juice by pressing one fruit: the liquor seemed to me not inferior to malmsey" (Léry 1578: 211–12).[13] Physician and explorer, Francisco Hernández de Toledo (1514–87) found the flavor to be "that of a peach [*melocotón*], but somewhat more acid … fresh fruits are cut into slices and steeped in brine for a short time" (Hernández 1651: 311).

The pineapple easily repaid the trouble of cultivation in Hispaniola, Oviedo observed, although a ripe fruit would keep for no more than fifteen or twenty days (Amador de los Rios 1851–5: 1:280). Acosta, well aware of its origin in Brazil, called it a "widely travelled fruit." He traced its successive transplanting to the West Indies and then the East Indies, where the price of single fruits had fallen in his time from ten ducats to two reales "not because of a decline in their aroma or flavour but because they are so widely available." He also offered evidence of the pineapple's acidity: "if a knife used to cut the fruit is left in it for a night or a day, the part of the knife that remained in the cut will be consumed and disappear" (Acosta 1578: 350–1).

Two fruits of the Solanaceae family mark a complete contrast with cashews and pineapples. Both tomatoes (*Solanum lycopersicum*) and chilies (*Capsicum* spp.) are at home in a warm temperate climate, though certain chilies like it hotter. Both were therefore obvious candidates for planting in southern Europe, especially since the seeds of these annual plants are easy to collect, have reasonably long viability, and are easy to germinate.

As it happened, they were also easy to fit into existing botanical categories, since the resemblance of the tomato, in particular, to mandragora and to the recently introduced aubergine was obvious. The first published description is in fact found in the 1544 commentary by Pietro Andrea Mattioli (1501–77) on Dioscorides' Greek *Materia medica*—where tomatoes could have no legitimate place (the plant being unknown to Dioscorides) except as one of a series of supplements to the chapter on mandragora: "Yet another kind is brought to Italy these days, flattened in shape like *rosa* apples, divided internally, green at first but golden when ripe" (later editions add that there are also red sorts) "whence they are called *pomi d'oro*, that is, golden apples. They can be eaten like the other kind [i.e. like aubergines], but they often cause nausea and vomiting" (Mattioli 1544: 327; 1574: 760). A fuller and more favorable estimate was given by Konrad Gessner (1516–65), who had seen tomatoes growing in German gardens:

> The golden apple or love-apple as it is called, or apple of the Other World, bush-sized, introduced, annual, apparently related to the solanums … with a heavy and fetid odour. The fruit is almost odourless and not unpleasant or harmful as food … round;

> but often larger, irregular and swollen; golden in colour or red in a second kind, white in a third. It grows easily with us and ripens in good time. It responds well to watering and to rich soil in yards or in pots.
>
> (Gesnerus 1561: f. 273v)

Equally untroubled by nausea, Clusius, in a supplement to his translation of Dodoens' herbal, observed that tomatoes were "eaten by some, dressed with pepper, salt and oil and cooked" (Clusius 1583b: 455).

There was, however, widespread reluctance in Europe to appreciate the tomato as a food. John Parkinson, writing in England in 1629 and calling it "love apple," treated it as a kind of ornamental in which the fruit, not the flower, was the eye-catching feature (so did Costanzo Felici, see Chapter 8 of this volume). Parkinson acutely observed that the small-fruited sort is quite distinct from the others and would in some years "rise of their own sowing in my garden." Those were cherry tomatoes, now distinguished by botanists as *S. lycopersicum* "var. *cerasiforme*," thought to have been important in the history of tomato domestication in Mexico. He went on to give his take on the tomato problem: "In the hot countries where they naturally growe, they are much eaten of the people, to coole and quench the heate and thirst of their hot stomaches" (Parkinson 1629: 379–80). In close accord with this view, the fullest early description written in sunny Mexico, that by Francisco Hernández, dealing with tomato and tomatillo together, treated them as a significant element in the normal diet without any caveat: "A very pleasant sauce is made from these fruits, crushed and mixed with chillies. It improves the flavour of almost all dishes and foodstuffs and revives a moribund appetite" (1651: 296).[14]

Thus Hernández brings us to the chilies, one of the first food plants that Europeans saw in the New World and among the most valuable of all that they have transplanted to the Old. "They use as seasoning a spice called *agi*, with which they season fish as well as fowl when they can get them," Diego Alvarez Chanca, one of Columbus' men, wrote in 1494 of Hispaniola (Fernandez de Ybarra 1907: 455; Navarrete 1858*:* 370). As Spanish settlement proceeded, every subsequent source confirms that Europeans in the New World used this new spice liberally. Garcilaso added that even those who returned to Spain from Mexico and Peru continued to "eat it regularly and prefer it to East Indian spices" (Garcilaso 1609: f. 210r; Livermore 1966: 1:505).

It is unexpected that a very early mention of the naturalization of chilies in the Old World relates to Britain. In William Turner's *Names of Herbes* the relevant entry is headed *piperitis*, to which the English name "Indishe peper" has been added: "The herbe groweth in certeyne gardines in Englande" (1548: 62; Britten 1881: 205). Continental authors soon gave more detail, with continuing enthusiasm. "Indian (or rather American) pepper" is what the punctilious Clusius called chilies. He expressed admiration for the chili gardens of Brno (now Czech Republic), observed in 1585, ancestral to the paprika farms of modern Hungary. Julius Caesar Scaliger (1484–1558), more concerned with natural philosophy than botany, vaunted the chilies grown and appreciated in cookery in Provence and the Basque country (1557: 200–1).[15] Clusius, again, recalled the chilies he had seen in 1564 on a journey through Old and New Castille, "grown by gardeners and also by women on their windowsills." He added a description and illustration of "Brazilian pepper," a new kind that he had first noticed on the same visit to the Iberian peninsula while traveling from Lisbon to Coimbra: he was told that it had come to Europe from the Pernambuco region in Portuguese Brazil, where it was widely grown.

FIGURE 2.6 The tomato [*center*], with its relative the mandrake [*top right*], amidst other decorative plants on a page of John Parkinson's *Paradisi in sole Paradisus terrestris* (1629: 381). Hand-colored copy. Image courtesy of Getty Research Institute.

This Brazilian kind seen by Clusius in Portugal might well have been the same observed by Léry in Brazil

> which our savages crush and pound with salt (which they are able to do by digging hollows to retain seawater). They call the mixture *jonquet*, and use it very much as

> we use table salt. Whether it be meat, fish, or other food, they season it to add relish, picking up each morsel and adding a pinch of *jonquet* before swallowing it.
>
> (Léry 1578: 216)

Among Clusius' posthumous notes is incorporated a short monograph on these admirable fruits, *De varietate capsicorum* (On the Variations of Capsicum), sent to him in manuscript by the Capuchin herbalist Gregorio da Reggio, who grew as many as twenty-five kinds in his monastery garden in Italy. Dried chili powder, he reported, "gives a good flavour to foods and even wine, but especially watery and windy foods. One scruple, taken daily in capon or veal broth, warms a cold stomach" (Clusius 1611: 104, 107; Egmond 2010).[16]

CONCLUSION

Transplanting may be regarded as a natural human activity: the Neolithic revolution depended on it. In this period transplanting was encouraged by the recent examples of Islamic agriculture in Spain and the settlement of the Atlantic islands. In parallel with those precedents should not be forgotten Bernal Díaz's observation: the Mexican priests at Rio de Tonalá tended his seedlings without knowing what they were, and they became the first orange trees of the American mainland. His report, like the reports of the first plantations in Madeira and Hispaniola, is unusual not because those transplantations took place but because they were reported. The sowing of a seed is barely memorable.

But why were some of the introduced foods of this period readily accepted, while others equally popular in their regions of origin were held at arm's length? Answers suggest themselves in quotations throughout this chapter. Local custom counts for something, as seen in the observations in Orta's dialogue that some food traditions favor cinnamon, others pepper or cardamom. In most cuisines, in the long run, melegueta pepper, Sichuan pepper, and molle or Peruvian pepper, have not been preferred to what was already available. Winter's bark and the other rivals to cinnamon that were optimistically identified in the New World have not found a place in the Old World repertoire. Other species were easy to accept because they were as good as, or even better than, an existing food in a definable category. New World cashews (*Anacardium occidentale*) were as good as Old World pistachios (*Pistacia vera*). Pineapple juice was scarcely inferior to malmsey wine, and pineapples themselves tasted like peaches but with more acidity.

These reflections help understanding of the contrasting reception of tomatoes and chilies. Columbus was looking for spices. He failed to find a West Indian equivalent of cinnamon or mastic, but he did find a pepper equivalent. Chilies were as hot as black pepper (some dieticians found them hot in the fourth degree, even hotter than pepper). They had beneficial dietary effects not identical with, but quite comparable to those of pepper, ginger, and other eastern spices. What was more, with a little loving care they could be grown in a temperate climate, as pepper never could. Chilies have now been welcomed into the culinary repertoire almost worldwide.

The first Europeans in the Caribbean were not looking for new fruits or vegetables. They nonetheless found some, and accepted some, but the tomato was a step too far. It was easy to grow, but where was it to be placed in the diet? If assimilated to ancient European plants it was most like mandragora and the nightshades, either poisonous or on the border of edibility. It was like aubergine, a recent introduction, but this black fruit with its sinister taste and suet-like texture had not found wide acceptance either. The tomato, though rarely rejected as poisonous, was neither sweet enough nor acid enough to fit easily among fruits. Mattioli found it nauseous. Parkinson's comment, already quoted, here repeated with my emphases, marked tomatoes as decidedly foreign: "In the *hot* countries *where they naturally growe*, they are much eaten of the people, to coole and quench the *heate and thirst of their hot stomaches.*" The tomato did not belong, and took many more years to earn even the half-hearted fruit-counted-as-vegetable acceptance that we accord it today.[17]

CHAPTER THREE

Trade and Exploration

Plant Hunting 1450–1650

DAVID MARSH

Sometimes the great dates in history do not really mark great changes. 1492 is an exception. The number and range of plants coming into Europe in the next two centuries, while not as great as happened in the following two centuries were, in other ways, more transformative. The opening up of new routes to India and the Orient, and the discovery by Europeans of a new continent with a completely different flora, opened up the inevitability of a rethinking of taxonomy and slowly created an atmosphere in which natural philosophy, or "science" as it was known by contemporaries, became the dominant way of thinking about the world. Led largely by inquisitive physicians and acquisitive merchants it also appealed to the curious and the consumerist elite. It happened at different speeds in different parts of Europe, and did not really take off in England until 1570s, but by the mid-seventeenth century plant collecting had joined the list of pursuits that formed part of the education and leisure activities of every gentleman and many gentlewomen. Plants went from being largely valued for their use, and only occasionally their beauty, to being valued for their own sake as much as anything else. It meant an intricate network of plant exchange and led in turn to a flourishing trade in plants.

This chapter will examine these changes using the example of plant enthusiasts—collectors, merchants, nurserymen, and gardeners—based in England, but because England was very much in the second rank of exploration and involvement in trade for most of the sixteenth century, the story will be set within a much wider European context.

"ALL PLANTS, HERBS, AND TREES, THAT GROW UPON THE EARTH"

Botanical knowledge at the time of the first European transoceanic voyages was, as has been shown in the earlier volumes of this series, based on the work of classical writers such as Dioscorides (first century CE), Theophrastus (*c.* 371–287 BCE), and Pliny the Elder (23/24–79 CE). It was not until the publication of William Turner's *A New Herball* (1551–68) that "a new Epoch in botanical literature in England [with] an English work written on a scientific basis" commenced (Henrey 1975: 1:21). But Turner, writing some sixty years after Columbus' first voyage, did not include a single reference to plants from the Americas. Although flora from the new world soon appeared in print in Europe, this did not occur in an English text until 1597 and the publication of John Gerard's *Herball, or Generall Historie of Plantes*.

The importance of "exotic" plants in the late sixteenth century is, however, encapsulated in a few lines from Christopher Marlowe's play *The Tragical History of the Life and Death of Dr Faustus*. It was first performed no later than 1592, although not first published until 1604, eleven years after Marlowe's death. In other words the play was written several years before Gerard's *Herball*, which is traditionally seen as the great turning point in the English understanding of plants. It suggests, too, that there was a much greater knowledge and appreciation of plants and their importance earlier than might have been expected given the lack of published material.

Marlowe had Faustus demand three things of Mephistopheles in return for his soul: "fain would I have a book wherein I might behold all spells and incantations, that I might raise up spirits when I please." This was granted immediately, as was the request for "a book where I might see all characters and planets of the heavens, that I might know their motions and dispositions." Finally Faustus requested "one book more,—and then I have done,—wherein I might see all plants, herbs, and trees, that grow upon the earth" (Act 2 scene 1 lines 161–71; Modlen 1912) (see Figure 3.1a, b, and c).

Marlowe derived his storyline from *The Book of Faust*, an English translation of a German book of 1587, which was itself based on the life of a real-life Faustus from Ingolstadt in Germany earlier in the century. Significantly, there are no references to a book of plants—or indeed astronomy—in the English *Book of Faust*, so why did Marlowe include plant knowledge as a crucial part of Faustus' price? What was it about plants that commanded a desire for complete knowledge of every single one in the world? Was it perhaps a recognition of their beauty, or perhaps of their economic worth as well as their more obvious medicinal values? Or was it perhaps something more abstract: the realization that the expansion of the known world had transformed existing systems of botanical understanding and opened up access to an almost infinite variety of new possibilities?

That Marlowe ranked plants alongside magic and astronomy presumably also reflects on the state of English botany and plant knowledge. But where did he get the idea that plant knowledge was not only useful and valuable but also a potential source of power? Perhaps, given the political state of Europe at the time, the answer lies in elsewhere in the text, where Faustus' friend Valdes says:

> Faustus, these books, thy wit, and our experience,
> Shall make all nations to canonise us
> As Indian Moors obey their Spanish lords ...
>
> (Act 1 scene 1 lines 117–19; Modlen 1912)

CABINETS OF CURIOSITIES

English State Papers for the mid- and late sixteenth century show that London was a cosmopolitan city, with the official returns of aliens and strangers suggesting around 5 percent of the city's population were foreigners, rising to perhaps 10 percent at the peak of the religious persecutions in the 1570s. Of these around 75 percent were from the Low Countries, with others from France, Spain, Italy, Portugal, Greece, Germany, Scandinavia, and even Turkey. The 1571 return included 4,850 names, probably mainly religious or

a

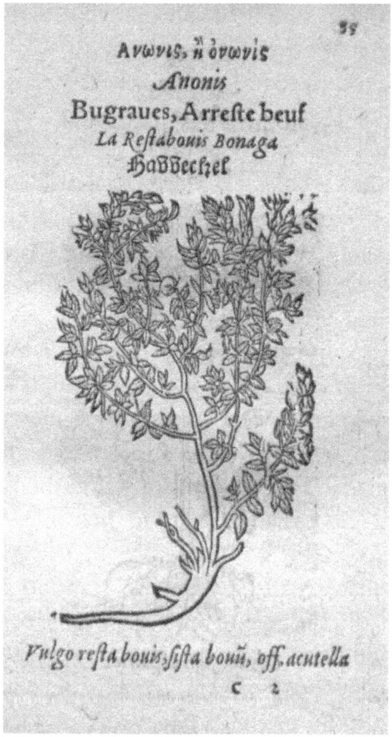

b

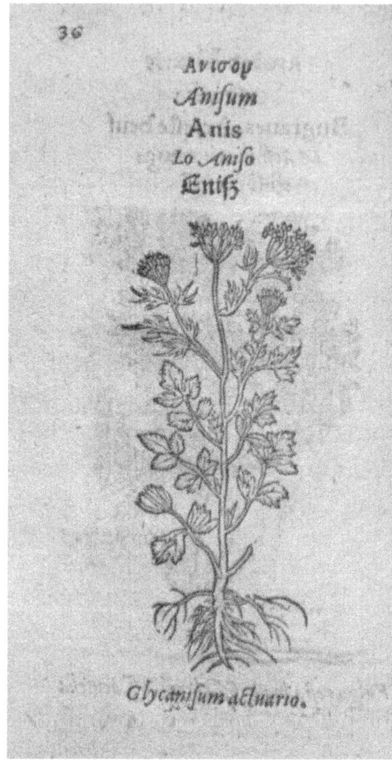

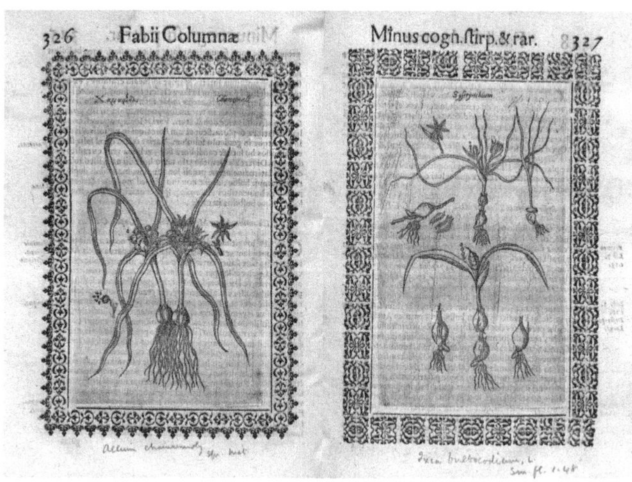

c

FIGURE 3.1a, b, and c Double-page spreads from the ideal plant book that Doctor Faustus demanded. Leonhart Fuchs, *Plantarum effigies* (Images of Plants, 1549: 35–6): a) restharrow (*Ononis spinosa*), and b) anise (*Pimpinella anisum*), labeled in Greek, Latin, Italian, French, and German. c) Fabio Colonna of the Accademia dei Lincei, *Minus cognitarum rariorumque nostro coelo orientium stirpium ekphrasis* (Description of Plants Less Known and Rarer Than Those Growing Beneath Our Skies, 1616: 326–7): two uncommon wild flowers, dwarf garlic (*Allium chamaemoly*), and a false crocus, *Romulea bulbocodium*. Images courtesy of Missouri Botanical Garden.

political exiles, from a wide range of professions and trades. Apart from the well-known groups of textile workers and merchants there were glassmakers, clockmakers, potters, and of particular interest, three apothecaries, twelve physicians, sixteen surgeons, and five gardeners. What the returns also show is that there were thousands more who did not congregate in London but were "encouraged" by the government to settle in areas around Canterbury, Sandwich, Colchester, Southampton, and Norwich, which were soon to be the centers of innovative market gardening.

Immigration on this scale and of this intellectual and skilled nature made London a vital and lively environment for scientific debate and exchange (Harkness 2007). Marlowe, with his links to courtiers like Sir Francis Walsingham, and his connections to publishers and booksellers, would have had been well placed to follow the course of intellectual advances amongst such "strangers" and their "curious" English counterparts. Marlowe would also have been aware that "nature" was for sale in shops and marketplaces in London. This was not just in the form of medicinal plants or culinary spices from the Mediterranean and Levant but increasingly from the new worlds—or rather, no longer quite so new worlds—of Asia and the Americas. Such goods were sold alongside other curiosities of the natural world: from dried strange fish, ostrich eggs, and fragments of Egyptian mummy to aromatic woods, unicorns' horns, basilisks, flying serpents, and hydras. Some of this material would have ended up in cabinets of curiosities which were well-established phenomena by Marlowe's day. Although primarily thought of as an elitist pursuit, as indeed it originally was in mainland Europe, in London the cabinet became as much the preserve of the wealthy commercial elite as that of the aristocrat. The best known, perhaps because the best recorded, was that of Sir Walter Cope (*c.* 1553–1614), who had "spent much time in the Indies," but Thomas Platter, the Swiss visitor who described the contents after a visit in 1599, mentioned that he had been to others around the city. Significantly, Platter was taken to Cope's Castle (now Holland Park House) in Kensington by Matthias de Lobel (1538–1616), the great Flemish botanist who was living in exile in London.

Some of the "marvels" that Platter listed—the unicorn's tail or the "thunder-bolt dug out of a mast which was hit at sea during a storm"—may appear amusing to us, but such "wonders" had a lengthy history in travelers' tales, and they were shown together with obviously "real" items from North America, China, Arabia, and Java. Surprisingly, no plants were listed, although there was a fan made from a single leaf, coral, and sea plants, as well as many other unspecified items in the collection (Williams 1937). However, dried plants are not particularly spectacular at first glance, so were perhaps were not in the front rank of what was shown to Platter from Cope's collection.

Did Marlowe's contemporaries share his knowledge or more pertinently Faustus' desire for plants as a potential source of power, knowledge, and wealth? If that was the case when did this process start, because there was little sign of it in the great herbals of William Turner (1509/10–68) and Henry Lyte (1529–1607), which surveyed indigenous plants, or in the earliest English gardening books written by Thomas Hill (b. *c.* 1528). Turner, considered with good reason to be the father of English botany, included long-known spices such as nutmeg, but had just one reference to "new" discoveries when in the entry on pepper he said rather vaguely: "Ther ar many thynges found out of late yeares by the saylyng of the Portugalles and diverse other adventurus travalers in far cuntres" (Turner 1562: f. 90r). This lack of more recently introduced plants was despite the fact that Turner was a keen and extremely accurate observer and was in correspondence with many continental scholars, notably Leonhard Fuchs (1501–66), a

German doctor and humanist, who had included several newly arrived plants in his own herbal, *De historia stirpium commentarii insignes* (Notable Commentaries on the History of Plants) published in 1542. Yet by 1587 there existed an account of "Gardens and Orchards" in Raphael Holinshed's *Chronicles* which clearly shows the impact of both new plants and "strangers" on the English garden and economy more generally. It was written by William Harrison (1534–93), who had contributed a lengthy and detailed "Description of England" to the 1577 edition of Holinshed, but without any real mention of gardens or plants. Harrison was a protestant cleric and in his other writings fiercely anti-Catholic, so "his despair at England's failure to emulate the social ideals of protestant humanism" (Parry 2004) may well have colored his historical judgment but did not affect the horticultural reality he depicted.

In it, having bemoaned "the excesse and vanitie" of his own countrymen, Harrison described how "such strangers also as dwelled here with us, perceiving our sluggishnesse, and espieng that this idlenesse of ours might redound to their great profit, foorthwith imploied their endevours to bring in the supplie of such things as we lacked, continuallie from forren countries." After claiming that vegetable and fruit production had been neglected until his own day, Harrison went on to explain how

> their use is not onelie resumed among the poore commons, I meane of melons, pompions, gourds, cucumbers, radishes, skirets, parsneps, carrets, cabbages, navewes, turneps, and all kinds of salad herbes, but also fed upon as deintie dishes at the tables of delicate merchants, gentlemen, and the nobilitie, who make their provision yearelie for new séeds out of strange countries, from whence they have them aboundantlie.
> (Harrison 1587: 208)

Nor was it just the kitchen garden that had been transformed. "If you looke into our gardens annexed to our houses, how woonderfullie is their beauty increased, not onelie with floures … but also with rare and medicinable herbs sought up in the land within these forty yeares: so that in comparison of this present, the ancient gardens were but dunghils and laistowes to such as did possesse them." Orchards, too, had benefited, for they

> were never furnished with so good fruit, nor with such varietie as at this present … in comparison of which most of the old trees are nothing woorth: so have we no lesse store of strange fruit, as abricotes, almonds, peaches, figges, cornetrees in noble mens orchards. I have seene capers, orenges, and lemmons, and heard of wild olives growing here, beside other strange trees, brought from far, whose names I know not.
> (209)

Harrison was clear that these changes had come about within the preceding forty years, and that they were continuing: "It is a world also to see, how manie strange hearbs, plants, and annuall fruits, are dailie brought unto us from the Indies, Americas, Taprobane [Sri Lanka], Canarie Iles, and all parts of the world." All of these places were under the control of either Spain or Portugal, and the implication must surely have been that there was considerable trade between them and the English.

Harrison went on to spell out the scale and effect of this trade in seed and plant imports:

> There is not almost one noble man, gentleman, or merchant, that hath not great store of these floures, which now also doo begin to wax so well acquainted with our soiles, that we may almost accompt of them as parcell of our owne commodities. They have

> no lesse regard in like sort to cherish medicinable hearbs fetched out of other regions neerer hand: insomuch that I haue seene in some one garden to the number of three hundred or foure hundred of them, if not more: of the halfe of whose names within fortie yeeres passed we had no maner knowledge.
>
> (209)

Two things appear clear in this, the first major description of the state of English horticulture. Harrison believed that the improvement of English gardening began around four decades before he was writing, and that the impetus for such improvement had come from overseas, in the form of knowledge and skills from "strangers" and new plants from Asia and the Americas.

NEW WORLDS AND THEIR PLANTS

One hundred years before Harrison was writing, and as the Tudors ascended the throne of England in 1485, Catholic Christendom was confined to western Europe by a ring of Muslim states that controlled the land routes to almost all the rest of the known world. From the straits of Gibraltar round to the midpoint of the Adriatic, the whole southern and eastern seaboard of the Mediterranean was also in Muslim hands. In 1487, when the Portuguese discovered that Africa had a Cape and that a sea-route to India was feasible, everything began to change. Vasco da Gama (*c.* 1460–1524) established a workable route to open up trade with India and beyond and this effectively broke the Muslim control of trade with eastern and southern Asia. But, important though that was, it was with the almost simultaneous stumbling across a new continent, and the subsequent Spanish conquest, that the worldwide expansion of Europe had begun and with it the contents of Western gardens.

Both Iberian countries were soon involved with the commodification of this new-found natural world. This is evident from the earliest days. Christopher Columbus (1451–1506) was not only interested in plants but understood that nature was a repository ripe for commercialization. His journal shows that he was captivated. On October 19, 1492, he noted: "As we neared this cape we were met by the soft, balmy smell of the trees and flowers ashore, the sweetest fragrance in the world." Nor was his sense of the economic potential far away: "I have no doubt there must be many plants and trees which would be valuable in Spain for tinctures and medicinal spices, but I am very sorry to say that I am unfamiliar with them." Two days later on October 21 he outlined his plans to benefit Spain: "The trees, too, are of a thousand kinds, each with its own fruit and pleasant fragrance; it grieves my heart not to recognise them, for I am sure they must all be useful. I am taking samples of them all, and of the plants" (Cummins 1992; Navarrete 1858: 186–8).

When Columbus returned triumphant to Spain, he met Spanish regents Ferdinand II and Isabella I at their court in Barcelona and presented them with his finds, which would have included seeds and plant specimens. In the audience was a fourteen-year-old page boy to the Crown Prince, Don Juan, named Gonzalo Fernandez de Oviedo. Also in Barcelona at the time was an Italian exile working for the Spanish crown on navigational and engineering matters, Giovanni Cabotto (better known as John Cabot, *c.* 1450–*c.* 1500) who realized that Columbus had not discovered Marco Polo's Cathay as he claimed, and so formulated his own plan to sail further to the west and reach Asia. He sought assistance initially from the small English merchant community in Seville.

There was quick action from the Spanish crown, too. Realizing the potential of what they had been told about by Columbus, they obtained a papal "bull" (decree) in May 1493 to monopolize their discovery and neutralize possible Portuguese claims. It was followed by diplomatic wrangling between the two Iberian powers that ended in 1494 by the Treaty of Tordesillas, which granted Spain and Portugal a duopoly over exploration and exploitation of the non-European world and dividing the Atlantic world into two hemispheres, along the modern meridian 46°37' W.

In those crucial last years of the fifteenth century it was, however, not just Spain and Portugal that were potentially involved in the expansion of the known world. England was, too. Seamen from Bristol had made at least two voyages of Atlantic exploration in 1480 and 1481, trying to find new fishing grounds, after the Hanseatic League had disrupted their use of Icelandic waters. The Cabot Project in Bristol believes that Cabot started exploring the American coastline from Newfoundland southwards in 1498 and actually encountered Columbus' fleet in the Caribbean in 1499, before returning to Bristol in 1500 where shortly afterwards he died.[1] It was a Bristol sailor, John Day, who was later to make the controversial claim to Columbus that Bristol men had discovered Brasil "as yr lordship well knows" (Williamson 1962: 212–14).

Ships connected with a fourth European power, France, were also in the region not long afterwards. This meant that in these early days of transatlantic and trans-Indian ocean voyaging, four countries had the opportunity to bring back accounts of the flora as well as seeds, bulbs, and specimens. They may well have done so, although there is little evidence of exactly what came back or how it was distributed.

PUMPKINS AND RHINOCEROS, MARVELS AND MONSTERS

However, the example of members of the cucurbit family—pumpkins etc.—shows that, under the right circumstances, new discoveries could be disseminated rapidly. While the first printed depiction in England of pumpkins appeared in Gerard's *Herball* of 1597, they had already featured in Fuchs' herbal of 1542, and it is clear that they were being widely grown in Europe decades earlier (see Figure 3.2). Research work on the murals of the Villa Farnesina in Rome—painted for the Sienese banker Agostino Chigi between 1515 and 1518—has revealed 130 species of plants painted by Giovanni da Udine probably from specimens in the garden of the palazzo itself. Included amongst the swags of fruit and vegetables are several sorts of curcubits (Janick and Paris 2006). Chigi was banker to the pope, Leo X, and is also thought to have been one of the financiers of Portuguese missions to India. Pumpkin seeds were almost certainly a gift to the pope from the Spanish crown. Given the number of seeds within any pumpkin, it would not have taken more than a few years to have them growing in any elite garden whose owner had the right connections. Elsewhere in the palazzo's murals are maize and beans, which are also indigenous to Central and South America. Such plants would have been considered marvels, although perhaps less spectacular than the pope's elephant which arrived from India in 1514 or his rhinoceros which arrived, albeit dead, the following year.

French and American scholars as recently as 2006 have identified an even earlier image of cucurbits in the Book of Hours of Anne de Bretagne, Queen of France (1477–1514). This was illustrated by Jean Bourdichon between 1503 and 1508 and depicts over three hundred plant species thought to have been drawn from the royal gardens at Blois and Tours. The margin to one page depicts an obviously living branch bearing flowers and fruits, which has been has identified as *Cucurbita pepo* subsp. *texana*, a plant indigenous

FIGURE 3.2 Four of the seven pumpkin or squash cultivars illustrated in John Gerard's *Herball* (1597: 774). Image courtesy of Missouri Botanical Garden.

to the Gulf Coast of what is now the United States (Paris et al. 2006). Another page depicts haricot beans also from Central America. It is no surprise that it should have been the personal prayer book of Anne de Bretagne (1477–1514) that carried images of these botanical novelties. As the ruler of the formerly independent Duchy of Brittany and the wife of His Most Christian Majesty (Maximilian I), Anne was a powerful woman in a central dynastic position, with family ties to the Spanish royal family, and strong diplomatic links with the papacy too. Unusual seeds, particularly ones that were so productive and ornamental and which came from hitherto unknown places, would have been sought-after gifts and presents. How Anne got the seeds is unknown but when these early ships returned from their Caribbean voyages there was, unsurprisingly, huge interest in the novelties they bought with them. A letter was sent as early as May 1494 to Cardinal Sforza, the pope's secretary of State by Queen Isabella's chaplain Peter Martyr D'Anghiera, mentioned more fully below. In it he enclosed seeds brought back by Columbus on his second voyage, including maize.[2] Perhaps others were sent by the Spanish crown to their French counterparts, or perhaps they came from a French ship ignoring the Treaty of Tordesillas.

Certainly the English initially turned a blind eye to the Treaty until after Cabot's return from his final voyage in 1500. Unfortunately he was virtually empty-handed, with no gold or other obviously useful commodities to impress his patron, so Henry VII (1457–1509), who was negotiating the marriage of Katherine of Aragon to his son Arthur and anxious to stay on the good side of the Spanish, withdrew his support for further voyages. This acceptance of Spanish control of most of the Americas by the other European seafaring powers lasted until about mid-way through the century. The result was that the two continents did not really figure in English culture for the rest of the sixteenth century. Instead they were to remain an unknown or mythical world like Cathay or Prester John's Africa, and the subject of travelers' tales full of "marvels" and "monsters." English references to the new world mainly concerned privateering such as Faustus' comment that it was "from America" that came "the golden fleece that yearly stuffs old Philips treasury" (Act 1 scene 1 lines 129–30; Modlen 1912).

Exploration vanished from the French agenda until the 1520s and 1530s—and from English agenda until the 1570s, even being only resumed unsuccessfully. This withdrawal had a profound effect on botanical knowledge and plant introductions because it meant that genuine first-hand information was entirely in the hands of Spain and thus at a premium. The Spanish monarchy quickly realized this and, because the Spain of Ferdinand and Isabella was dynamic and innovative, unlike the Spain of later centuries, it moved to take control. As early as 1503, seven years before the establishment of the first permanent settlement on the American mainland, the Spanish established the Casa de Contratación, a customs warehouse in Seville specifically for handling imports from the Americas. They began to govern their growing trade and later empire through the Council of the Indies: a term in use by 1519, and before the conquest of Mexico. They started a systematic compilation of eyewitness accounts using this knowledge and experience as part of the imperial strategy for controlling, exploring, exploiting, collecting, studying the nature of the New World.

This activity led in turn to the publication of a series of accounts edited by the royal chaplain, Peter Martyr d'Anghiera, who was appointed Chronicler of the Indies in 1520. The first in 1511 is based on Columbus' journal and includes the letter of 1494 cited above. Over the next fifteen years Peter Martyr published a further nine short reports on Spanish exploration and conquests. This coordinated approach and openness was continued by Charles V from his accession in 1519. A royal decree in 1525 instructs ship masters to bring animals and plants back to Spain, and the following year saw the publication of the first part of the *Natural History of the Indies* published by Gonzalo de Oviedo y Valdés (1478–1557), who had witnessed Columbus' return. Oviedo had access to all the official papers of the Council of the Indies, he knew many of the important people involved in the exploration and early exploitation of the Americas, including Columbus, and he had worked with closely with the royal family. Crucially, he was the also first writer who had lived and worked in the Americas.

AUTHENTICATING THE DOUBTFUL AND ESTABLISHING THE NEW

Oviedo's book was the first attempt at a systematic recording of what was being discovered in the Caribbean and beyond. It also shows clearly how Europeans, viewing these new phenomena and trying to fit them into accepted classical norms, gradually realized the impossibility of this. Oviedo wrote, "As Pliny said with his own pen ... it is hard to make old things new: to fit authority on what is new and authenticity on what is doubtful."[3] He went on to say that there were many plants that "not even the Indians had a name for them much less the Christians." Of the seventy-six illustrations in his text, twenty-three are of plants, many of them named in native languages rather than invented Spanish equivalents, which suggests that Oviedo, too, was beginning to question classical approaches.

This questioning can be seen in his account of the pineapple (*Ananas comosus*) written around 1535 (see Figure 3.3). Given that almost no one in Europe would ever have seen a pineapple, and that probably only King Ferdinand had tasted the one that survived the transatlantic crossing (Peter Martyr 1530: f. 33v, quoted in the Introductory Chapter of this volume), Oviedo found himself with the task of describing a completely unknown object to a European reader. Even the name was problematic. He explained that the first Christians who saw it thought it resembled a pinecone and so called it just that: a *piña*. However, he went on,

> a person could enquire why since this is a thistle is this fruit not called an artichoke ... I maintain it was the responsibility of the first Christians who came here to give one name or another and it seems to me the better name would have been artichoke with respect to the thistle and spines amongst which it grows, although it does not resemble a pine cone more than an artichoke. The truth is that is not totally unlike an artichoke ...
>
> (Myers 2007: 75)

His solution was to call it both: *piña o alcarchopha*—pinecone or artichoke. He went on to name several kinds of pineapples, each with different native names—*yayama*, *boniama*, and *yayagua*—but struggled to describe their differences and qualities. This would be hard enough for a modern author writing for a readership with experience of the fruit, but for Oviedo it was almost impossible since his audience had no such knowledge. To compensate he added a drawing, so that "in some manner the readers eye will be able to share the truth." His simple woodcut became the stereotype for all images of pineapples until the eighteenth century and helped the fruit become an object of such intense interest that collectors wanted it in their gardens.

In the earliest days of the Spanish conquest primary knowledge and experience like this informed the Council of the Indies, but gradually the early openness about information begins to dry up, with only a few sporadic examples of drawing on the knowledge of the indigenous people for commercial exploitation. In 1528, for example, there was a report to the Council on the cultivation of balsam (the so-called Hispaniola balsam) in Santo Domingo by Spaniard Antonio de Villasante (1477–1536), who had married a native chieftainess. But within a few decades very little material or even knowledge left the

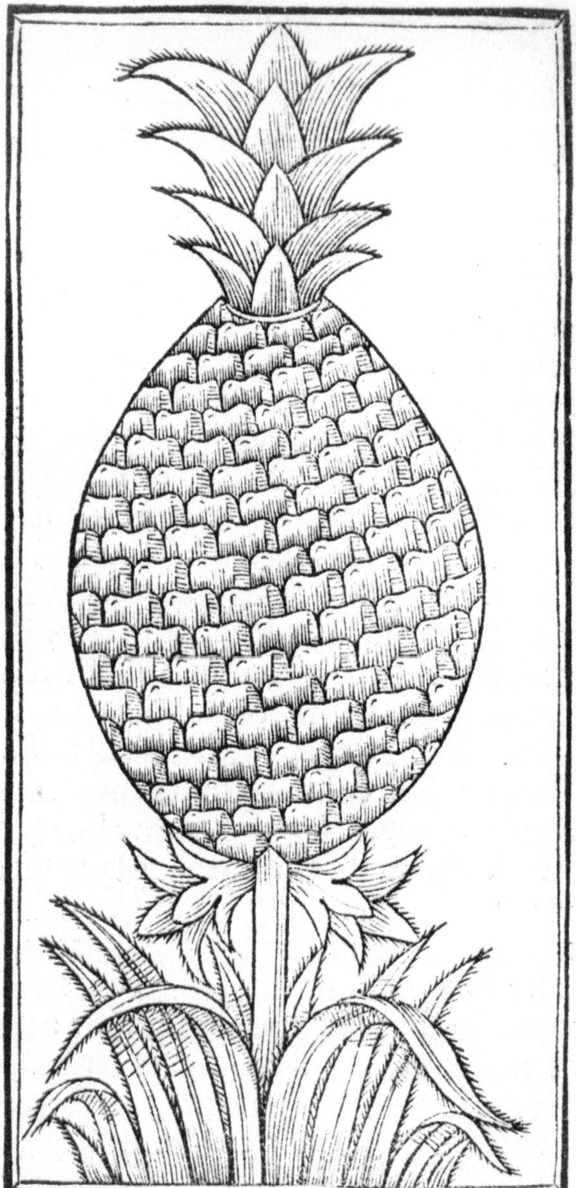

FIGURE 3.3 A pineapple, illustration from *De la natural hystoria de las Indias* (Natural History of the Indies) by Gonzalo Fernández de Oviedo (1478–1557), 1526 (woodcut), earliest representation of this fruit. Image courtesy of Bridgeman Images.

Spanish court. The post of chronicler of the Indies held by Oviedo and intended to act as focus for information-gathering and further-published history was abandoned on his death and left in abeyance for fifteen years.

Even at the highest levels there seems to have been widespread ignorance of basic things, as witnessed by what is usually assumed to be the earliest portrayal of the landscape of America by a European artist. Jan Mostaert (*c.* 1475–1552/3), a court painter for Margaret of Austria, the aunt of Charles V and governor of The Netherlands, painted a strange *New World Landscape* probably in the mid-1540s. Although the story it depicts is unclear, it was hardly an accurate depiction of Caribbean topography, people, fauna, or flora.[4] Although Mostaert had never crossed the Atlantic, his inaccuracy was not because of a lack of skill or available knowledge. Margaret's court would have been well informed by letters, books, and visitors who had actually been to the new Spanish colonies. Instead, perhaps the painting marked a new indifference, which went hand in hand with curtailing of publication of first-hand accounts, and the willful and widespread destruction or ignoring of indigenous knowledge.

Spain had good knowledge, too, of successful agricultural systems of the South and Central American civilizations. The empires they had conquered utilized plants from well outside their geographical origins, showing that plant migration and exchange networks were already established. There were gardens of medicinal plants, while many members of the upper class had ornamental gardens. Bernal Diaz (*c.* 1496–1584), one of the conquistadores, was one of the earliest to see Montezuma's gardens: "with their many varieties of flowers and sweet-scented trees ... his gardens were a wonderful sight" (Carasco 2008: 172; Serés n.d.: 289). But, unsurprisingly, it was the "useful" plants that were the first to attract Spanish attention, and there was an initial wave of introductions to Europe: maize, pumpkins, and chili all appear in Fuchs' herbal in 1542. However, with the exception of tobacco, new plants of medicinal use, which might also reasonably be expected to have been of great interest, hardly appeared in a European herbal for a century after Columbus' first voyage, and this might also imply that no attempts were being made to grow them. It was not until Gerard's *Herball* of 1597 that plants of potential medical use begin to creep in to Western texts. While this may have been because the "foreign" drugs had not been accepted by European medicine, it was echoed by either confusion or a lack of information about many other plants and their origins.

The very word "Indies" was so ambiguous that it led to the dissemination of all sorts of errors and misunderstandings. Maize, one of the staples of the indigenous diet, was, for example, at first associated with Pliny's "millet of India" by Oviedo, an error accepted in the first reference in a European herbal by Hieronymus Bock in 1539 who thought maize came from India. Fuchs, in 1542, thought it had been bought from Asia by the Turks and called it *frumentum Turcicum*. Jacobus Theodorus (Jacob Diether, 1588–91)—known as "Tabernaemontanus" and the so-called father of German botany—thought there were two kinds: *frumentum Turcicum* and *frumentum Indicum*, the latter, correctly, traced from America via Spain. Gerard agreed with him but still called it Turkey corn or Turkey wheat, while Parkinson thought that *frumentum Indicum* came from both West and East Indies but was "not found natural in any place."[5]

Nevertheless, there were two important non-herbal texts about new medicinal plants since, despite the slow atrophy of the Spanish state, there remained a strong culture of a more open humanism amongst its scholarly community. The great French botanist and Imperial gardener, Charles de l'Écluse (1526–1609)—better known as Clusius—saw this for himself when he visited Seville in 1564–5 and met physicians, botanists, and plant collectors, and heard how they had an established an exchange mechanism with correspondents in Mexico. One of those that Clusius visited in Seville was Nicolas Monardes (1493–1588), who was in the first generation of those growing up with the existence of a new continent largely under Spanish control as part of his everyday experience. In the very year of Clusius' visit he published the first part of his *Historia medicinal de las cosas que se traen de nuestras Indias Occidentales* (Medical Study of the Products Imported from our West Indian Possessions). Gradually enlarged, it appeared in further editions under varying titles between 1569 and 1580. It was based on his investigations in Seville, where ships returning from the Americas would dock and unload, as well his correspondence with Europeans in Peru and Mexico, and his commissioning of travelers to bring him seeds and samples on their return. He grew, observed, classified, and experimented with them in his garden in Seville. It would be surprising if he did not also share and exchange them through his network of correspondents.

Unfortunately, Clusius did not have the opportunity to meet the other pioneer author about non-European medicinal plants, since Garcia de Orta (1501?–68), a Portuguese physician, had gone to Goa in 1534 and remained there until his death in 1568. Instead Clusius heard about Orta's *Colóquios dos simples e drogas he cousas medicinais da Índia* (Conversations on the simples, drugs and medicinal substances of India), published in Goa in 1563, by chance when he was in Lisbon in December 1564. Soon afterwards he translated it into Latin (Clusius 1567). Apart from its insights into Indian medicine, Orta was scornful of Portuguese attitudes to plants:

> The Portuguese, who navigate over a great part of the world, only procure a knowledge of how best to dispose of their merchandize, of what they bring there, and what they shall take back. They are not desirous of knowing anything about the things in the countries they visit. If they know a product, they do not seek to learn from what tree it comes, and if they see it they do not compare it with one of our trees, nor ask about its fruit, nor what it is like.
>
> (Markham 1913: 86–7; Orta 1563: ch. 12)

He also gave some useful insight into the plant trade in the Persian Gulf and Indian Ocean, explaining for example that there was a flourishing trade in aloes, particularly from Socotra to Turkey, Persia, Egypt, and thence to Europe (Markham 1913: 6; Mathew 1997; Orta 1563: ch. 2).

Monardes' work was soon well known across Europe. In addition to Clusius' Latin an Italian version came out in 1576, a French one followed, and it appeared in English in 1577 as John Frampton's *Joyfull Newes out of the New Founde Worlde: wherein is declared the rare and singuler vertues of diverse and sundrie hearbes, trees, oyles, plantes and stones, with their aplications, aswell for phisicke as chirurgerie*. Frampton updated it in 1580 based on Monardes' revised edition of 1574. He highlighted the importance of the plant trade, particularly for medicine: "since the aforesaid medicines mentioned in the same work of Doctor Monardes are now by merchants and others brought out of the West Indias into Spain, and from Spain hither into England by such as doth daily traffic thither" and are known to the English to be precious remedies for all manner of illnesses incurable by "the old order and manner of physic, as also proven by their use in Spain and other countries ... " (Frampton 1580: 4).

Frampton (1559–81) was a Bristol merchant, who had been based in Seville until increasing hostility between Spain and England led him to be being imprisoned and having his goods confiscated before he returned to England. Once back, apart from Monardes, he also translated recent Spanish books on navigation, and traveled to the West Indies and Asia. It is significant that most of these works were dedicated to Sir Edward Dyer, who was secretary to Elizabeth's favorite, the Earl of Leicester. They found a receptive audience since the political and economic mood in England was changing rapidly as the country began to emerge as a mercantile marine power, eager to return to exploration, expand overseas trade and begin colonisation.

"HOW TO DRESSE, SOWE AND SET A GARDEN"

This all coincided with a new interest in gardens and garden-making. The first real gardening book in English had only been published in 1558. Thomas Hyll's *A most briefe and pleasaunte treatise teaching how to dresse, sowe and set a garden* was only forty-two pages long and based on extracts from classical writers which he had "Englished," but it must have been relatively successful because it was extended to sixty-eight pages and reprinted in 1563. It was then enlarged again and renamed in 1568, when it appeared as *The proffitable arte of gardening now the third tyme set fourth*. Why had Hyll (b. *c*. 1528), who was the author of other popular works on everything from astrology and physiognomy to dreams and beekeeping, switched his attention to gardening?

His second gardening text, *The Gardeners Labyrinth* [sic], which appeared posthumously in 1577 under the pseudonym of Didymus Mountain, was also largely based on classical sources and says little about plants and nothing about new introductions. However, its dedication to Sir William Cecil (1520–98), Elizabeth's chief minister and himself an extremely keen and innovatory gardener, speaks volumes. Cecil already had impressive gardens at Theobalds (in what is now Cedars Park in the parish of Cheshunt in the English county of Hertfordshire), and by 1567, he had just completed the building of his new house on the Strand in London, complete with an expensively ornamented garden which John Gerard (1545–1612) took charge of around 1577 (Husselby and Henderson 2002). But Cecil was not alone in constructing new gardens around this very time.

Christopher Hatton (1540–91), the Lord Chancellor, was building a vast palace with grounds to match at Holdenby (in Northamptonshire) in the 1570s, whilst in 1573–5 Robert Dudley (1532–88), Earl of Leicester, created the splendid garden at Kenilworth (in Northamptonshire) for a state visit by Queen Elizabeth I. The latter had an aviary of exotic birds, and like Cecil's gardens, was ornamented with Italian marble. There is no mention of exotic plants at any of these locations, in any of the surviving plans or descriptions, but if they were anywhere in England at this time they would have been in these prestigious settings. Yet within a few years such exotics were not uncommon.

FRANCISCO HERNÁNDEZ DE TOLEDO AND SPANISH BOTANIZING

In the Americas, plant exploration continued. In 1570, around the same time that Monardes and Orta were writing, Philip II of Spain sent Francisco Hernández (1514–87), one of his personal physicians, and his son Juan to carry out empirical research into Mexican flora and fauna and medicinal plants. The latter spent seven years collecting and collating indigenous knowledge, and the results were little short of astounding. Whereas Theophrastus had identified around 350 plants, Dioscorides and early Muslim botanists around 500–600, Hernández identified more than 3,000. He wrote up his findings in Latin in twenty-two manuscript volumes and began to translate them into Spanish and Nahuatl, but on his return to Spain it was clear that however much Philip might have wanted the information, he was going to do very little with it. Not only had the European religious and political situation altered the crown's priorities, but it is quite possible that the sheer volume of what Hernández found made it controversial, its richness and variety threatening Philip's more orthodox world view (Weiner 2002). Alternatively, the information may have been viewed as too potentially advantageous to disseminate from a commercial perspective. In any event, like the final section of Oviedo's work, that of Hernández remained unpublished on the library shelf. The drawings and pages of the manuscript were even given away as "curious" presents. The manuscript was lost in a fire at the Escorial in 1671, but fortunately Hernández had made copies, including some left in Mexico, and from 1615 onwards some parts were published. It was translated into Dutch in the 1630s and into English in 1659.

Yet despite the restrictions imposed by the Spanish, from the 1570s onwards, some exotic plants and presumably knowledge about them, their origins and properties, still managed to reach European collectors and their gardens. There is little direct evidence of how this happened, although the answer is fairly obvious. From 1492 onwards, ships were criss-crossing the Atlantic and Indian oceans in ever-increasing numbers, and bringing back not just the recognized economically important cargoes of gold, silver, and spices, but anything and everything that took the voyagers' interest. Although plants themselves would not have survived the long crossings, they could be bought back as seeds or bulbs. These required very little effort to transport since they were small, dry, and easily packed not just by the owners and masters of the ships but every member of their crew. There are no entries for such things in any of the surviving port books, which record ships, their masters, and cargoes on which duties were payable, although even the data for commercial imports of plant material such as spices or raw medicinal products was limited before the early seventeenth century (Wallis n.d.). Nevertheless, this must have been how most plant specimens originally arrived.

"THE SWOLLEN COLEWOORTE OF AL OTHERS IS THE STRANGEST"

The sack of Antwerp by the Spanish in 1576 shifted the economic fulcrum of northern Europe very quickly to London and Amsterdam. But the continuing wars in The Netherlands and the continuing influx of religious refugees to London cemented its position as the new entrepôt for horticulture and much else besides. London now contained a community of knowledgeable apothecaries, physicians, and botanists from a mix of nationalities including many from France, Flanders, and Holland. They included Petrus Pena and Matthias de Lobel, authors of *Stirpium adversaria nova* in 1571, a pioneering book which began the process of classifying plants. Nor was it just an intellectual community but a physical one too, since many of these newcomers lived on and around Lime Street in the City of London. The Flemish diaspora stayed connected. Families such the Coels (anglicized to Cole), the Garets, and van Meterens maintained personal links with family and friends in Europe as well as having trade links. These networks were crucial in the introduction and exchange of plants across the whole continent.

From surviving letters, books, and friendship albums (*alba amicorum*) it is possible to construct a list of friends and connections showing the extent of this community. In addition to those in and around Lime Street, this community included Edward, Lord Zouche, a diplomat and courtier who had a botanic garden in Hackney supervised by Lobel; William Camden the antiquary; Edmund Spencer the poet; Richard Hakluyt the writer; and Hugh Morgan the Queen's apothecary. Lobel's *Stirpium adversaria nova* lists another dozen or so doctors and apothecaries from whom he had obtained information, and in the "*Observationes*" forming part of *Plantarum seu stirpium historia* (1576) another series of plant collectors and gardeners including Dr. Hector Nonnius, a Portuguese living in London; Sir Nicholas Pointz (knight of Iron Acton in Gloucestershire); Dame Catherine Killigrew (who was William Cecil's sister-in-law); and Sir Thomas Throckmorton. The circle even numbered explorers and privateers including Francis Drake, Walter Raleigh, and Thomas Cavendish.[6]

It was against this background that John Gerard published his *Herball or Generall Historie of Plantes* in 1597. He drew on the enormous reservoirs of information and support from the wider scholarly community in compiling it, and the fact that Gerard fell out with these acquaintances over errors in his *Herball* does not stop it providing specifics about the trade in plants. One of the reasons Gerard's *Herball* has become more famous than its predecessors, notably the work of Lobel, is because Gerard wrote in English rather than Latin and had a more conversational style. Equally important for tracking the history of plants, he sometimes detailed how, when, and where plants were found or who gave them to him. This is obviously particularly important in the case of "exotics." Apart from the Lime Street community there were overseas acquaintances such as the French royal gardener Jean Robin, who had sent him seeds of nasturtium (native to South America). There were a series of professional gardeners: Vincent Pointer of Twickenham, Henry Banbury of Tothill Street, Master Fowle the keeper of the Queens House in St James, and Master Huggens Keeper of Hampton Court. His work reveals that Cecil and Zouch were not the only aristocrats with interests in gardening: for example, the Earl of Sussex grew fine melons in Bermondsey, London whilst Gerard also noted particular plants growing in the gardens of Sir Francis Walsingham's at Barnes (or, Barn Elms, London) and at Sir Francis Carew's house at Beddington, Surrey. Some of the other plant lovers mentioned—William Martin, a surgeon and later master of the company of Barber Surgeons, and William Gooderus, the queen's chief physician—came from the usual medical backgrounds but others came from the commercial elite of the City. These included Sir James Harvey, one-time Master of the Ironmongers' Company, who lived on Lime Street and was mentioned for growing "mad apples" or aubergines; John de Franqueville, originally from Cambrai, who knew both Pena and Lobel but who was trading from London by 1582; and Nicholas Leate.

Leate (1565/6–1631) was a prominent and well-connected member of all the major trading companies, including the East India, Levant, Turkey, and Virginia. He was deeply involved, too, in civic and national politics, being a Deputy Alderman, three times master of the Ironmongers' Company and an advisor to the Privy Council. He was also a very assiduous plant collector. Gerard reported, for instance:

> The swollen Colewoorte of al others is the strangest, which I received from a worshipfull marchant of London master Nicholas Lete, who bought the seeds thereof out of France; who is greatly in love with rare and faire flowers and plants, for which he doth carefully send into Syria, having a servant there at Aleppo and in many other countries; for the which my self and likewise the whole land are much bound unto.

Gerard grew a novel kind of red beetroot from seed obtained by Leate "from beyond the seas," as well as "a gilliflower with a yellow flower. The which [Leate] procured from Poland and gave me thereof for my garden which before that time was never seene or heard of in these countries" (Gerard 1597: 246, 251). A further link is also possible since amongst his portfolio of investments Leate had shares in many ships, including one of 150 tons named the *Hercules*, and it was in a ship of that name that Gerard's servant William Marshall secured a post as surgeon to go plant hunting in the Mediterranean.

GARDENING'S "CREEPE INTO ENGLAND"

Gerard's *Herball* is one sign of how English horticulture had been transformed by the time of Elizabeth's death in 1603. This process had begun in the early years of her reign, speeded up as exploration and trade routes increased access to the wider world, and was helped by the presence of so many European intellectuals and entrepreneurs in London. With the accession of James I in 1603 there was another step change in the general economy and particularly in horticulture. It was summed up by Sir Julius Caesar, James' Chancellor of the Exchequer and himself the son of an early Venetian immigrant, who commented proudly in 1608 that while during the previous reign "Englishmen were not so skilful in trades to make all kinds of wares ... now ... the people had mightily increased both in numbers ... and in all good skill ... and [are] skilful in all kind and manner of trades" (British Library, Lansdowne MS 152/64/237).

That horticulture was one of these trades that were transformed is reflected in remarks made in a long letter sent to Samuel Hartlib, an emigre from Danzig (Gdansk), and published in *His Legacy of Husbandry*. "Gardening"—by which was probably meant productive horticulture—"is but of few years standing in England, and therefore not deeply rooted, nor well understood. About 50 years ago, about which time ingenuities first began to flourish in England, this art of gardening, began to creepe into England" (Hartlib 1655: 8–9).

One of James' many interests as king was in the wider "public health" of his people, as seen in his establishment of the Society of Apothecaries, and interventions in the affairs of the Grocers' and Barber-Surgeons' companies and the Royal College of Physicians. But, conscious of the great European famine during the 1590s, he also took an interest in improving food production. On September 18, 1605, a group of gardeners "inhabiting within the city of London and within six miles compass thereof" were granted a royal charter which gave their new company near-exclusive rights over "gardening, grafting, setting, sowing, cutting, arboring, rocking, mounting, covering, fencing and removing of plants, herbs, seeds, fruits, trees, stocks, sets and contriving the conveyances to the same" (Guildhall Library, MS 3396, Appendix 1, Letters Patent September 18, 1605). Although in its early days the company represented the whole range of the trade, from the aristocracy's gardeners—like the earlier-mentioned professional gardeners Banbury and Fowle, who were amongst Gerard's contributors—to common laborers, most were just still growing basic field crops. The range of skills and specializations outlined in the charter should not be taken literally to mean that all the members were capable of doing them: indeed, there was very little specialization until after the Restoration.

TERRESTRIAL PARADISES: PRESAGING THE AGE OF COMMISSIONED COLLECTION

In the early years of the seventeenth century ornamental plants and planting were of little commercial value, and the concept of a botanical garden, although established on the Continent in the sixteenth century, was still in its infancy in England, with the first not being established until 1621 at Oxford. Exotics were still largely reserved for the wealthy "amateur" and they still obtained their specimens largely through the exchange networks discussed above rather than purchase. But valuable information and exotic books about their exotic plants *could* be purchased. John Parkinson's lavishly illustrated folio volume *Paradisi in sole Paradisus terrestris* (1629) was not aimed at the man who *must* garden, as perhaps Hyll's early books or those of Gervase Markham were, but at those who *wanted to* garden: "gentlemen of the better sort and quality" who were seeking "the most choicest" plants. This is almost the beginning of gardening of the "U and non-U"

FIGURE 3.4 Yucca. From John Parkinson, *Paradisi in sole Paradisus terrestris* (1629: 435). Hand-colored copy. Image courtesy of Getty Research Institute.

sort (upper class and aspiring middle class, respectively) with plants ranked into the fit and unfit, aimed at "establishing gardening as a fit pursuit for gentlemen and women—a better sort of gardener" (Francis 2008). It was a luxury book bringing the rare and exotic to the attention of a wider public. It shows the extraordinary range of plants that had been introduced, most of them comparatively recently, including "the Marvaile of the world," "the Indian Cresses, or yellow Larkes heeles [nasturniums]," "the Flower of the Sunne [sunflower]," and the "Indian Daffodil [*Sprekelia formosissima*]" with a red flower.

Parkinson (1567–1650), like Gerard, often recorded the plant's provenance, and from that one can reconstruct the network of exchange that must have existed. The case of the yucca is a good exemplar (see Figure 3.4). Parkinson's own plant came from John de Franqueville, who in turn had received it from Vespasian Robin at the Jardin du Roi in Paris. It was Vespasian's father Jean who had been sent a plant by John Gerard, and he had obtained his from an Exeter apothecary whose servant had originally bought it back from the Caribbean.

It also shows the difficulty that plant collectors faced with new introductions. The original yucca "perished with him that got it from his widow" while Parkinson knew nothing of "the vertues" of the yucca of which he was so proud. "Wee haue not heard of any, that hath either read, heard, or experimented the faculties hereof, nor yet whether it hath good or evill taste; for being rare, and possessed but by a few, they that have it are loth to cut any thereof, for feare of spoiling and losing the whole roote." And little was known of growing conditions, so that plants failed: "This naturally groweth in the West Indies, from whence it was brought into Spaine, where it bore both in June and July … and hath beene sent from Spaine, unto divers lovers of plants, into severall parts of Christendome, but have not thrived long in these transalpine colder countries, so far as I can heare" (Parkinson 1629: 434).

Parkinson reported, too, that new plants were still coming from merchants like Leate, who sent him a double yellow rose from

> Constantinople, which (as wee heare) was first brought thither from Syria; but perished quickly both with him, and with all other to whom hee imparted it: yet afterwards it was sent to Master John de Franqueville, a merchant also of London, and a great lover of all rare plants, as well as flowers, from which is sprung the greatest store, that is now flourishing in this kingdome.
>
> (Parkinson 1629: 420)

But they were also coming from a much wider range of people, because crossing the Atlantic or Indian Oceans, while still dangerous, was by the mid-seventeenth century almost routine. It now became possible for the wealthiest collectors no longer to rely on the arbitrary arrivals from merchants such as Leate but to safely commission their servants or agents to go abroad and collect plants for them and their friends. The age of deliberate commissioned plant hunting and collecting was beginning.

CHAPTER FOUR

Plant Technology and Science

Frondi tenere e belle

INGRID D. ROWLAND

Frondi tenere e belle
del mio platano amato
per voi risplenda il fato.
Tuoni, lampi, e procelle
non v'oltraggino mai la cara pace,
né giunga a profanarvi austro rapace.
Ombra mai fu.
di vegetabile,
cara ed amabile,
soave più.

Tender and beautiful fronds
of my beloved plane tree,
let Fate shine for you.
May thunder, lightning, and storms
never offend your cherished peace,
nor the fierce south wind come to disturb you.
Never was the shadow
of a dear and lovely vegetable
so pleasant.

(opening aria, Georg Friedrich Händel's 1738 opera *Serse*)

Quid *πλατανῶν* opacissimus? How about that shadiest of plane trees?
(Pliny the Younger, quoted by George Seferis, *Mythistorêma*)

INTRODUCTION

The opening aria for Georg Friedrich Händel's 1738 opera *Serse* (*Xerxes*) may have been one of the most beautiful pieces of music he ever composed, but, like the opera it heralded, "Ombra mai fu" left its first London audience in a state of hopeless confusion. After an overture decked out with all the flourishes appropriate to an *opera seria* (on a serious theme), the legendary Persian king Xerxes, sung by a soprano castrato, strode onstage in his plumed finery to spin out a ravishing love song—to a tree. The lyrics were as absurd as the music was sublime: "Never was the shadow of a dear and lovely vegetable so pleasant." Furthermore, the aria ended abruptly, without the customary ornamented repetition, giving the public approximately three minutes or so to register the crazy lyrics, the poignant melody, and the weird effect produced by the two in concert (Briant 2002: 234–5). The rest of *Serse's* mixture of high tragedy and low comedy, conveyed in short, ballad-like arias, resulted in an almost complete failure of communication between composer and audience; only the learned Fourth Earl of Shaftesbury (1711–71) would later write, "My own judgment is that it is a capital opera notwithstanding tis called a ballad one."[1]

Yet for the same reasons that it perplexed Enlightenment London, *Serse*'s seventeenth-century libretto, with its tree-loving Persian king, had captivated Baroque Italy in earlier decades, in earlier musical settings. The original libretto, including "Ombra mai fu," was drafted (as *Xersé*) by the Venetian lawyer Nicolò Minato in 1654, at a time when tragedy and comedy could intermingle easily onstage, and no one, in the aftermath of the Dutch tulip crisis of 1637, doubted the lengths to which a person might go for a plant.

Xerxes' love song to the plane tree appealed to Baroque sensibilities on a variety of levels, all of them with direct bearing on the way Europeans of the time regarded vegetation: a regard conditioned by their renewed interest in classical Greek and Roman literature, a new world of intercontinental commerce, new developments in natural philosophy, and, not least, a way of life, and of international diplomacy, centered on villas and gardens, evoking ancient Roman leisure while furthering the far-from-leisurely pursuit of contemporary political and business interests. By 1694, when a new musical setting of *Xersé* was presented to Italian audiences (again including "Ombra mai fu"), a glorious Oriental plane tree (*Platanus orientalis*), the same species as the famous tree of Xerxes, had been spreading its *frondi tenere e belle* over the botanical garden of Padua for fourteen years. Today, hollowed out after a long-ago lightning strike, that majestic "vegetable" is still thriving, as is Händel's glorious aria to *Platanus orientalis*, now at last assured of a devoted audience.

The popularity of this story of Xerxes and his plane tree provides a perfect example of the Early Modern fascination with classical antiquity: the tale came from the Greek historian Herodotus (*c.* 484–*c.* 425 BCE), who recorded that in 480 BCE, as the Persian monarch and his troops marched across Asia Minor in hopes of conquering Greece, they paused just before reaching Sardis, the endpoint of the Persian Royal Road, and there Xerxes "found a plane-tree, which he adorned with gold because of its beauty, and he assigned one of his Immortals [the Persian imperial guard] to guard it. On the next day he reached the city of the Lydians."[2]

Planes were apparently venerated as holy on the arid highlands of ancient Persia, and are still dearly loved for their shade all over the sunny Mediterranean; at an earlier stage of Xerxes' journey, his counsellors reminded him that Pythius, one of his hosts along the royal progress, had given his father, Darius I, a golden statue of a plane tree. And trees certainly figure prominently in the reliefs at Persepolis—capital of the Persian Achaemenid Empire from the reign of Darius I (the Great, r. 522–486 BCE) until its destruction in 330 BCE—where vast armies pay tribute to Darius (see Figure 4.1).

Not everyone in the ancient world shared Xerxes' arboreal enthusiasms. In the early third century CE, the Roman writer Aelian [Claudius Aelianus, *c.* 175–*c.* 235 CE] scoffed:

> Xerxes deserves justly to be laughed at, who ... fell in love with a Plane, and doted upon the Tree: for seeing (as it is reported) in Lydia a tall Plane-tree, there he stayed a whole day, no necessity requiring, and pitched his Tents in the Wilderness about the Plane-tree: he also hung upon it many rich ornaments, honouring the boughs with chains and bracelets, and left it a Keeper, as the Guardian and Protectour of a Mistress. But what did this profit the Tree?
>
> (*Varia historia* 2.14; Stanley 1665: 39)

FIGURE 4.1 Cypress tree and a Persian soldier, Eastern Staircase, Apadana Palace, Persepolis, Iran, late sixth–early fifth century BCE. Photo by Stefano Politi Markovina/Alamy Stock Photo.

Aelian's second-century countryman, Gaius Plinius Caecilius Secundus (Pliny the Younger, 61–*c*. 113 CE), felt almost as passionately about *Platanus orientalis* as the Persian king; in a letter to a friend in northern Italy, he asked:

> What's happening in our beloved Como? How about your marvelous suburban house? How about that portico where it's always spring? How about that shadiest of plane trees?
>
> (Pliny, *Letters* 1.3 to Caninius Rufus)

In 1935, the Greek poet George Seferis (who won the Nobel Prize in Literature in 1962) would adopt Pliny's question about the plane tree, *quid platanon opacissimus* (How about that shadiest of plane trees?), as the epigraph for one of the poems in his collection *Mythistorêma* (Mythical Narrative), for the Greeks have always been as passionate about plane trees as the ancient Persians; according to Pliny's uncle, tourists to Delphi in the first century CE were shown an exemplar planted by none other than the legendary King Agamemnon, leader of the vast Greek contingent that initiated the Trojan War.[3] In broad strokes, this chapter, in which the plane tree will be a recurrent motif, traces the acquisition and practical application of plant knowledge, botanical "science," in Early Modern Italy. Insofar as it is possible, the focus of this story will be on plant knowledge and technology in contexts other than medicine and pharmaceutics, disciplines that had historically resided at the core of botanical studies.

THE CLASSICAL TRADITION AND EARLY MODERN PLANT KNOWLEDGE

The tale of Xerxes and the plane tree may have ranked as one of the most picturesque bits of plant lore from classical antiquity, but most Early Modern readers consulted ancient authors for more serious purposes, seeking out practical advice about medicine and agriculture in their own day. Botany, the scientific investigation of plants as an end in itself, only began to emerge as an independent discipline towards the end of the modern period, but everyone, from the Middle Ages to the Scientific Revolution, revered the accumulated plant knowledge of ancient writers and their later commentators as an incomparable, authoritative, and evergreen source of information about the cultivation and use of plants, as applicable to modern times as to antiquity. Among the most popular of these ancient botanical writers, at least among Early Modern readers, was surely Pedanius Dioscorides of Anazarbus (*c.* 40–*c.* 90 CE), a Greek physician assigned to accompany the Roman army in the last years of the emperor Nero's reign. His treatise "on medicinal materials," *De materia medica*, filled five papyrus scrolls with information about over six hundred medicinal plants, with such authority that Romans of the imperial age continued to copy the text for generation after generation. Dioscorides is a Greek name, and he wrote in Greek, the official language of the eastern half of the Empire. As the Roman world began to break apart, scholars translated Dioscorides into their own languages: Latin (eighth century), Arabic (ninth century), and Syrian (ninth century); no matter the language of its presentation, the information contained in *De materia medica* retained its timeless validity. Later scribes often added copious notes to the text, culled from Arab and Indian medical books, as well as illustrations, creating beautiful manuscripts packed with a wealth of information, each one of them unique. In effect, every pharmacist, every physician, every chemist and alchemist in the Early Modern Mediterranean world referred to Dioscorides as an indispensable source, but before the age of printing, each copy of Dioscorides' text differed from every other to at least some degree (see Chapter 5 of this volume for a detailed discussion of plants and medicine) (see Figure 4.2).

Some of the most spectacular surviving manuscripts of this essential text arrived in Venice, the principal crossroads between Constantinople and western Europe, and between the Mediterranean and Germany. Here the first printed edition of Dioscorides was published in 1478, a Latin translation based on an eighth-century (Latin) manuscript now in Munich.[4] In 1499 the Venetian printer Aldus Manutius (1449/52–1515), responsible for so many editions of classical texts, published the first Greek Dioscorides. Advances in classical scholarship alerted readers to the shortcomings of the medieval texts on which these earliest editions were based, full of mistranslations and later insertions, and in 1544, the Sienese physician Pietro Andrea Mattioli (1501–*c.* 1577) tried to remedy the situation by publishing a new, scholarly translation of Dioscorides into Tuscan vernacular, together with brief comments about the individual plants. By writing in vernacular, Mattioli hoped to attract a wider circle of readers in his own region. He may also have been testing the waters for a far more ambitious project, first published in 1554: a carefully revised Latin Dioscorides with over 560 woodcut illustrations and his own extensive commentaries, aimed at an international audience. Mattioli continued to improve this grand Latin text for the rest of his career: his revised Venetian edition of 1566 ran to more than 1,600 pages and exerted a European-wide influence.

Other important ancient botanical works, on the other hand, were only recovered from obscurity in the fifteenth century, part of the transformation contemporaries called the

FIGURE 4.2 Sweet violets, miniature from the *Tractatus de herbis*, attributed to Dioscorides Pedanius (*c.* 40–*c.* 90), Latin manuscript, f. 138r, 1458. Fifteenth century. Modena, Biblioteca Estense. Photo by DEA/DeAgostini/Getty Images.

"rebirth of letters" and we call the Renaissance. On a book-hunting trip to the monastery of St. Gall in Switzerland in 1417, the Florentine cleric Poggio Bracciolini (1380–1459) found, among other treasures, a ninth-century copy of *De re rustica* (On Rural Matters), an agricultural treatise composed in the first century CE by Lucius Junius Moderatus Columella (4–*c.* 70 CE). During the later Middle Ages, Columella's name was well known, but his work had survived only in tantalizing fragments. *De re rustica* appeared in the venerable manuscript together with a work specifically devoted to trees, *De arboribus* (On Trees), transmitted under Columella's name but probably written by another author, alongside several other works by Roman agricultural writers, known collectively as the *Scriptores de re rustica* (writers on rural matters). These *scriptores*, Marcus Porcius Cato the Elder (234–149 BCE), Marcus Terentius Varro (116–27 BCE), and Palladius Rutilius Taurus (fourth or fifth century CE), may have been wealthy men who delegated the day-to-day running of their farms to slaves, but they nonetheless maintained a lively interest in the details of country life. The earliest Romans had been humble farmers and herdsmen, as later Romans remembered with reverence. Even the cosmopolitan citizens of the late Republic and the Empire continued to identify their own values with the frugal traditions of their agrarian ancestors, consciously modeling themselves on Cincinnatus, the fifth-century BCE Roman consul who after serving two terms as emergency dictator returned "to the same bullocks, and his small ancestral inheritance of four *iugera*."[5] Columella's tenth book (of twelve) is composed in dactylic hexameter verse in tribute to the most popular agricultural writer of all, the poet Vergil (70–21 BCE), who survived the Middle Ages without ever falling out of fashion. In his second major work, the *Georgics* (Agricultural Matters) of 29 BCE, he discussed everything from beekeeping to constructing a plow—all in gloriously convincing hexameter poetry. Vergil may have been the supreme Latin poet, but he also knew how it felt to have dirt under his fingernails, and so did many of his Renaissance readers.

Although Early Modern Italy's political structures and patterns of land use had changed drastically from the slave farms (*latifundia*) of late antiquity to the sharecropping system known as *mezzadria* (in which peasants gave half their produce to property holders in exchange for housing and employment), these ancient agricultural writings struck a chord with Early Modern landowners, who identified their responsibilities within their own city-states with the civic virtues of the ancient Roman Republic, and their country life with the ancient Roman patrician idea of cultivated leisure, *otium cum dignitate*.[6] Poggio took his identification with the ancient Romans still further: mistaking the Carolingian minuscule script of his Columella and other ninth-century manuscripts for genuinely ancient Roman handwriting, he altered his own penmanship to match, thereby creating modern Italic lettering. Other Early Modern Italians adopted the Latin name "Palladius" to bring them closer to Palladius Rutilius Taurus, the ancient agricultural writer, including the Roman humanist, papal secretary, and avid gardener Biagio Pallai (d. 1554), who wrote as Blosio Palladio, and the great designer of Venetian villas, Andrea di Pietro della Gondola of Vicenza (1508–80), renamed Andrea Palladio in 1538.

Technology on the farm had changed little from ancient times to the fifteenth century, and in any event the *scriptores de re rustica* and Vergil wrote for the pleasure of landowners rather than to instruct illiterate laborers. Farming techniques were handed down in fields, gardens, vineyards, and orchards by immemorial tradition, adapted to changes in climate as they occurred. The Earth's temperature in antiquity was cooler than the Medieval Warm Period around 1000 CE, but much warmer than the period now known as the Little Ice Age, extending from roughly 1400 to 1700.[7] Villas in Roman Britain included vineyards, but the Elizabethans were compelled to import their wine; grapes no longer survived in the chill climate and sixteenth-century English farmers turned to growing hops instead. Tales abound in the Renaissance of northern Europeans, from King Arthur to a German cardinal to foreign students at the University of Padua, surrendering their inhibitions, and often their dignity, to Italian wine (Brown et al. 2001).

VILLA LIFE IN ANCIENT ROME AND THE RENAISSANCE

The focal point of the rural life celebrated by the *scriptores de re rustica* was the country house or villa (e.g. Columella, *De re rustica* 1.4). Originally the word *villa* simply denoted a farmhouse, but by the time of the later Roman Republic that definition was changing rapidly. One of Columella's fellow *scriptores*, Marcus Terentius Varro (116–27 BCE), devoted the third book of his treatise *De re rustica* to this transformation. If villas had begun as rustic working farms, by Varro's day many of them were turning into showcases for the wealthy, much to the dismay of traditionalists, including, to a certain extent, Varro himself. Most lavish of all were the properties of the famous general and gourmand Lucius Licinius Lucullus (118–57/56 BCE), who decorated his villas with the fabulous booty he brought back in 66 BCE from nearly a decade of successful campaigning in Pontus (a region roughly equivalent to modern Turkey). Lucullus also established a famous suburban garden on the outskirts of Rome, at the top of the modern Via Veneto, decorated with the first glass mosaics known to Roman art, as well as paintings, statues, terraces, and rare plants, including cherries, apricots, and perhaps what we call Swiss chard, all gathered in the course of his foreign conquests and subsequently brought as spoils of war to Italy.[8] Plutarch's *Life of Lucullus* (39.3) reports that one Stoic philosopher described him as "Xerxes in a toga" for his profligacy, but the parallel may run deeper: the Gardens may have been inspired by Persian gardens, like the grove of Daphne outside Antioch, which Lucullus had seen during his military campaigns.

The retired general's lavish retreats were not simply a way to show off to his contemporaries. Lucullus had returned from the East in 66 to find a deadly political situation in Rome. Undercut at every opportunity by his ruthlessly ambitious rival Pompey the Great, he was forced to spend three years outside the *pomerium*, Rome's ritual boundary, waiting for the Senate to award him the triumph for which he was fully qualified (his friend the statesman Cicero finally managed to procure it in 63 BCE). For Lucullus, in effect, withdrawal was the only alternative to assassination, and he took his near-exile philosophically, stocking his villas with libraries and pursuing his lifelong interest in the philosophy of Plato and the Academy. Meanwhile, the opulence of his banquets drew even the devious Pompey to his dinner table. Lucullus and his contemporaries may have called this kind of existence *otium cum dignitate* (dignified leisure), but they all knew that villa life in the late Roman Republic was no longer a rustic retreat. It had become intensely political (see Dewar 2013).

Some 1,500 years later, in 1512, another forced retiree, the Florentine Niccolò Machiavelli (1469–1527), withdrew to his own villa near San Casciano Val di Pesa, tortured three times by the Medici, his career in ruins. Beneath the pines and live oaks, he walked in the woods, hunted birds, and idled in the local tavern.[9] Machiavelli's evenings were not only spent in reading the ancient authors. He also wrote *The Prince* and the *Discourses on Livy*, the works that guaranteed his place in history. He may have retreated from direct participation in the running of early-sixteenth-century Florence, but he exchanged that temporary influence for lasting authority in the world of political ideas.

Both Lucullus and Machiavelli understood that retiring from civic life into a verdant retreat can be the most powerful political move of all. They had learned their lesson from a master: Plato of Athens (428/427 or 424/423–348/347 BCE), whose Academy, taking its name from the famous sacred grove and plane tree-shaded gymnasium on the outskirts of his native city, only pretended to renounce the world. In fact, the Academy, a perfect stopping point for anyone approaching Athens from the southwest, figured on every traveler's itinerary for the eight hundred years of its existence, and fixed its grip on political action all over the ancient world, not to mention later eras, by shaping political thought. It was no accident that one of Plato's most enthusiastic readers in the fifteenth century was the wily old power broker Cosimo de' Medici (1389–1464), who had used his own string of country villas to secure his family's grip on Florence by hatching secret plans for action among his allies, and cultivating the ideas of the scholars he supported as well as his vineyards. It was always safer to confide a secret among the incidental noises of a garden than it was to conspire indoors. Further, knowledge of one's garden's plants and their products, ranging from comestibles to medicaments, dyes, and poisons, had become integral to the nobility's political "toolkit."

In 1419, Renaissance readers recovered access to what is still the most informative textual source of information on ancient Roman villa life: the letters of Pliny the Younger (61–*c.* 113 CE). A native of Comum (modern Como) in northern Italy, Pliny belonged to a successful provincial family that made its fortunes in Rome. His uncle, Pliny the Elder, had risen to the office of admiral, and famously died in 79 CE while attempting to rescue victims of the eruption of Vesuvius. But the elder Pliny had also been a famous naturalist, who wrote a massive *Natural History* that considered virtually everything animal, vegetable, mineral, and artistic. Pliny the Younger was eventually made governor of Bithynia, the province adjacent to Constantinople, but he wrote of villas on the shores of Lake Como, at Lavinium on the coast south of Rome, and in the upper valley of the Tiber river, near present-day Città di Castello. His loving descriptions of these properties guided Renaissance readers, like Cosimo de' Medici, in the design of their own country retreats, where they hoped to feel something of the same joy Pliny expressed when he could return to his own fields and gardens. The vagueness of his descriptions—he was a lawyer, not an architect—allowed Renaissance designers to give free rein to their imaginations in recreating the shape of buildings and gardens, but there was no question about identifying the plants; despite the cooling trend of the Little Ice Age, the trees and flowers that delighted Vergil and Pliny still dominated in Renaissance gardens, as they did in contemporary paintings. Cypresses (*Cypressus sempervirens*) and umbrella pines (syn. stone pines, *Pinus pinea*) lined the roadways, and still do today; groves of live oak (*Quercus ilex*) still provided—and provide—shade year-round. Some of the plants that had counted for Pliny as exotic imports had long since become fixtures in the landscape: the cherries, peaches, and apricots first imported by Lucullus, and citrus fruits, which in Roman times included only citron and lemon. Oranges arrived with the Arab conquerors of Sicily in 850 (Langgut 2017: 818).

Renaissance villas' gardens and their plantings, which evidenced a novel union or interplay of nature and art in which the designers and their patrons took both pride and delight, resulted in a growing interest in (and knowledge of) plants—both in and of themselves and in their cultivation—based less on their pharmacological utility than on their aesthetic qualities and their rarity (Bellorini 2016: 51). In the case of Cosimo de' Medici's garden at Castello, its plantings and their intermingling with architectural forms was described by philologist Claudio Tolomei (1492–1556) as having the appearance of "natural artifice" and "artificial nature" at once (Lazzaro 1990: 61). And, as his letters reveal, Cosimo had keen interest in the procurement, placement, and propagation both ornamental plants and those planted in gardens of simples. Walls and grottoes at Castello were draped with living veils of vegetation, while shrubs and trees of interesting form or imposing size served as architectural elements or artless topiary. There was a wondrous allée of mulberry, a faux woodland ("*salvatico*") of cypress, ilex, and fir; and a labyrinth of box, fragrant myrtle and bay laurel, and tapering cypress. The garden included plants rare and novel in Italian gardens, too, such as oleander, a native Mediterranean plant that had not yet found its place in Early Modern gardens; Indian fig (*Ficus Benghalis*); and a pair of imported plane trees.

The opening of the New World to commerce in 1492 brought a whole new repertory of plants to Renaissance gardens and to Renaissance tables, too. The first tomatoes were small and bitter; they entered Italian cooking, and bonded with their soul mate, the basil plant, only after several generations of selective breeding (Gentilcore 2010; see Chapter 2 of this volume). Potatoes, originally restricted to the remote highlands of the Andes, made a comparatively late arrival in Europe, only becoming a staple food in the seventeenth century (Hawkes and Francisco-Ortega 1993; Nabhan 1993: 100–10). Squash, however, was already praised as a filling for stuffed pasta in a treatise of 1584, Giovanni Battista Rossetti's *Dello scalco* (On Stewardship).[10]

EARLY MODERN PLANT REALITIES

The legacy of ancient Greece and Rome may have been crucial in shaping the perception of plants, gardens, and villas in the Early Modern era, but what made that classical reverie so continually compelling, ironically, was its increasing remoteness from daily life. Between the fifteenth and eighteenth centuries, expanding populations and the development of global commerce brought a whole series of rapid changes to European society, ideas, and technology. People who might daydream about classical antiquity had no intention of giving up modern innovations like the magnetic compass, paper, printed books, or (although the thought is terrifying now) "modern" medicine. Renaissance patrons may have built villas inspired by ancient texts, but they swiftly began to work their land using new methods and new tools. As expanding populations in the fifteenth and sixteenth centuries created a pressing need for food, farmers and landlords experimented with intensive agriculture and the development of new staple grains. Rice, for example, had been known to the Mediterranean world since the conquests of Alexander the Great (356–323 BCE), but the Romans used it only as a medicine. The Arabs cultivated rice as a staple food and created delicious recipes, while medieval Europeans continued to classify rice among the spices and use it one pinch at a time. Europe's first rice field was planted in 1482, by the Milanese warlord Ludovico Sforza, who in that same year also hired a Florentine engineer named Leonardo da Vinci (1452–1519). The immense, swampy valley of the River Po provided a perfect environment for the flooded paddies (as well as a large, hungry population), and Leonardo, as one of Ludovico's engineers, must have helped to design the complex irrigation systems that fostered the expansion of this new agricultural industry. Risotto alla Milanese, rice spiced with saffron, made its first triumphant appearance on Lombard menus. Sforza's wife Beatrice d'Este is credited with commissioning the first recipe for what is now a staple Italian sweet, *budino di riso* (rice pudding), in 1491.[11]

Other areas of Italy quickly followed Milan's example, including the islands of Sardinia and Sicily, despite the fact that rice fields clearly helped to spread the "bad air" thought to cause malaria (Majori 2012; Snowden 2006). Ludovico also cultivated mulberry trees and fostered the local production of silk. His father had nicknamed him "The Moor," "il Moro," because of his dark complexion, but "*moro*" was also the word for the purple-black mulberry, which "il Moro" adopted as one of his personal symbols.[12]

Improved methods of irrigation increased the yields of conventional grain, now harvested by a modern tool, the scythe, a large-scale, infinitely more efficient version of the traditional sickle.[13] Sickles forced reapers to bend uncomfortably over their work, cutting only a few ears at a time; the scythe permitted its user to stand upright and mow down swathes of grain or grass. Modern eyes are used to thinking of the ancient god Saturn and the personification of Death as a Grim Reaper with scythe in hand, but both originally wielded sickles; hence the scythe-wielding Grim Reaper in Buonamico Buffalmacco's *Triumph of Death* from the monumental cemetery of Pisa, painted between 1336 and 1341, would have been a startling image in its own time, an ultra-modern Grim Reaper showing off the latest in agricultural technology to one of the most advanced cities of the Mediterranean (see Figure 4.3). In 1348, bubonic plague struck Pisa and brought a new urgency to this nightmare vision of reaping on an industrial scale.

Early Modern farmers developed their own new techniques to increase agricultural productivity: replacing the traditional system of fallowing with a "three-field" system of crop rotation, and the use of cross-plowing to break up soil more finely for cultivation (Sider 2005: 213; see also Chapter 1 of this volume). Rather than allow intensively cultivated fields to recuperate by lying totally fallow, European farmers adopted an Arab technique from Spain, sowing them instead with forage crops like alfalfa and clover, plants that replaced nitrogen in the soil while also providing fodder for farm animals (hence the word "alfalfa" derives from Arabic "*al-fasfasa*" (fresh fodder), though in the Renaissance the plant was simply known as "Spanish grass").[14] Animals, in turn, produced manure that could be used as fertilizer, further increasing crop yields. (Unlike their contemporaries in Asia, Europeans hesitated to use human waste as a fertilizer: Ferguson 2014). By rotating crops on a continuous cycle that alternated nitrogen-consuming with

FIGURE 4.3 The Grim Reaper with the latest in agricultural tools, from Buonamico Buffalmacco, *Triumph of Death*, Pisa, Campo Santo, 1336–41, restored fresco. Image by SaskiaS/Alamy Stock Photo.

nitrogen-fixing plants, farmers could sometimes increase their yield to three harvests per year, with irrigation and manuring further enhancing the richness of the soil.

Early Modern irrigation technology developed with particular sophistication in Italy, with its strong legacy of ancient hydraulic engineering; in Spain, with its own complement of Roman ruins and brilliant Arab waterworks; and in The Netherlands, Europe's most densely populated region, patiently reclaimed from the sea (Russo 2004; Shulman 2018: 29–98). Not surprisingly, the technological miracle of irrigation served a variety of purposes in Early Modern Europe, sustaining the practical needs of agriculture, propelling water mills for Early Modern industries (such as flour, paper, and cloth), but also underpinning the creation of elaborate ornamental gardens that mixed modern technological wonders with plantings meant to heal the body and soothe the spirit. In the new global economy, the search for new medicines and a spirit of competitive display contributed to an extraordinary trade in exotic vegetation.

As a general trend, Early Modern Europeans shifted noticeably from medieval timber or wattle-and-daub construction to buildings of brick and masonry (Tussenbroek 2017). They did so for the same reason as the ancient Romans: fire safety. Roman author and architect Vitruvius (80/70–after *c.* 15 BCE), who must have seen how swiftly flames swept through the flimsy houses of Rome's slum, the Suburra, declared: "However advantageous [wattle-and-daub] buildings may be in terms of speed and for covering broad expanses, they are still a greater source of disaster, and on a large scale, because they are as good as torches when it comes to catching fire" (Vitruvius, *De architectura* 2.8.20; Rowland 2009). Seventeen centuries later, London's Great Fire of 1666 would afford dramatic proof.

Even brick and masonry buildings, however, were roofed with wooden joists and erected with the help of wooden scaffoldings, using different types of timber carefully selected for each separate task (just as Vitruvius had recommended in his own day). The gracefully curved wooden ceilings of Venetian churches like Santo Stefano, and the Basilica in neighboring Vicenza, were actually constructed by shipbuilders, already skilled in shaping curved hulls.

The burgeoning trade networks of the Age of Exploration fed a taste for wood from faraway places. The same cedars of Lebanon (*Cedrus libani*) that had roofed the Temple of Jerusalem and built the Phoenician navy still thrived on the slopes of the Shouf Mountains, as well as in Cyprus and Turkey, but the wood itself remained an exotic import in Europe until the end of the eighteenth century, when cedars were transplanted to Italy with great success. Ebony, an African tree, began to appear in the sixteenth century, its hard, black wood used above all for the elaborate cabinetry that became a universal fashion, especially when contrasted with ivory ornament. Not surprisingly, perhaps, one of the first examples of ebony and ivory furniture in the Renaissance was the four-poster bed that belonged to prominent Italian banker and merchant Agostino Chigi (1466–1520), specifically mentioned in his will and immortalized in the fresco that decorated his bedroom.

The most ingenious use of timber, however, may have been the creation of Venice from a maritime swamp starting in the fifth century CE. Only two parts of the city, Rialto (the name means "high bank") and Dorsoduro ("hardback"), rest on land. All the other tiny islands that compose this unique city are made begin with densely packed pilings of water-resistant Slovenian alderwood or larch, driven through the sand until they sink into a layer of harder clay. The salt water of the lagoon and a lack of oxygen slow the wood's decay, and most of the city's original pilings still survive to this day. Above the pilings, the

Venetians placed slabs of Istrian limestone or wooden planks, and on these foundations they erected their buildings, as bold and magnificent as the very idea of creating a city on the shifting sands of a maritime lagoon. Over a million alder logs, for example, support Baldassare Longhena's magnificent church, Santa Maria della Salute, begun in 1631 and completed fifty years later (Macchioni et al. 2016).

MADONNAS IN TREES AND THE FOREST OF MATTER

As plants were considered elements of nature that formed part of a dense and mysterious web of "being," knowledge of their place in nature as well as of their properties and uses demanded consideration of their symbolic associations. In the case of symbolism, too, antiquity provided a solid foundation on which to build (see also Chapter 6 of this volume). The Roman writer Aelian may have ridiculed the Persian king Xerxes' passion for his plane tree, but the ancient Greeks and Romans had their own tree cults, from the whispering oaks at the oracle of Zeus in Dodona to the sacred trees of Rome. The Romans also attributed supernatural powers to barren trees, termed *arbores infelices*, "unfortunate trees." According to Pliny the Elder, these were "condemned by religion" (*damnatae religione*: *Naturalis historia* 16.45) and used to produce the crosses used in crucifixions. Romans found guilty of treason within the city were either thrown from the Tarpeian Rock on the Capitoline Hill or hanged from an *arbor infelix*. Among the unfortunate species, he numbered the tamarisk, poplar, alder, the Atinian elm (*Ulmus minor* 'Atinia'), and the Italian buckthorn (*Rhamnus alaternus*) (*NH* 16.45).

Even Aelian, despite his harsh words about Xerxes, regarded trees with a certain veneration:

> The beauty of a tree consists of fine branches, abundant leaves, a sturdy trunk, deep roots, movement in the wind, shadow spreading all around, change in accordance with the passing of seasons, with irrigation channels to support it and rain water to sustain it. Xerxes' robes, barbarian gold, and the other offerings did not ennoble the plane or any other tree.
>
> (*Varia historia* 2.14; Wilson 1997: 87)

Roman tradition is filled with stories about trees, none more famous than the tree in Vergil's *Aeneid* that will admit the hero Aeneas to the Underworld. In a prophetic trance, the Sibyl of Cumae tells Aeneas to seek the shores of Lake Avernus, an ancient volcanic crater north of Naples, surrounded by a dense forest, probably then as now consisting of live oak and chestnut trees:

> Receive my counsel. In the neighb'ring grove
> There stands a tree; the queen of Stygian Jove
> Claims it her own; thick woods and gloomy night
> Conceal the happy plant from human sight.
> One bough it bears; but (wondrous to behold!)
> The ductile rind and leaves of radiant gold:
> This from the vulgar branches must be torn,
> And to fair Proserpine the present borne,
> Ere leave be giv'n to tempt the nether skies.
>
> (*Aeneid* 6.136–41; Dryden 1697)

Remarkably, Vergil, who wrote so precisely about trees in his pastoral poems, the *Eclogues*, and his agronomic didactic epic, the *Georgics*, never revealed which species bore the Golden Bough.

Some remnant of these arboreal traditions survived from late antiquity through the Middle Ages. When Saint Zenobius of Florence died in 429, his coffin was said to have brushed against a barren elm tree, an *arbor infelix* that suddenly sprang to verdant life. A freestanding column next to the Florentine Baptistery supposedly marks the spot where that tree grew; when it finally died, many years later, its wood, appropriately for a redeemed *arbor infelix*, was carved into a crucifix rather than a cross.

In Rome, pious legend held that Pope Paschal II had founded the church of Santa Maria del Popolo in 1099 on the site of another *arbor infelix*, an enormous walnut tree haunted by the restless ghost of the Emperor Nero. This ghost so disturbed travelers entering the city that Paschal himself took an axe to the tree's roots, liberating a cloud of devils and uncovering the Roman Emperor Nero's corpse, which the resourceful pope promptly threw into the Tiber (Bentivoglio and Valtieri 1976: 9–10). The church of Santa Maria del Popolo certainly existed in the late eleventh century, but the earliest reports connecting its foundation to the ghost in the walnut tree date from the fifteenth century, a century in which central Italy experienced a whole series of miraculous events connecting trees, especially various kinds of oaks, with churches dedicated to the Madonna. In 1417, for example, a local artist from the city of Viterbo painted an icon of the Virgin Mary and hung it on an oak tree in the countryside—which happened to grow along one of the main routes out of the city. When the image began to work miracles during an outbreak of plague in 1467, Pope Paul II commissioned first an altar, and then a protective chapel. Paul's successor, Sixtus IV, greatly expanded the church and its associated convent, and the Madonna della Quercia still continues today as an important regional pilgrim site. Another painted Madonna hanging in an oak tree, this one in Lucignano south of Siena, performed similar miracles in 1467, and like her Viterbese sister earned her own church, Santa Maria delle Querce—the plural *querce* at Lucignano suggests that this lucky tree formed part of an entire oak grove. In 1501, an image of the Madonna appeared spontaneously in a cork oak near Tolfa, the site of important alum mines. The banker who leased the mines, a wily, wealthy Sienese named Agostino Chigi (mentioned above for his ebony bed), was delighted to endow an appropriate church to the Madonna of the Cork Oak, *Madonna della Sughera*; the clever Chigi probably commissioned the Madonna as well, as an encouragement to the alum miners. The countryside between Rome and Florence, the ancient land of the Etruscans as well as the Romans, harbors a multitude of similar stories, many of them featuring Madonnas hung on trees by some rural wayside. Perhaps these figures were as common in the early Renaissance as the wooden crucifixes that still dot alpine trails in the present day.

Groves and forests exerted the same pull on Renaissance imaginations as they did on ancient and medieval fancy, from Shakespeare's forests: the fairies' realm in *A Midsummer Night's Dream* and the Forest of Arden from *As You Like It*, places fertile with love, transformation, and mystery, to the magic forests of Italian writers like Dante (*c.* 1265–1321), Boccaccio (1313–75), and Ariosto (1474–1533). Dante began his *Divine Comedy* envisioning himself as a middle-aged wanderer lost "in a dark forest," his way blocked by three allegorical beasts: a wolf, a lion, and a dog. He would emerge from this *selva oscura* completely transformed, the darkness replaced by a lucid vision ranging from the depths of Hell to the pinnacle of Heaven. Boccaccio's early poem, *La Caccia di Diana* (Diana's Hunt) perhaps composed around 1334, took up a favorite setting of medieval romance, the magical forest, but transformed his woods into the countryside of Naples (where Boccaccio lived at the time), presided over by the ancient Roman goddess Diana. Therein, Diana's nymphs were chaste huntresses, all but one, who argued that love is the highest human calling of all. Boccaccio described their lair, at the poem's opening, as a narrow valley,

> ... encircled by
> four low-lying mountains, bountiful with green
> grasses and flowers, and in its flowering midst there
> stood a clear fountain, fair, large, and with abundant
> water; and the water that overflows spreads out and
> makes a stream that bathes all the grasses; then it runs
> off to one side. Each mountain is covered with trees
> with fronds so thick that the sun can scarcely pass
> through them to the fields; upon them various birds
> perch to pipe their carols, and a slight breeze stirs the
> boughs there ....
>
> (Canto 2. 1–15, Cassell and Kirkham 1991: 101)

In Ludovico Ariosto's bestselling and outrageously hilarious epic, *Orlando Furioso*, written two centuries later, the chivalric forest had become so crowded with wandering knights and lovelorn ladies that traffic jams occasionally occur; Orlando spends most of the poem streaking stark naked through the trees, driven mad by love. In Ariosto's imagination, the sacred grove of ancient classical tradition has become the kind of topsy-turvy world as the forest outside Athens in which Shakespeare's Bottom spends much of his Midsummer Night's Dream wearing a donkey's head.

One of the most influential Renaissance transformations of the forest, however, took place in a theological treatise, composed in Rome between 1506 and 1512 with the aim of harmonizing the classical learning of the Renaissance with Christian spirituality. Its author, Giles of Viterbo, headed the Augustinian order, the largest religious order of the era, at a time of urgent calls to reform the Christian church—Giles was, in fact, Martin Luther's superior. Taking the standard theological textbook of the day, the twelfth-century *Sentences* of Peter Lombard, Giles drafted a commentary "according to the mind of Plato" in order to show how the ancients had prepared humanity for Christian revelation. As a central image, he took Dante's *selva oscura*, the "dark woods," and explained that this forest symbolized the material world, where the human soul, like the huntress Diana, tracked down traces of God's presence just as the ancient goddess tracked her prey. The soul's prize, however, was not a mortal animal, but eternal knowledge, the knowledge that could inspire Diana, the soul, to lift her gaze above the forest canopy, the world, to contemplate the moon—by which Giles signified the transcendent region that Plato described as the realm of Forms or Ideas (Rowland 1997). The whole story of the Forest of Matter captivated early-sixteenth-century scholars, religious, and the lay audiences who heard Giles' immensely popular sermons. Thus when Renaissance artists painted Diana hunting in the woods, they might be conveying a profoundly Christian message, and the forest itself might be an image drawn from Plato to signify the material world. Sometimes, of course, as in the sixteenth-century painter Parmigianino's fresco of *Diana and Actaeon*, decorating a villa near Parma, they were simply enjoying the opportunity to paint beautiful figures with scant clothing, romping in a verdant forest that stood for nothing more than a wilderness remote from civilization.

SYMBOLIC GARDENS AND THEIR PLANTS

Like the woods that surrounded them, Renaissance gardens were often dense with symbolism. Some meanings survived from the ancient world: bay laurel, the plant sacred to the god Apollo, remained the sign of poetic inspiration. Vergil's image of a mighty oak buffeted by storms made the association of the oak (*Quercus robur*) with fortitude memorable beyond the inherent strength of oaken timbers:

> As, when the winds their airy quarrel try
> Justling from ev'ry quarter of the sky,
> This way and that the mountain oak they bend
> His boughs they shatter, and his branches rend.
> With leaves and falling mast they spread the ground,
> The hollow valleys echo to the sound:
> Unmov'd the royal plant their fury mocks
> Or, shaken, clings more closely to the rocks
> Far as he shoots his tow'ring head on high,
> So deep in earth his fix'd foundations lie.
>
> (*Aeneid* 4.441–9; Dryden 1697)

With this passage of Vergil in mind, Raphael painted a figure of Fortitude grasping an oak sapling for the private apartment of Pope Julius II in 1511—well aware that the pope's family name, Della Rovere, meant "of the oak," and a cleverly twined specimen of *Quercus robur* appeared on his coat of arms.

Vergil praised the roses of Paestum for blooming twice a year, but the medieval association of roses with the Virgin Mary proved especially durable. Dante's final vision of heaven was an enormous celestial rose—the Virgin—and roses often appeared in the gardens of convents and monasteries with precisely the same significance. Legends abound of saintly figures like Saint Benedict and the Blessed Giovanni Chigi (a relative of the banker Agostino) who defeated the impulse to violate their vows of celibacy by rolling in a thorny rosebush.

But whole gardens could also carry a symbolic meaning, as is the case with the famous villa garden of Cardinal Ippolito D'Este at Tivoli, outside Rome, which sits on a precipitous slope. From its vantage, on clear days, the dome of St. Peter's basilica is visible on the distant horizon, framed by the sea behind (see Figure 4.4). The dome was incomplete in the cardinal's own day, but construction was well underway on the gigantic church, and its rising silhouette must already have been visible across the plain. Inspired by and reliant on this vista, the cardinal's garden develops an overarching theme of "Tivoli and Rome." A fountain on the villa's panoramic terrace recreates a miniature version of the Eternal City, complete with a tiny river Tiber, while the uphill reaches of the garden evoke the wooded slopes of Tivoli (Beneš 2013). This theme, however, is layered and must be understood as part of a dense complex of motifs rooted and intertwined in this symbolically rich garden and its plantings. Unsurprisingly, classical mythology, transplanted into and vivified by a living Renaissance landscape, underpins the garden's narrative as it unfolds for visitors progressing from the Via del Colle entrance up through four ascending terraces. The first and lowest terrace garden, which featured espaliered citron, establishes the identity of the garden as a whole as a second Garden of the Hesperides, mythological nymphs who tended the trees yielding golden apples that were sought by the hero Hercules in the course of his famed Twelve Labors; and, importantly, just as Hercules introduced Hesperidean apples to the Mediterranean from the edges of the known world, so Cardinal D'Este could be presented as a new Hercules, an adventurer bringing exotic plants, like the jasmine and citron that were planted here, to Italian soil (Bay 2019). Diana, the embodiment of Nature, appears in her most imposing form, as the colossal Diana of Ephesus, spurting streams of water from no fewer than twenty-two breasts as she presides over pools and fountains that simultaneously represented the life-giving as well as the threatening aspects of the local river Aniene. Just beyond lay the dense and awe-inspiring *boschi* (woods) of the second terrace, hiding the dragon who guarded the Hesperides' precious fruit. Contemporary purchase records suggest that this mythological dragon's wooded lair was planted with elms (*Ulmus glabra* and *Ulmus minor*), fir trees (*Abies alba*), and sweet chestnuts (*Castanea sativa*), the predominant arboreal species of sixteenth-century Tivoli. Chestnuts were particularly important to the local economy, both for their tasty nuts (which were also ground into flour), and for their timber, especially prized for making wine casks in this important wine-growing region. A colossal statue of the Albunean Sibyl, the prophetic goddess associated with Tivoli and its sulfurous hot springs, dominates the third terrace of Villa d'Este, surrounded by a sacred grove of plane trees. The villa's owner, Cardinal Ippolito d'Este undertook engineering projects to ensure the region's supply of drinking water (and the functioning of his villa's fountains); to safeguard the healing springs; and to protect against the Aniene's damaging floods. With the Albunean Sibyl as patron deity, this program of symbols and garden plantings cast Tivoli and the Villa as a specifically Latin, regional version of Mount Parnassus, a home to Muses who rivaled those of Rome (who in turn rivaled the Muses of ancient Greece in their fabled Vale of Tempe). The connection with Parnassus became explicit for visitors when

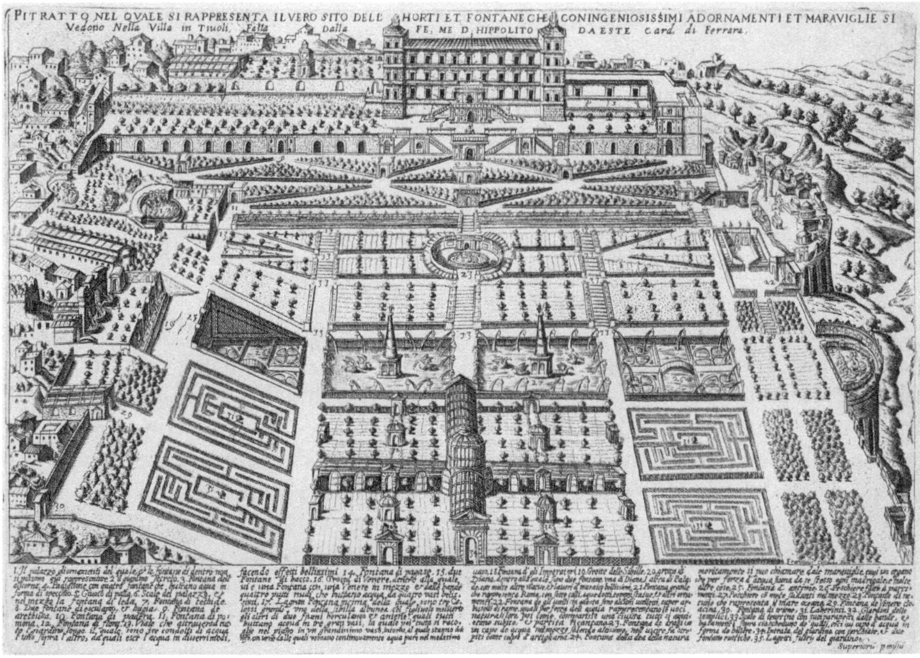

FIGURE 4.4 Villa d'Este, Tivoli, Italy, engraving from *Palazzi di Roma de piu celebri architetti* (Palaces in Rome by the Most Famous Architects), by Pietro Ferrerio, 1655. Photo by Icas94/De Agostini/Getty Images.

they finally reached a fountain topped by a statue of the winged horse Pegasus, who had created the Hippocrene, the original fountain of the Muses on Mount Helicon in Greece, by striking the ground with his immortal hoof. Symbolically, the steep ascent through the grounds of Villa d'Este toward the villa proper, following in the footsteps of Hercules and Cardinal Ippolito, became a pilgrimage toward Virtue, a journey crowned by arriving at the grotto of Diana, now presented as the lithe, chaste, girlish huntress of virtue rather than an elemental Earth Mother, in a grotto suitably studded with stucco apples of the Hesprides.[15]

About 100 km (60 mi) to the northwest of the Villa D'Este lies the Villa Lante, yet another example of a symbolically pregnant garden created in the same spirit, at virtually the same time, and by another member of a sophisticated circle within the Vatican. This villa, created by Cardinal Gianfrancesco Gambara in the town of Bagnaia outside Viterbo in 1566, provided the busy churchman with his own place for contemplative pilgrimage, again conceived as an uphill climb toward virtue. An ardent proponent of strict discipline within the Catholic Church (he would be appointed Grand Inquisitor in 1567), Gambara had participated in the closing sessions of the reforming Council of Trent in 1562–3. Rather than evoking ancient Roman sensuality, the plans for his villa intentionally pointed to prayerful Christian retreat—but in a recognizably Renaissance style. As Gambara and his guests made a physical ascent by mounting a steep slope behind the villa's main pavilion, they could perform spiritual exercises by contemplating the sculptures and fountains they encountered on their uphill journey (as well as a well-stocked aviary). At the summit, however, body and spirit met again, at picnic tables loaded with food, drinks cooling in the uppermost set of fountains, and cool breezes to reward the pilgrim's progress (Benocci 2010). At the same time, this hallowed retreat rested on pagan mythological foundations as firm as those of Villa D'Este. Visitors entering the garden through its side gate would find themselves in another Parnassian landscape, wandering through a *bosco* densely planted with groves of cypress, oak, beech, fir, and chestnut as well as with fruiting "orchard" trees. On one level, this woodland, focused on an Acorn Fountain, signified the paradisiacal Golden Age established by the primeval god Saturn in Latium; for acorns had served as the staple food for humble, innocent humanity during that pious era (Lazzaro-Bruno 1977). At the *bosco*'s summit, visitors entered a more manicured, less "wild" landscape representing the Age of Industry and Art under the auspices of Jupiter (a symbol for the Pope) and his earthly agent, the Cardinal.

When he drew up the plans for his villa, Cardinal Gambara must have delighted in the contrast it would make with Rome's most famous secular paradise: the suburban "greenhouse" (*Viridario*) commissioned in 1509 by the Sienese banker Agostino Chigi, the same Chigi of the ebony bed, who endowed the church of the Madonna of the Cork Oak at the alum mines of Tolfa, and probably commissioned the tree-loving paintings of the Madonna as well. The Tolfa mines formed the basis of Chigi's financial empire, which extended from London to Constantinople from its base in the banking district of Rome. By 1509, Agostino had become so powerful that he no longer wanted to live among his fellow bankers. He purchased a riverside lot in the semi-rural zone between the Vatican and the dense neighborhood of Trastevere, where several cardinals maintained garden properties called *vigne* (vineyards), simple houses with grape arbors and gardens where they could withdraw for lunch and a siesta. Chigi and his architect, Baldassare Peruzzi, had more ambitious ideas about how to develop what would become the banker's chief residence in Rome: with the guidance of the ancient architectural writer Vitruvius, they recreated a specific ancient type of ancient dwelling: a suburban villa for a Roman patrician. Chigi belonged to the nobility in Siena, and dearly hoped that Rome would accord him the same honors, but in the meantime, his new home would proclaim the status he hoped to achieve. In the sixth of his *Ten Books on Architecture*, Vitruvius detailed a carefully calibrated social hierarchy for houses, beginning with the rustic dwellings of farmers, and moving on to the homes of professional men:

> For bankers and tax collectors [houses] should be larger and more impressive and secure from danger. For lawyers and orators they should be more elegant still, and more spacious, in order to accommodate gatherings. For the leaders [*nobiles*] who must dedicate their service to the citizens by holding offices and magistracies, their vestibules must be lofty and regal, their atria and peristyles as expansive as possible, their woodlands and garden paths laid out on a broad scale in keeping with the dignity of divine majesty.
>
> (Vitruvius, *De architectura* 6.5.2; Rowland 2009)

By profession, Chigi belonged to the category of "bankers and tax collectors," but he spent the last decade of his life moving slowly away from these businesses in order to ensure his children full membership in the Roman aristocracy. Furthermore, the very fact of choosing a property among cardinals on the Via della Lungara compelled him to build in keeping with their style, not with the muted scale of the close-packed banking district. His *Viridario* was not exceptionally large, but its interior opulence more than made up for that exterior modesty (Barbieri 2014). Indeed, the garden of the Chigi villa was, if anything, more ambitious than Baldassare Peruzzi's innovative house, which contemporaries, including Chigi, immediately recognized as a full-fledged palazzo (and it was regal enough to inspire the plan of Versailles). The greenery of the *Viridario* seems to have drawn significant inspiration from the royal gardens of Poggioreale ("Royal Hill") in Naples, where Chigi maintained a dense network of business connections (Benocci 2005: 73; Rowland 2005 a and b). Giuliano da Maiano's original design of the complex at Poggioreale may have drawn its own inspiration from one of Lorenzo de' Medici's Florentine villas (de Divitiis 2015; Giuliano was Florentine himself), but the climate on the Bay of Naples was far kinder to vegetation than the hills of Tuscany, especially during the rigors of the Little Ice Age. By all accounts, the garden at Poggioreale was an incomparably lush paradise, nurtured on rich volcanic soil beneath a beaming Mediterranean sun (Visone 2016; Zecchino 2002). Chigi endeavored to repeat the miracle by ordering bare-root plants for his *Viridario* from a Neapolitan nobleman.[16]

According to the poets who celebrated the *Viridario* in verse, the garden's presiding deity was Venus, who served a variety of symbolic purposes: ensuring, as a goddess of fertility, that Chigi's business ventures would thrive as fruitfully as his greenery, but Venus was also the ancestress of the Emperor Augustus, these same poets compared Agostino, a merchant by calling, but "a king in spirit."[17] In keeping with that kingly largesse, the floral frescoes that decorated Agostino Chigi's *Viridario*, executed by Raphael's talented colleague Giovanni da Udine, provide one of the earliest and most sumptuous displays of exotic plants, imported from the New World, Africa, and Asia (Caneva 1992; Caneva and Carpaneto 2011; Sgammellotti and Caneva 2017; see also Chapter 8 of this volume and Figure 4.5). By the early seventeenth century, international botanical commerce had become big business, with The Netherlands at its center. Two plants native to Turkey, tulips and Crown Imperials (also known as imperial fritillaries), were particularly prized by Dutch gardeners and thrived on Holland's sandy soil. In those days, as trophies, they were planted in splendid isolation rather than in the vast fields that have become so

FIGURE 4.5 Exotic fruits and vegetables, Loggia of Cupid and Psyche, Villa Farnesina (formerly Chigi), Rome, 1518. Fresco by Raphael and his workshop. Photo by David Silverman/Getty Images.

famous a feature of the western Netherlands. For a brief period in the 1630s, tulip prices fluctuated wildly before settling into the equilibrium that transformed the cultivation of tulips into an important Dutch industry. Although historians generally conclude that the Dutch "tulip crisis" was less cataclysmic than we might think from the satirical tracts and allegorical paintings it inspired around the critical year of 1636, the disturbance nonetheless shows that by the seventeenth century, the plantings in most European gardens had become international in their scope (Goldgar 2007; Pavord 1999; see also Rowland 2009 and Chapter 7 of this volume).

NATURAL PHILOSOPHY AND THE KNOWLEDGE OF PLANTS

In 1492, the remarkable (and remarkably long-lived) classicist, chemist, botanist, and physician Niccolò Leoniceno (Nicolaus Leonicenus in Latin), Professor at the University of Ferrara, penned a scathing attack on Pliny's *Natural History*, *De Plinii et aliorum in medicina erroribus* (On the Mistakes of Pliny and Others About Medicine), criticizing not only the Roman writer's grasp of his Greek sources, but also his first-hand knowledge of nature, evident especially in his inaccurate treatment of medicinal plants.[18] "Many things that he writes about in his books on Natural History have not been tested and explored," Leoniceno declared, asking rhetorically "whether Pliny concurs in these matters with Dioscorides, Galen, Paul [of Aegina], and the other outstanding physicians, and with experience, the mistress of all things, or whether, as they say, even Homer, despite his greatness, occasionally falls asleep" (Leonicenus 1492: a3 recto–verso). We already know from the title of his booklet that Leoniceno would catch Pliny napping. As he observed, the ancient Roman wrote as an orator, not as a practicing physician who desperately needed to know which plant is which.

Some of Leoniceno's readers, including the dedicatee of his incendiary little book, the Florentine scholar Angelo Poliziano, were shocked at the bold way it stripped an ancient author of his authority, holding Pliny to the same standards of reasoned argument and empirical testing that were developing for contemporary natural philosophy. Poliziano himself had done pioneering work in subjecting literary texts, classical and vernacular, to critical scrutiny, but Leoniceno applied this critical stance much more broadly, to texts like Pliny's *Natural History*; to Pliny himself as an author, to his sources, and to his medieval and Arab commentators considering them all as actors in history; to the examination of plants in nature; to the examination of plants in medical practice, to the study that he would have called natural philosophy and that we would eventually call science.[19] Often acknowledged as the father of modern botany, he was also a significant figure for the development of chemistry, of medicine, and of classical scholarship—and *De Plinii ... erroribus* also makes some pungent comments about astronomy.

Sixteenth-century scholars, like the herbalist Gherardo Cibo (1512–1600), began to collect dried specimens of plants. His herbarium, alphabetically ordered, is still preserved today in Rome's Biblioteca Angelica (Celani 1902; Findlen 1994: 167–70; Tongiorgi Tomasi 1989; see also Chapter 5 of this volume). At the same time, the old medieval medicinal gardens in three Italian university cities were reorganized to reflect the increasingly empirical guiding principles of Early Modern medicine (in which Leoniceno's explorations of plant taxonomy were important milestones). The Botanical Garden of Pisa was founded in 1543 to serve the chief university of the Duchy of Tuscany, sponsored by Duke Cosimo de' Medici and laid out by the botanist and physician Luca Ghini, who created a teaching garden rather than simply a source for medicinal herbs. To keep students informed about the entire lifecycle of the plants on display, he assembled an extensive herbarium of dried specimens, and worked on creating credible classification systems. Two years later, in 1545, Duke Cosimo asked Ghini to set up a second botanical garden in Florence, transforming the old medicinal garden of the Dominican convent of San Marco into a "Giardino dei Semplici" (Garden of Simples) run on the same educational principles as the garden in Pisa. ("Simples" were plants used as they were for medication rather than mixed into compounds.) A few months earlier, however, the University of Padua, the official university of the Republic of Venice, had founded its own Botanical Garden—not to be outdone by the Tuscans. Appropriately, the Botanical Garden of Padua was designed by the Venetian scholar Daniele Barbaro, the great friend and collaborator of a figure we have already met briefly: the architect Andrea di Pietro della Gondola of Vicenza (1508–80), who in 1538 took the name Andrea Palladio to honor the ancient agricultural writer Palladius Rutilius Taurus, and who, inspired by Palladius, created some of the most enchanting villas ever designed—including one for Daniele Barbaro. Because the Botanical Garden of Padua can boast that, unlike its Pisan counterpart, it has remained in the same place for its entire history (the garden in Pisa has been moved twice), its 1997 UNESCO World Heritage citation calls it "the original of all botanical gardens throughout the world, [which] represents the birth of science, of scientific exchanges, and understanding of the relationship between nature and culture. It has made a profound contribution to the development of many modern scientific disciplines, notably botany, medicine, chemistry, ecology, and pharmacy." And here, in 1680, a splendid example of *Platanus orientalis* was planted, still majestically spreading its branches today.[20]

CHAPTER FIVE

Plants and Medicine

FLORIKE EGMOND

INTRODUCTION

Probably speaking a mixture of English, French, and Spanish or Portuguese, a few European men and three boys from Japan and the Philippines sat down in a house in Lime Street, London, to discuss various East Asian plants, their indigenous names, what they were used for, and how to prepare or administer them. Were they food, spices, stimulants, herbal remedies, poisons, or something else entirely? The ship's surgeon among them gave their host some seeds of star anise (a tree native to parts of China and Vietnam) that he had personally brought back from Asia. This host, the apothecary, drugs and spices merchant, and naturalist James Garet, noted its aromatic flavor and tried, unsuccessfully, to grow star anise in his London garden. The meeting took place on July 28, 1589, but even without the precisely dated source there are several historical reasons why this event could not possibly have taken place earlier than the 1570s–1580s (Egmond 2010: 191–4) (see Figure 5.1).

Those historical reasons—to which we will return in what follows—help to highlight some of the major changes in European knowledge of medicinal plants between the early fifteenth and the early seventeenth centuries. We will first take a look at some important developments in this period, then turn to changes in the media, and finally inspect contacts between European herbal medicine, the New World, and the Far East. The main focus remains, however, on plants and their role in Early Modern European medicine.

Though impressive, the changes of these centuries should be seen against the background of some long-term continuities. The first and most important is the predominance of plants as the principal ingredients of medicaments. It lasted, unbroken, from antiquity until the Early Modern period. Paracelsian medicine with its stronger emphasis on mineral therapies became prominent in the late sixteenth century, but never replaced plant-based medicine. The importance of phytotherapy and plant derivatives in modern pharmaceutics and cosmetics shows that the medicinal lives of plants have in many ways never ended. In fact, a large part of the modern star anise crop serves the production of anti-influenza drugs. Two further important continuities were the influx of practice-based ("folk") knowledge about medicinal plants into "learned" medicine, about which more will be said below; and the non-existence of a boundary between the categories of food and medicine. Galenic-Hippocratic medicine with its humoral theory attributed key characteristics of warm, cold, dry, and moist to every plant, with a specification of the degrees of intensity. Irrespective of whether a modern person would have classified it as food, medicine, or stimulant, *all* plants were deemed to have characteristics (*properties*)

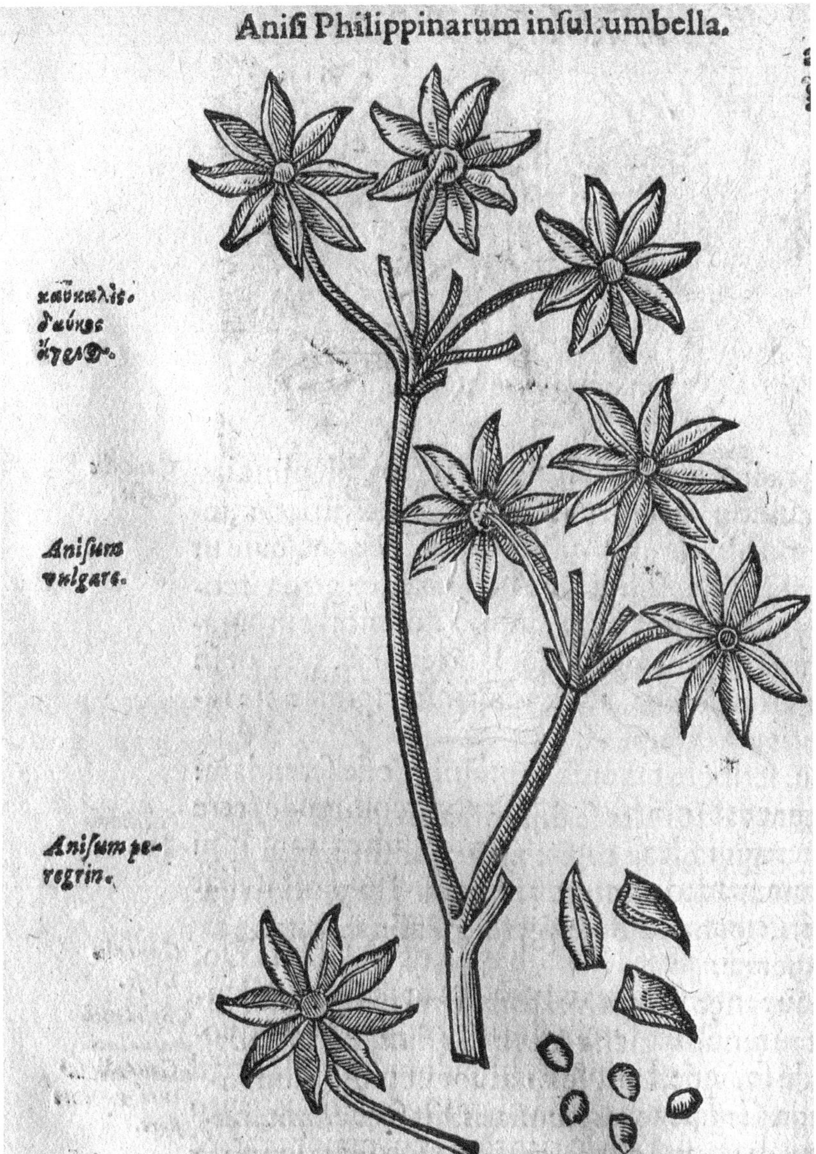

FIGURE 5.1 Star anise (named "Anisus Philippinarum insularum," woodcut illustration based on Garet's information in Carolus Clusius, *Rariorum plantarum historia* (History of Rare Plants, Antwerp: Plantin-Moretus, 1601: ccii). Image courtesy of Missouri Botanical Garden.

that could affect human health and illness positively or negatively according to their interaction with a person's constitution and the balance or imbalance of humors (Nutton 1993). Finally, during most of the period discussed here knowledge of plants was rarely separated from knowledge of their medicinal uses. Botany and herbal medicine were part of a single domain of knowledge. They only began to split into two disciplines in the second half of the sixteenth century.

THE CHANGING PANORAMA

The first reason why the London meeting of 1589 could not have taken place much earlier is straightforward: no British explorations had reached parts of the world where star anise grew before Francis Drake's circumnavigation of the world (1577–80). Two of the men present at James Garet's house were, in fact, the British explorer Thomas Cavendish (1560–92) and his ship's surgeon; they had returned only recently from their circumnavigation of the world (1586–8). In these years Garet was one of the major experts on exotic *naturalia* in London and an important dealer in both European and exotic medicinal drugs and spices. Lime Street was the location of a circle of highly active naturalists. James' brother Pieter too was a pharmacist and merchant in spices and sugar. Like James in London, Pieter in Amsterdam kept a close eye on all exotic naturalia that arrived from the 1580s to the early 1600s (Egmond 2010: 192–207; Harkness 2007: 15–56).

The lives of the Garet brothers, émigrés from Antwerp, coincided with a crucial phase (*c.* 1570s–1610s) in explorations by the British and the Dutch. They were entering parts of the world where European access had largely been limited to the Portuguese and Spanish, in so far as sea routes were concerned, and to the Italians for the overland route via the Middle East to India and China. The latter route had been used with fluctuating intensity ever since antiquity for the transfer to Europe of Middle Eastern and Asian natural products, including a range of vegetable substances for use as medicine or food. Venetian trade with the Far East met with new forms of competition once the Portuguese reached India via the sea route around the Cape. The geographical center of gravity of the importation of Asian drugs and spices shifted in the late fifteenth through the early sixteenth century from the Mediterranean to the Atlantic. Lisbon, Seville, and Antwerp, all in the Habsburg Empire, functioned as key ports on the Continent. But changes in the balance of power between the Habsburgs and the north European nations led to the fall of Antwerp (1585) and the closure of its port by the emerging Dutch Republic. British and Dutch explorers, such as Drake and Cavendish, entered on the world stage, and their northern nations broke the Portuguese monopoly on the sea routes to the Far East in the early decades of the seventeenth century. They gained a foothold even in the New World, most of which remained under the dominion of the Spanish, however.

These changing relations between Europe and the rest of the world greatly stimulated a series of developments that affected all branches of nature study or "natural history" (Ogilvie 2006; Reeds 1991; Zemanek 1998). The main changes in botany and plant medicine can be summed up under the labels of innovation, professionalization, and discipline formation, but we need to look at people, places, and practices to know what that actually means.

Many different medical practitioners whose expertise was mainly shaped in practice and transmitted via apprenticeship were involved in plant medicine in Early Modern Europe. They ranged from herb women, midwives, folk healers, alchemists, and apothecaries to aristocratic ladies and (male and female) members of religious communities who were responsible for the healthcare of their extended households. Most of these practitioners used vernacular languages, but many apothecaries also had working Latin, while some apothecaries and plant experts in religious institutions were university trained and had excellent Latin. Several of the great sixteenth-century plant experts came from exactly this religious institutional background. The famous German herbalist Otto Brunfels (1488–1534), for instance, turned to Protestantism but had been a Carthusian monk. The Italian Evangelista Quattrami (b. 1527), who acted as chief herbalist, plant hunter, and distiller for the noble Este family in Ferrara and Rome in the 1570s–80s, had a doctorate in theology and was an Augustinian friar from Gubbio. The Este garden on Monte Cavallo (Quirinal) in Rome and the gardens of Villa d'Este in Tivoli—planted and supervised by Quattrami—were among the most prestigious in the whole of Italy (Egmond 2010: 91–3).

Academic plant study was largely the domain of physicians. In the course of the sixteenth century the study of plants came to occupy an increasingly important space in their university training. From the 1540s, special university chairs were founded in Europe for what we might call medicinal botany: examples are Padua, Bologna, Valencia, Montpellier, Wittenberg, Basel, and Leiden. Often professors of herbal medicine and their assistants also specialized in anatomy. They practiced botany and did fieldwork in spring and summer, and used the cold months for anatomical investigation. From the 1530s–40s specific locations were created for the study of plants: the first university botanical gardens in Europe originated as places of both teaching (medical students), research, and experimentation. University-trained plant specialists never gained a monopoly on high quality knowledge of medicinal plants. Their expertise was permanently fed by practice-based knowledge, and at the same time deeply influenced by what might be labeled "green fashion": a fascination with living nature that spread widely among the European elites. It manifested itself in collecting rare naturalia, the creation of *Kunst- und Wunderkammern* (cabinets of art and curiosities) and private gardens that functioned as both living collections and sites of plant experimentation, the use of naturalia in decorative arts, and so on. Patronage by elite collectors also resulted in the rise of the new phenomenon of the plant hunter, and in expeditions to discover new plants that could be interesting for medicine and collectors (Egmond 2010, 2018; Ogilvie 2006; Reeds 1991).

In a crucial methodological development that can be traced throughout the fifteenth and sixteenth centuries the authority of the rediscovered classical works on living nature and plant-based medicine (especially those of Dioscorides, Theophrastus, and Pliny) was increasingly challenged by and confronted with new information based on direct observation and eyewitnessing (*autopsy*). James Garet's explicit reliance on the personal experience of his Japanese-Philippine informants and of those Europeans who had actually seen tropical plants in their natural habitat is a good example of this trend. Classical texts by no means lost their authority, nor did these confrontations lead to a subversion of Galenic medicine or Aristotelianism, but personal experience became a more important source of authority. The critical, at first mainly philological analysis by humanists led from textual study to the confrontation of ancient texts with the surrounding reality of living nature and to more testing of the actual medicinal effects of plants (Ogilvie 2006).

A good example is the vast amount of experimentation, testing, and chemical and botanical research done by naturalists-physicians and apothecaries all over Europe in their efforts to re-create the classical wonder drug *theriac*. This compound mentioned by Galen was reputed to preserve health, act as antidote against poisons, and cure those who were ill. Throughout the Early Modern period debate never finished as to the identification of the many *simples* (medicinal herbs) that should be used to create it, and their correct mix (Findlen 1994: 241–5, 267–86). The potential effects of *theriac* in cases of contagious disease became even more topical in the years 1570–1650 with their severe outbreaks of the plague. Theriac is by no means forgotten even now: in 2017 the pink garlic of Sulmona was advertised at a Roman market as "Galen's theriac for the poor."

While their involvement with the classical tradition required impressive linguistic and philological abilities, many Early Modern plant experts were not only scholars but also men of action. From the very early sixteenth century onwards they undertook journeys and expeditions in order to discover new naturalia and potentially new medicines, and *re*-discover ones that had been famous in antiquity. Naturally, they first of all focused on the southeast Mediterranean and the Middle East: the core areas of classical knowledge. The search for the supposedly extinct plant *silphium* (also *laserpicium*), known from the works of Theophrastus, Dioscorides, and Pliny, illustrates some of these continuities. In antiquity its juice formed a basic ingredient in pharmacy and was used as an aromatic in cooking. The best silphium came from Cyrenaica, the eastern coastal zone of present-day Libya, but the plant may have become extinct already in Roman times. By the mid-sixteenth century it was no longer known what a silphium plant looked like, and naturalists suggested various possibilities. The Italian physician-naturalist Pietro Andrea Mattioli (1501–77), for instance, first identified its juice as a benzoin resin produced by styrax trees, but later revised his opinion. In the early seventeenth century Prince Federico Cesi (1585–1630) of the Roman Accademia dei Lincei regarded laserpicium, like manna, as a miraculous rain. Even today silphium continues to fascinate: the debate on the correct identification and the question of why silphium may have become extinct are still undecided (Amigues 2004; Guerrini 2006: 28–31; Hardy and Totelin 2016).

Field explorations in search of new plants soon extended to many high mountain zones in Europe with their varied native flora, and to Iberia, central and northern Europe. The above-mentioned Mattioli was one of the first to undertake systematic fieldtrips, for instance in the Dolomites, Central Italy, and the Bohemian mountains (1520s–50s). The physician-naturalist Conrad Gessner (1516–65) explored the Swiss Alps in the 1550s. And during the 1540s–50s Luigi Anguillara (1512–70), who is best known as director of the Padua university botanical garden, explored nearly the whole of Italy as well as parts of the Adriatic Greek coast, Crete, and parts of the Turkish empire, reaching perhaps as far as Aleppo and Tunisia (Egmond 2018). Results of the long journey of the German physician-naturalist Leonhart Rauwolf (1535–96) to Palestine, Lebanon, Syria, and Iraq (1573–6) can be found in his extant herbarium, which contains some nine hundred dried plant specimens. He also returned with seeds of medicinal plants whose effects he had studied on the spot while acting as a traveling physician in the Middle East (Herde and Walter 2010; Rauwolf 1582–3).[1]

Journeys and fieldwork thus added to the corpus of medicinal plants transmitted since antiquity, but at the same time raised many questions about the medicinal use of those "new" plants. They also held out the promise that even more could be discovered in other parts of the world.

CHANGING MEDIA AND READERSHIP

Book printing with moveable type was probably the most crucial change in the media of the Early Modern period—media that are at the same time *our* historical sources. Before the introduction of book printing in the second half of the fifteenth century, botanical and plant medicinal knowledge was mainly transmitted orally, via handwritten texts, and via colored drawings. Drawings and texts that identify the plant, describe its uses in food and (Galenic) medicine, and sometimes give recipes, figure in a large variety of manuscript albums or codices that are usually known as herbals (Arber 1986). Somewhat confusingly, most of the Early Modern printed works on such topics are known as herbals, too. Here we will speak, therefore, of *manuscript* and *printed herbals* to distinguish the two. Within the large and diffuse category of herbals finer genre distinctions are often made between compendia on plants and plant medicine, recipe books, books of health, and natural history encyclopaedias. Their contents often overlapped, however, and examples of all subgenres can be found both before and after print.

Generally, the core of plant medicinal information in the manuscript herbals goes back to classical and Arabic authorities, but nearly all of them also contain Early Modern comments, corrections, and further additions. The colored drawings usually have various functions: they always facilitate plant identification, but also serve as decoration and sometimes have heraldic and symbolic meanings. Two important developments date from the late fourteenth and early fifteenth centuries—well before print, therefore. First, the number of manuscript herbals increased exponentially: Collins mentions a total of 193 surviving fifteenth-century herbals as compared to 136 for the whole of the preceding nine centuries (Collins 2000: 278).[2] Geographically, most of the extant herbals originated in Italy (in particular its northern regions), Germany, France, and the Burgundian area. Most of them survey a wide range of naturalia and health practices, but some show signs of specialization. A few (mainly Italian) herbals concentrate on plants used in alchemy, for instance; others include only medicinal plants. Some herbals focus on therapeutic uses and give recipes; others pay more attention to the plants themselves. Parts of the famous and still undeciphered Voynich manuscript (presumably early fifteenth century, if not after all a modern or historical fake; Clemens 2016; Touwaide 2015) may have been meant as a pharmacopoeia, manual on perfumes, herbal, or book of health.

A second important development that occurred well before print is visual innovation. Comparison of many fifteenth-century herbals and their predecessors has led to the identification of chains of copying—both over time and within the same period—and thus to that of "families" of manuscript herbals. Repeated copying tended to result in loss of detail in the plant drawings and, in some cases, in an increasingly schematic character, although this did not necessarily impede plant recognition. However, three North Italian manuscript herbals from the 1390s–1440s are regarded as the first in post-classical European history to show signs of turning away from schematic (or perhaps we should say, strongly stylized) representation and towards direct observation of real plants as the source of a more naturalistic representation (Baumann 1974; Givens et al. 2006; Kyle 2017; Olariu 2015; Pächt 1950).[3] Modern botanists have ascertained, for example, that several of the drawings in the *Codex Bellunensis* (Belluno Codex, 1420s) depict local wild plants and must be based on investigations in the mountains, presumably by herb gatherers, physicians, or apothecaries (Mariani Canova et al. 2006). Though these early fifteenth-century painted herbals by no means completely broke with schematic representation, it is hard not to see this partial turn to lifelike representation as a visual strategy connected with the above-mentioned methodological shift towards direct observation.

Interestingly, the same *Codex Bellunensis* also contains early visual evidence of the presence of exotic (in the sense of non-European) medicinal plants and substances in the 1420s Veneto. One of these is tamarind, depicted here not as a bush or tree but as a small chunk of fruit pulp, and described as coming from overseas (see Chapter 2 of this volume). Another example in the *Bellunensis* documents how inhabitants of the Veneto attempted to replace expensive exotic drugs such as "acacia" (made from the green pods of the Egyptian acacia) with an extract made from the immature fruits of the local wild plum tree.[4]

Printed herbals with woodcut illustrations began to appear in the late fifteenth century, but print never replaced handwriting. Considerable numbers of manuscript herbals continued to be produced throughout the late fifteenth and sixteenth centuries. A late-fifteenth-century Italian manuscript herbal with the signature of Federico Patella, for instance, depicts some two hundred medicinal plants and presents short descriptions of their medicinal properties.[5] And two albums with beautiful plant drawings by the Italian naturalist-painter-aristocrat Gherardo Cibo from *c.* 1564–84 show full-page colored drawings on the right-hand page with a description of its medicinal properties on the facing left-hand page (see Figure 5.2). The manuscript herbals of the sixteenth century show signs of specialization and innovation as well. Cibo's albums, for instance, combine

FIGURE 5.2 Mid-sixteenth-century colored drawing of "Daphnoide" by Gherardo Cibo. Image courtesy of Alamy.

a survey of medicinal plants with a regional *flora* and have a highly original visual format: most drawings show the strongly enlarged plant against a backdrop of its habitat (Mangani and Tongiorgi Tomasi 2013).[6]

The manuscript herbals of the sixteenth century did not merely co-exist with the printed ones. They had different characteristics and quite possibly different functions (Blunt and Raphael 1979; Egmond 2017). Precisely the fact that manuscript herbals were unique collections on paper gave their makers freedom to add, correct, update, and repeatedly reorganize botanical and therapeutic information. That was also true for *herbarium* collections in which dried plants were pasted on paper and arranged in large albums according to alphabetical name, plant shape (morphology) or, occasionally, medicinal function. All of the earliest extant ones are connected with Italy: a herbarium created by the above-mentioned Gherardo Cibo (begun in the early 1530s); the anonymous Italian *En tibi* herbarium (*En tibi perpetuis ridentem floribus hortum* [Here for You a Smiling Garden with Everlasting Flowers], 1540s–50s); a herbarium attributed to Michele Merini, a priest from Lucca, with dried plants that may have been collected in the Pisa botanical garden (1540–5); and the Hurtado de Mendoza herbarium (1576) that may actually have been compiled in Italy in the 1540s.[7] Crossovers between print and manuscript also occurred. Some manuscript herbals made for home use copied drawings, plant descriptions, and medicinal uses from printed works, but also added further information, plants, or recipes based on personal experience or local tradition.[8]

In the course of the sixteenth century large *visual* collections that consisted mainly of original, colored plant drawings and brief handwritten descriptions evolved into plant "databases" and surveys. The three largest extant collections of this kind—all from the mid-sixteenth century—were created by Conrad Gessner (close to 700 folios with several drawings per folio); the German physician-naturalist Leonhart Fuchs (*c.* 1,540 pages); and the Venetian aristocrat-naturalist Pietro Antonio Michiel (990 folios).[9] Although most plant image collections of this period remained closely connected with medicine, they aimed at encyclopaedic coverage of all known plants rather than at a function as books of health. The manuscript herbal of the Nuremberg apothecary Georg Oellinger with its *c.* 650 pages of plant drawings is a good example. Its title *Magnarum medicinae partium herbariae et zoographiae imagines* (Images of a Large Number of Medicinal Plants and Animals) (1553) explicitly refers to medicine, but the herbal actually contains little information about the plants' medicinal properties. The so-called *Camerarius Florilegium* (Camerarius' Flower-book, *c.* 1589) took this development a step further: this florilegium is a luxurious collection of plant images, a kind of painted garden in which medicinal uses have more or less disappeared from sight.[10] One of the most beautiful, but *printed*, examples of this latter genre is the *Hortus Eystettensis* (Eichstätt Garden) by Basilius Besler (published in Nuremberg, 1613) with its stunning hand-colored copper engravings (Egmond 2017; Givens 2006; Hoeniger 2006; Kusukawa 2012; Ogilvie 2006).

Book printing had become established in numerous European cities by the last two decades of the fifteenth century, with important consequences for the distribution of information. It is indicative of the high esteem for the classical works by Dioscorides (on plant medicine) and Pliny (on the natural world) that these were among the very early texts to be divulged in print in Europe (Funk 2016; Rozzo 2011). Some herbals made the transition from manuscript into print: the so-called *Circa instans* (*Circa instans negotium*, Concerning the Present [Topic of Simple Medicines] or The Book of Simple Medicines), for instance, originated in the twelfth century as a southern Italian illustrated work with almost three hundred medicinal substances, organized alphabetically. It is preserved in more than two hundred manuscripts, in Latin and in many vernacular languages. It was first printed in Venice in 1497, and remained an influential source for other herbals and pharmacological collections until the sixteenth century. The first printed herbals with woodcut illustrations reached a much wider European readership than the manuscript ones. Among the most influential were the German *Gart der Gesundheit* (Garden of Health, 1485), which discusses some 380 plants and their medicinal uses, and the Latin *(H)ortus sanitatis* (Garden of Health, 1491), which covers the medicinal uses of animals and stones as well as plants (Baumann and Baumann 2010; Collins 2000). Like the manuscript herbals, nearly all printed herbals from the late fifteenth to the early seventeenth century present some kind of plant survey and an evaluation, update, and correction of botanical and plant medicinal knowledge dating back to antiquity.

The long sixteenth century, *c.* 1480 to *c.* 1620, was the heyday of large printed, often richly illustrated encyclopaedic works on plants that contained both botanical and plant medicinal information. Among the most famous early ones are: Otto Brunfels' herbal of the 1530s in both Latin and German editions with woodcuts based on the magnificent lifelike drawings by Hans Weiditz (Brunfels 1530–6, 1532–7); Hieronymus Bock's German *Kreutterbuch* (Plant-book) of 1539 (unillustrated) and 1546 (illustrated) with some seven hundred plants; and Leonhart Fuchs' Latin *De historia stirpium commentarii insignes* (Notable Commentaries on the History of Plants, Basel 1542), of which German, Dutch, and English adaptations appeared almost immediately. Fuchs' herbal was alphabetically organized and discusses some four hundred wild and one hundred domesticated plants and their medicinal uses.

The works of three naturalists from the southern Netherlands dominated a large part of the second half of the sixteenth century. Rembert Dodoens' (1517–85) *Cruijdeboeck* (Plant-book, 1st edn. 1554; 2nd expanded edn. 1563) with its 715 woodcut images was translated from the original Dutch into various other vernacular languages and into Latin (Clusius 1583b). It paid great attention especially to medicinal herbs and organized the plants not according to the alphabet, but in six groups, based on properties and affinities. In contrast, Matthias de Lobel's (1538–1616) *Icones stirpium* (Images of Plants, 1591) is sometimes regarded as an early attempt to classify plants, numbering about 1,200, not according to their medicinal properties but on the basis of the shape of their leaves (see Figure 5.3). That kind of effort was in the air at the time. In his principal publication *De plantis libri XVI* (Sixteen Books on Plants, 1583)—one of the very few herbals of this period to have no illustrations at all—the Italian physician-naturalist Andrea Cesalpino (1519–1603), too, was looking for a way to classify plants not by what could be regarded as accidental properties, such as medicinal effects, but by formal and inherent characteristics (Greene 1983: 807–31).[11] In terms of contents one of the most influential British publications on medicinal botany, John Gerard's (*c.* 1545–1612) *Herball or Generall Historie of Plantes* of 1597 relied heavily on works by Dodoens and Lobel.

Carolus Clusius (1526–1609) was one of the first naturalists to produce both general plant compendiums and specialized works. Many of his publications of the period 1566–1605 were brought together in *Rariorum plantarum historia* (History of Rare Plants, 1601) and *Exoticorum libri decem* (Ten Books of Exotica, 1605), the latter of which focuses mainly on *exotic* naturalia, a term that was coming into more general use exactly in this period. In his works on the plants of Iberia and of Pannonia (1576, 1583a) Clusius invented a new subgenre in botanical writing, the regional or national flora that covers as fully as possible the plants that grow in a particular location. In terms of genre the botanical works by the Swiss brothers Jean (1541–1613) and Gaspard Bauhin

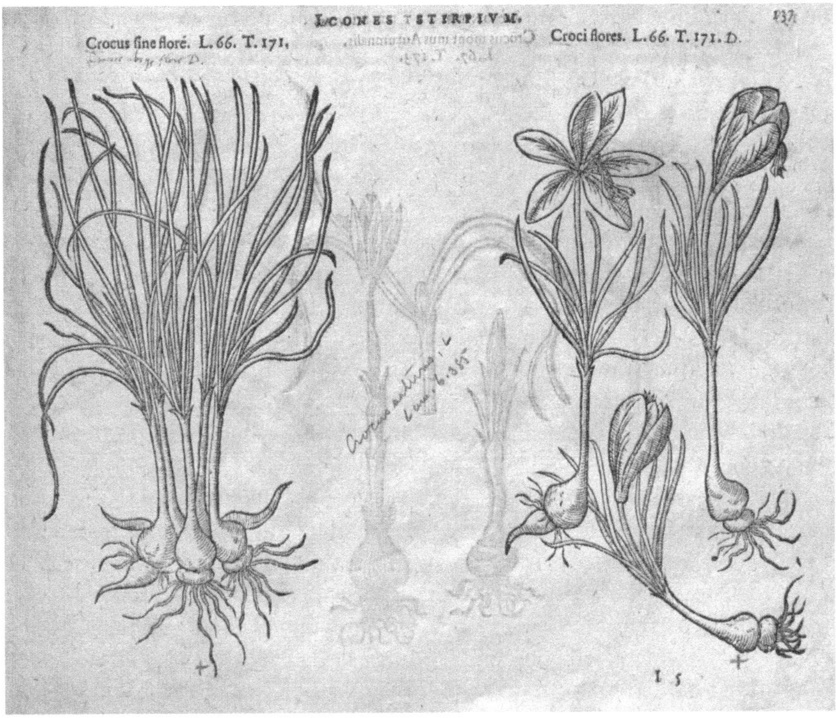

FIGURE 5.3 The saffron crocus in leaf and in flower. From Matthias de Lobel, *Icones stirpium* (Images of Plants, 1591: 137). Image courtesy of Missouri Botanical Garden.

(1560–1624)—especially the latter's (unillustrated) *Pinax theatri botanici* (Theater of Botanical Nature) (Bauhinus 1623) which describes and orders some six thousand species—can be regarded as the tail end of the long sixteenth century. But they also embodied trends that became stronger in the course of the seventeenth century: more botanical systematics, less emphasis on illustration, and a growing split between (plant) medicine and botany (Egmond 2010; Ogilvie 2006; Reeds 1991).

In the 1560s–80s three important works by Iberian physician-naturalists introduced the medicinal plants of Asia and the New World to a European public (see also below): *Historia medicinal de las cosas que se traen de nuestras Indias Occidentales* (Medical History of the Things that are Brought from our West Indies, 1565; augmented in 1569 and 1574) by the Spanish physician Nicolás Monardes (*c.* 1500–88); the *Colóquios dos simples e drogas da Índia* (Conversations on the Simples, Drugs and Medicinal Substances of India, 1563) by the Portuguese physician in Goa García de Orta (*c.* 1501–68); and the *Tractado de las drogas y medicinas de las Indias Orientales* (Treatise on the Drugs and Medicines of the East Indies, 1578) by the Portuguese physician Cristóbal Acosta (1515–94). These works reached a large non-Iberian European public via newly edited and expanded Latin translations by Clusius (Bleichmar 2005; Fontes da Costa 2015; Fontes da Costa and Leitão 2008; López Piñero et al. 1992).

Specialization occurred not only on the botanical but also on the pharmaceutical side of the slowly increasing disciplinary divide between botany and medicine. From the late fifteenth century printed pharmacopoeias began to appear that contained instructions and regulations for the preparation of compound medicines. They were usually issued by a professional organization of pharmacists or by local authorities. The earliest known printed one, from Florence, is dated 1498: *Nuovo receptario composto dal famossisimo chollegio degli eximii doctori della arte et medicina della inclita cipta di Firenze* (New Formulary Compiled by the Most Renowned College of the Distinguished Professors of Art and Medicine of the Magnificent City of Florence). It contains lists of plants and vegetable substances with information about how to collect and preserve them, besides detailed recipes for compound medicines.

What characterized this impressive series of printed herbals in the long sixteenth century more than anything were the multiple editions in both vernacular languages and Latin; the abundant use of woodcut illustrations; and the continuous updating with new information concerning European and non-European plants, which resulted in yet more editions and the increasing size of the volumes. Nothing comparable had happened earlier; nothing similar happened in the seventeenth century.

The printed works of the Italian naturalist Pietro Andrea Mattioli epitomize this process and at the same time give us some insight in distribution and readership. Mattioli was the first to translate Dioscorides from the original Greek into Italian. From the 1520s on, he devoted a large part of his life to rectifying errors in the transmitted Dioscorides texts and adding new information. He identified plants from antiquity, described new plants that he discovered during field research, and examined their medicinal effects—sometimes by experimenting on himself. His investigations resulted in a series of publications that began with his first augmented Dioscorides edition in Italian (1544) and an expanded and richly illustrated Latin edition of 1554. By then Mattioli had added some six hundred "new" plants to Dioscorides' original *c.* six hundred. In new Czech, German, French, and Spanish editions local editors and translators added large quantities of further material. The series concluded, at least during his lifetime, with the even further expanded Latin edition of 1565.

It is hard to exaggerate the reach of Mattioli's works. In the 1560s his principal (Venetian) publisher estimated that some 32,000 copies had been printed of the Italian edition alone. Geographically, Mattioli's works circulated in the whole of Europe. The floral embroideries created by Mary Queen of Scots (1570s), for instance, were mainly based on the woodcut illustrations in Mattioli's work (Bath 2008: 103–7). A colleague of Mattioli's wrote that copies of his work could be found in Syria, Persia, and Egypt, and that a Hebrew translation had been produced in Greece. The long-term influence of both Mattioli and Dioscorides is also evident from the newly updated and once more expanded Mattioli-Dioscorides edition of 1586 by Joachim Camerarius Jr., which itself was translated into many languages and published in several editions. A decade later, Gaspard Bauhin republished selected writings by Mattioli (Bauhinus 1598), and Mattioli's own *Commentarii* were republished once more in Venice in 1744, two centuries after their first appearance (Ciancio 2016; Fausti 2004, 2010; Findlen 1999).

Such huge sales were exceptional, but not completely out of line with other publications about plants and medicine. Assisted by Plantin's effective distribution, Clusius' heavily edited Latin edition of Garcia de Orta's work on Asian plants and medicine reached a wide readership both in Europe and outside it—helped also by the newly introduced illustrations. The first print run of 1,250 copies was sold out within one year (April 1567–April 1568). Italian and French editions were based on this Latin version, and via Clusius' many and regularly updated Latin editions Garcia de Orta continued to be an important source of information on Asian medicine in the Portuguese empire until the nineteenth century (Egmond 2015; Walker 2008: 251–2).

It is no coincidence that many sixteenth-century herbals first appeared in vernacular languages, nor that vernacular translations often appeared soon after the Latin editions. Not only highly educated Latinized circles but a much wider European public took an interest in herbals that functioned both as regularly updated natural history publications and as handbooks of plant-based medicine. That huge readership in itself may well have stimulated the formation of new (sub)genres in print, such as the flora, the florilegium, and the pharmacopoeia. The wave of sixteenth-century publications certainly went hand in hand with the increase of visual plant collections (gardens, herbaria, and albums with drawings) in the course of that same century.

EUROPE AND THE WORLD

Clusius was one of the greatest sixteenth-century divulgers of new information about exotic plants and their medicinal uses, yet he never set foot outside Europe. He collected information about non-European plants and *materia medica* via a wide network of informants. In this network the Garet brothers in London and Amsterdam were key figures as regards exotic plants and medicinal substances. As merchant-pharmacists, "*curiosi*," and information brokers they also were important figures in their own right, however, in the European circulation of new botanical and medicinal knowledge during the 1590s and early 1600s. Precisely during those decades, as mentioned earlier, contacts between the northern European nations and the Far East and South and Central America multiplied. Clusius systematically incorporated the Garets' visual and textual information in his printed works, especially in his expanded Latin editions of Garcia de Orta, Nicolás Monardes, and Cristóbal Acosta (Egmond 2010: 175–207; Harkness 2007).[12]

The Garets' letters to Clusius give us some idea of the variety of exotic plant substances that they received in London and Amsterdam during the 1590s–1610s. Many of these were curiosa, food, and medicinal drugs at the same time. From East Asia and the Pacific, they received beans; a branch of lavender tree; bamboo; roots of dracaena; *Piper longum* (Indian long pepper); an aromatic fruit like a clove; a fruit from the Pacific; a Maldive coconut (*Lodoicea maldivica*, actually from Seychelles); and various medicinal woods, besides star anise. A baobab fruit was shipped from East Africa. From the New World came the potato; Mexican prickly poppy; guaiacum; sassafras; a grain; a leaf of the paper tree; fruit of the Virginian calabash or gourd tree; further fruits, seedpods and roots from Virginia; fruits including soap berries from Guyana; Brazilian beans and seed pods; and dried "herba mimosa" from Puerto Rico. The American potato that James Garet successfully planted in his London garden in 1588–9, and ate, was one of the very first grown in Europe north of the Alps.[13]

By the late sixteenth century exchanges between Europe and parts of Asia went back many centuries. In contrast, the history of European contact with the plants and indigenous medicine of the New World was as yet barely a century old. Everything indicates that Europeans developed a strong interest in the medicinal plants of the New World almost immediately after Columbus and his men set foot in America in 1492. For example, the guaiacum (*Guaiacum officinale*) that the Spanish encountered during the first decade of the sixteenth century—probably in the Santo Domingo area—relatively quickly became a famous, though very expensive, medicine in Europe. There it was regarded as a new wonder drug for the treatment of the "*morbo gallico*" (now usually identified as syphilis) and as an herbal alternative to the more common treatment with mercury. Some regarded this great plague of the sixteenth century itself as an introduction from the New World, which made the efficacy of a New World drug even more plausible (Arrizabalaga et al. 1997). Another famous American medicine—known as cinchona or Jesuit bark or fever tree bark—arrived much later in Europe, during the 1630s, and was a source of quinine effective against malaria (Anagnostou 2000: 182–7; Crawford 2016).

Already by the early 1520s Europeans had begun to exploit the New World as a source of commercially interesting plant substances and drugs. The south German merchant family of the Welsers (allied with the Fuggers), for instance, exported not only sugar but also balsam, guaiacum wood, and canafistula from their domains in the area of Venezuela in the 1520s–30s (Denzer 2005). Once the potential of a few substances had been acknowledged, the European need for surveys of New World medicinal plants that listed their uses in indigenous herbal medicine only increased. Different types of surveys were created in the course of the sixteenth century. A famous example is an illustrated manuscript herbal, *Libellus de medicinalibus Indorum herbis* (Little Book of the Medicinal Herbs of the Indians also known as *Codex de la Cruz Badianus*), that describes and depicts plants used by the Aztecs and their curative properties. An Indian convert priest, Juan Badiano (1484–after 1552), compiled and translated it into Latin on the basis of an original text in Nahuatl by another Indian convert, Martín de la Cruz, at a school of higher learning established by the Franciscans in Central Mexico during the 1530s (Gimmel 2008; Gruzinski 1992: 189–92; Pardo Tomás 2013a: 28–32; Slater, López Terrada, and Pardo Tomás 2014). The resulting bicultural codex "combined Renaissance representation with pre-Hispanic canons" (Mason 2001: 129). The herbal had been commissioned for a very practical purpose by the viceregal Mendoza family. They presented it in 1552 to the Spanish king as part of "a shipment of botanical specimens that exemplified the agricultural/medicinal wealth of the New World" and with the intention of encouraging him "to support pharmaceutical cultivation and trade, and to put the Mendoza family in charge of that trade" (Gimmel 2008: 175).

In the early 1570s the Spanish King Philip II played a key role in a different attempt at surveying the medicinal flora in the New World. It took the shape of a full-fledged scientific expedition (1571–7) to New Spain (Mexico), led by Francisco Hernández, Philip's *protomédico* (royal physician). It was, in fact, the first ever European state-sponsored scientific expedition. Plans for such an enterprise had been circulating in both Italy and Spain since the late 1560s. One of its main goals was to investigate the potential use of New Spain and its plants for Spanish commercial-medicinal purposes. It involved systematic exploration on the spot, and combined direct observation of plants (and animals), collecting, describing and depicting, with the gathering of botanical and medicinal information by interviewing local indigenous and Spanish inhabitants. Several native medical experts took part and advised on curative properties of plants (Barrera Osorio 2006: 101–27; cf. Varey, Chabran, and Weiner 2000).[14]

Shortly afterwards (1579–85) and once more on the command of Philip II, the *Relaciones geográficas* (long questionnaires) were distributed in New Spain in order to collect information about many aspects of life, including health, medicine, and the natural environment. The answers to these questionnaires offer a mass of textual and visual information about food and medicinal plants and their uses in the provinces of New Spain (Barrera Osorio 2006: 81–100; Pardo Tomás 2013b, 2014). Meanwhile, from about the 1560s onwards, Jesuit missionaries and Portuguese settlers began to compile information in Portuguese Brazil about its flora and fauna in yet other written surveys, although much of this information did not become publicly known in Europe until well into the following century (Veracini and Teixeira 2017; Walker 2013).[15] All of these surveys mediated information between indigenous and European botanical-medicinal cultures.

On the other side of the world, the Portuguese first established Goa as their "hub for operations in all of Asia" in the 1510s (Walker 2015: 221–2). Here various forms of European and Asian medical knowledge and documentation merged. One example is an early-seventeenth-century European-style manuscript herbal that depicts Indian medicinal plants. It was created by a man who participated in multiple cultures of herbal medicine, just like Juan Badiano in Mexico. Manuel Godinho de Erédia (1558/60–1623) was born to a Portuguese soldier and a Malay mother. He trained at the Jesuit college in Goa, and was a talented artist, cosmographer, mathematician, navigator-explorer, engineer, and naturalist. His *Suma de árvores e plantas da Índia intra Ganges* (Summary of the Trees and Plants of India intra Ganges, 1612) has seventy-two pages of naturalistic colored drawings of Indian plants and fruit (Everaert et al. 2001). Its short texts have been described as a combination of an "apothecary's manual and pictorial herbarium. The choice of plants shows that he was copying from nature the plants that grew in the gardens in and around Goa" (Županov and Barreto Xavier 2014: 523–4). Although his album is unusual as a combined European-Asian herbal, Godinho de Erédia stood in a by then almost century-long tradition of merchants, apothecaries, and physicians in the Goa area who gathered information about medicinal plants and transmitted it across cultural and linguistic boundaries (Walker 2008, 2015, 2016; Županov 2002; Županov and Barreto Xavier 2014).

The best-known representatives of this tradition are the earlier mentioned Garcia de Orta, who spent a large part of his life in Goa, and the Dutchman Johannes Huygen van Linschoten (1563–1611). The latter worked as official in Portuguese service in Goa (1583–88) and after his return to the Dutch Republic published the internationally successful *Itinerario* (1596), in which he had incorporated texts on plants by the Dutch naturalist-physician Bernardus Paludanus. The *Itinerario* contains not only nautical and mercantile information that was crucially important to the Dutch and English in their competition with the Portuguese for access to East Asia and the spice routes, but also highlights medicinal and other *naturalia*.[16] It emphasizes once more how closely intercontinental commerce and plant medicine intertwined (Cook 2007; Cook and Walker 2013; Walker 2016).

While these examples give us some idea of European attempts to gather information in a context that was largely alien to them, they tell us little about how Europeans responded to herbal medicine in other continents; how quickly (or slowly) Europeans *in* Europe came to know about new medicinal plants from America or the Far East; or how and to what extent they actually began to *use* them in European medicine. European responses differed enormously. Pardo Tomás and López Terrada (1993) divide medicinal plants encountered by Europeans during their first fifty–seventy years in the New World into two categories according to the European response. The first category comprises new plant medicine under the guise of the old. In such cases Europeans either attempted to deduce (or "read off") medicinal properties from the similarities of American plants (in shape, taste, or smell) to their European counterparts; or they recognized certain cures in the New World, such as purging or wound treatment, and assumed that the American plants used in these cures would act in much the same way as their European counterparts. Based on this approach, American sarsaparilla—which belongs to the genus *Smilax* that also occurs in Europe—was successfully introduced into Europe from the mid-sixteenth century as a multifunctional medicine and potential remedy against syphilis (Pardo Tomás and López Terrada 1993: 217–18).

The second category includes all plants whose nature and use were unfamiliar to Europeans, such as tobacco, cacao, guaiacum, potatoes, coca, tomatoes, and maize. Guaiacum was introduced quickly in Europe as an anti-syphilitic, but in many other cases the various uses of such plants only emerged years, decades, or even centuries after their first introduction into Europe. Tobacco was another quickly incorporated drug: it arrived in Europe north of the Alps by the middle of the sixteenth century, and Europeans soon learned to smoke, masticate, and grow it. Its precise properties and functions (medicine, stimulant, element of ritual) and its beneficial or harmful effects on health long remained the subject of debate, however. Interestingly, in Central America itself, the harmful effects of smoking were already becoming clear to physicians before the end of the sixteenth century. On the orders of the Mayor of Oaxaca (Mexico) an autopsy was conducted in 1592 on a person who had been "a great devotee of tobacco"; his lungs were found to be "dry and dark" (Pardo Tomás 2012: 198). Sunflowers, on the other hand (see the Introductory Chapter of this volume), are documented from *c.* 1585 in northern Europe, but they were for a long time grown as purely decorative plants and not for oil. Potatoes, though already tentatively eaten by Garet in 1589, took until the eighteenth century to become a staple in Europe.

Names, properties, and places of origin of exotic naturalia were regularly lost in transport and translation. Of many plant parts that arrived in Europe it was even uncertain whether they were edible, medicinal, a stimulant, poisonous, a wrapping material, or simply useless. The detailed descriptions of exotic gums, resins, nuts, roots, and branches by European naturalists such as Mattioli, Dodoens, or Clusius were very often an essential part of a huge, collective effort at establishing first the identity of a plant, and secondly its possible medicinal and other uses (Gonzalez Bueno 2007). It is not a coincidence, therefore that herbals, image collections, and herbaria of the second half of the sixteenth century often contain comparative visual material and descriptions, which helped Europeans distinguish between, for instance, true and false costus, cinnamon, balsam, turmeric and so forth (see Figure 5.4). The hundreds of items in the German Ratzenberger herbarium (created from the mid-1550s to the early 1590s), for instance,

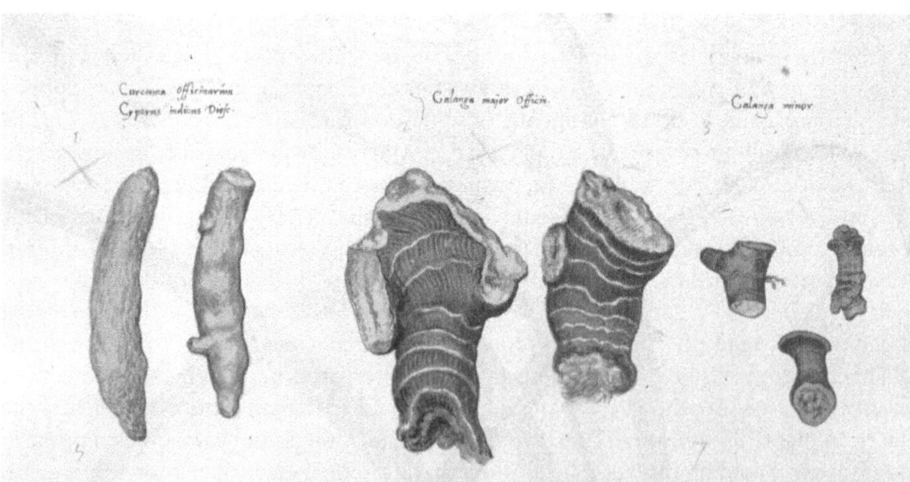

FIGURE 5.4 Exotic medicinal plant substances depicted in the mid-sixteenth-century Flemish *Libri picturati*. Detail. *Libri picturati* A19, f. 25. Image courtesy of Jagiellon Library, Krakow.

include many dried exotic medicinal plants, such as cassia, cinnamon, cardamom, cubeb pepper, turmeric, costus, and nutmeg from Asia and the Middle East, and American cashew, copal, liquidambar, tacamahaca resin, sassafras, guaiacum, jalapa root, besides nux vomica and balsam wood (Schaffrath 2012; Zahn 1902).[17]

Even their inclusion in text and image in European compendia and some knowledge of their uses did not necessarily entail a quick or easy integration of exotic medical plants into a European medicinal discourse, let alone wide distribution in Europe (Čermáková and Černá 2017; Schmölz-Häberlein 2013). Finding or creating a market for exotic drugs also involved a cultural adaptation: physicians, surgeons, apothecaries—and their patients—had to be convinced that new drugs could be effective and could somehow be fitted into the current cultural-medical framework. The success stories of guaiacum, tobacco, and cinchona bark may well have been the exceptions to a much more general pattern in which plants from remote parts of the world only very slowly made their way into the practice of European medicine and health care.[18]

Sources that inform us about the actual inclusion of exotic drugs in therapy and their presence in pharmacies or in the medicine chests or stocks of European households are not abundant and as yet only partly explored (Guerra 1966; Roberts 1965; Varey and Chabrán 1995; Wallis 2006; Worth Estes 2000). Detailed inventories of medicine chests are usually only available for members of the elite. Among the medicines in stock at the court of Philip II of Spain were, for instance, cinnamon water, rhubarb, unicorn and rhinoceros horn, elk's hoof, bezoar stones (some from Peru), terra sigillata, various precious and semi-precious stones, metals, corals, balsams and resins such as liquidambar, and Maldive coconut (Puerto 2003: 282–92). The early-seventeenth-century inventory of Count Anton of Arenberg—son of one of the highest noblemen of the Low Countries, Prince Charles d'Arenberg—comprises a rather similar list of *his* domestic stock. It included several exotic products and compound medicines that probably contained non-European ingredients: cinnamon water, terra sigillata, the horn of an elk, three horns of unicorns, mithridatium, rhubarb, the gum tragacanth, agaricus mushrooms, "mirchoacum" (jalap root), Venetian theriac, a green ointment against sprains, various substances to ward off the plague, syrups against kidney and bladder stones, herbal laxatives, manna, etc. (Muylle 2016). The Early Modern distribution of exotic medicinal plant substances in Europe—over time, geographically and socially—thus remains fairly hazy.

It is no less complicated to evaluate how and to what extent European plant medicine reached other continents and influenced their medicinal practices—not to mention the influence of European medicinal ideas on those of other cultures. Here, we will only briefly outline some of the ways in which European herbal medicine could reach the Far East and the New World. The first and most immediate form of contact was via seafarers and traveling merchants who carried European medicines for their own use. In 1589 James Garet in London, for instance, supplied the drugs for the medicine chest of the British fleet commanded by Paul Wheele and John Chidley, which sailed for the Pacific with a crew of some four hundred men. Garet's friend and colleague, the royal druggist Hugh Morgan, who received drugs from the Englishmen returning from the explorations of Virginia, in his turn supplied the contents of the medicine chest that accompanied Martin Frobisher in 1576 on his voyage to discover the Northwest Passage (Egmond 2010: 179, 191–2). European apothecaries were, therefore, not merely on the receiving end of non-European medicinal herbs, but also acted as intercontinental intermediaries.

Far more important than ships' chests were the medicinal provisions transported by the Spanish armies in the New World, the Portuguese who created mercantile footholds and settlements in Brazil and East Asia, and the missionaries and religious orders that followed them. In their wake European-style hospitals sprang up very quickly in the New World, for instance on Hispaniola in 1503, and in Mexico City by Hernan Cortés in 1524 (the Hospital de Jesús Nazareno only treated Europeans). Yet more hospitals, mainly for the indigenous population, were founded in many parts of New Spain in the course of the sixteenth century (Muriel 1990). In India the Royal Military Hospital of Goa dates back to *c.* 1520; a garden on the premises ensured supplies of "common local and imported healing herbs" (Walker 2015: 221–2). From at least the mid-sixteenth century medicinal plants that originated in different continents could be found next to each other in colonial hospital gardens in Asia and America (Walker 2016).

From the early seventeenth century onwards—though this is taking us somewhat beyond the period discussed here—the exportation of European drugs and herbal medicinal information took on a more systematic character, especially via the worldwide network of the Jesuits, which helped to spread the use of, for instance, (European) theriac as well as (American) fever bark and *ipecacuanha* in several continents. The Collegio Romano in Rome with its central library and herb garden functioned as a nerve centre in this sophisticated network of exchange of both objects and information. During the seventeenth and eighteenth centuries the Jesuits in particular united missionary work with health care and the creation of pharmacies, hospitals, and schools with libraries that included European works on medicine. Jesuits founded pharmacies in the major cities of the mission countries, such as Santiago de Chile, Lima, Buenos Aires, Mexico City, Pernambuco, Rio de Janeiro, Manila, Goa, and Macao. They stocked traditional European drugs as well as remedies that were indigenous to these countries (Anagnostou 2000, 2005, 2007; Harris 1996, 2005).

European knowledge about herbal medicine traveled far outside Europe in the form of texts, especially printed ones. It is only rarely known which books were carried to the Far East or America by individual pharmacists and physicians, but the inventories of the libraries linked with hospitals, missions, and in particular Jesuit colleges regularly mention sixteenth-century European publications. Gessner's works on natural history, for instance, could be found in China, and Clusius' publications were studied in the seventeenth-century missions of Spanish America and the Philippines as well as in China. Woodcut illustrations in these publications could even reach across major linguistic barriers, as is shown by the influence of Dodoens' mid-sixteenth-century herbal in seventeenth- and eighteenth-century Japan (Anagnostou 2007; Vande Walle 2001; Walravens 1973).

Relations between Europe and the rest of the world have so far been discussed here mainly in terms of a bilateral (and Europe-centric) process. Actual circulation patterns of medicinal herbs and related knowledge were far more complex (Cook and Walker 2013). That already emerged from the examples of Hugh Morgan and the Jesuit worldwide circulation, which was anything but one-directional, but is also shown in, for instance, environmental historian Richard Grove's ground-breaking research (1991, 1995, 1996). Almost from the very beginning of European contact with the New World, American medicinal and edible plants began to travel around the world via European intermediaries, but without necessarily arriving in Europe. The South American papaya is said, for example, to have been carried to the East Indies via the Philippines by the Spanish; the Portuguese carried manioc from South America (especially Brazil) to Africa in the sixteenth century, from where the plant traveled to East Asia.[19] In the other direction spice seeds or plants were carried long distance from the East Indies to New Spain during the 1550s with the intention of creating plantations of Eastern spices in the New World (De Vos 2006: 417–18; see Chapter 2 of this volume). Islands such as the Azores, Madeira, the Canary Islands, the Cape Verde islands, and São Tomé off the African west coast played a very important role in this worldwide circulation of plants. In the course of the seventeenth and eighteenth centuries botanical gardens in colonial settlements became increasingly important in this respect. The garden of the Dutch East India Company in the Cape Colony (South Africa), for instance, was first established in 1652 to provide fresh vegetables and fruit for the East Indiamen en route between Europe and the East Indies. The garden soon became a centre of plant acclimatization and intercontinental plant distribution (Fraser and Fraser 2011; Grove 1995).

Worldwide circulation of plant-based drugs and the medicinal plants themselves had effects that went far beyond their material presence. Exotic drugs, spices, and plants influenced European perceptions of what drugs were, or could be, and how medicinal or food plants from totally different climate zones could be related to European ones. The effects could range from confusion—was tobacco healthy, poisonous, pleasurable, a medicine, a noxious substance?—to increased botanical and medicinal knowledge based on comparison. In 1589, for instance, James Garet—when writing to Clusius—confirmed that his potatoes were the real thing (now known as *Solanum tuberosum*) and not sweet potatoes; he also remarked that they grew just like a small solanum. Garet's (correct) reference to the solanum and his comparison in the same year between the potato and the American tomato (which modern experts even put in the same genus) are an indication of his aptitude for comparative botany where exotic plants were concerned.[20] It is quite possible that Garet, as apothecary and drugs merchant, was particularly sensitive to resemblances in the solanum family.[21] Not only the edible potato, tomato, aubergine, and capsicums belong to it, but also poisonous or narcotic plants like belladonna (deadly nightshade), datura, and nicotiana (tobacco). An apothecary incapable of distinguishing the edible from the poisonous ones would soon have had little reputation and few customers left.

European notions of drugs and medicinal plant substances based on intercontinental comparison were, indeed, operative from the very first moment that Europeans set foot on land in the New World. Columbus took specimens of Southeast Asian cinnamon and pepper with him from Spain, and showed them on November 4, 1492, to the local population on Cuba in order to find out whether such spices grew there too—which they confirmed (see the Introductory Chapter of this volume). Of course, Columbus' use of Asian spices was completely understandable, because he was looking for Asia, and thought that he had arrived there (Varela 1984: 50–1). By 1579 Spanish officials in South America were well aware that Asia was a very long way away, yet even then they still used spices and drugs from the East Indies as a means of discovering potentially valuable American equivalents. The Spanish Viceroy of Peru, Francisco de Toledo, instructed a Spanish naval expedition that set out from Peru down the Pacific coast to the Strait of Magellan to find "populations of *Indios*" and discover "whether there are spices, or any kind of drugs or aromatic substances, which is why you are taking some types of spices, such as pepper and cloves, cinnamon, ginger, nutmeg and other things to show them in order that you may teach them and ask them about that."[22] Thus, in an imaginative example of "reading off," Europeans were to show East Indian drugs and spices to *Indios* in Patagonia in order to "discover" the latter's indigenous sources of herbal medicine. Whether the *Indios* could make head or tail of such requests is unknown. The example does reveal something about the Europeans, however. Plant substances from the East Indies (and not European medicinal plants) had apparently become a European model for all exotic simples.

FIGURE 5.5 Title page of the posthumous edition of Rembert Dodoens' herbal by Balthasar II Moretus (1644). Image courtesy of Missouri Botanical Garden.

European mental openness with respect to exotic plant medicine and a clear awareness of medicinal expertise outside Europe also emerge in the preface to the final and largest (posthumous) edition of Dodoens' *Cruydeboeck* by Balthasar II Moretus (1644: 1) in which many of Clusius' descriptions of non-European plants had been integrated: "… the science and knowledge of herbs and plants … which is even preserved and practised by wild and barbaric people; because one notices daily that they (however wild they are) know and use certain herbs and other similar things to heal their wounds and cure their illnesses" (see Figure 5.5).

CONCLUSION

Inevitably this chapter has only briefly discussed a number of very important developments that occurred in plant medicine during almost two centuries, *c.* 1420–*c.* 1620. While the overall Galenic approach remained predominant in this period, and plants continued to be the core ingredients of medicinal preparations, many other things changed. Printing had major effects on the distribution and circulation of information. A vast increase in the number of documented plants, and the gradual establishment of patterns in the description, visual representation, naming, and classification of plants resulted in growing specialization and, by the late sixteenth century, in a beginning disciplinary split between botany and plant medicine. Specialization resulted in new genres in print, such as the flora on the botanical side and the pharmacopoeia in the domain of pharmacy and medicine. Until well into the seventeenth century, however, nearly all works on plants—whether handwritten and illuminated or printed and illustrated—continued to pay attention to their effects on the health of human beings.

The long sixteenth century can also be regarded as the first period in history in which large amounts of new information about non-European medicinal plants and herbal medicine began to reach Europe. Except in the case of a small group of relatively famous medicines and plants, the reception and actual application of these exotic plants and medicines in Europe is as yet hard to estimate. Many seem to have been incorporated only slowly in European medicinal culture. During the same century European herbal medicine spread via seafarers, merchants, armies, missionaries, and especially hospitals created in colonial contexts both in the New World and in East Asia. This period also saw the start of a global circulation of medicinal plants and knowledge of their uses—a process that, like the medicinal lives of plants, has by no means ended.

Acknowledgments

I would like to thank Peter Mason, José Pardo Tomás, Holger Funk, Sarah Kyle, Dominic Olariu, and Andrew Dalby for their many helpful remarks and suggestions. The research for this article was done as part of the NWO-funded projects *Clusius Project*, *Re-reading the Book of Nature* and *A New History of Fish* at the University of Leiden.

CHAPTER SIX

Plants in Culture

LUKE MORGAN AND ELIZABETH HYDE

> Plants are, I find, named for different reasons. Some take their names from whoever first discovered, understood, demonstrated, cultivated, domesticated and acclimatized them: thus, herb-mercury, from Mercury ... Other plants have kept the names of the regions from which they were introduced elsewhere, as *malum medicum* for a citrus fruit [*poncire*] from Media where it was first found ... Others take their names by antiphrasis and contradiction; hence absinthe, which is the opposite of *pinthe*—tipple—because it is *not* pleasant to drink ... Others are named for their virtues and effects, as *aristolochia*, which helps women with their lochial fluids in childbirth ... others from the remarkable features seen in them, as the heliotrope (that is, the marigold), which follows the sun, opening at sunrise, stretching upwards as the sun ascends, drooping when it sets and closing up when the sun hides itself ... others from the metamorphoses of men and women with similar names, as daphne (that is, the laurel) from Daphne ... others, from similarities: as *hippuris* (that is, horse-tail), because it does resemble a horse's tail ... And, by naming the other way round, the Fabii were called after *faba* (the bean) ... others, from the shapes, as the trefoil, which has three leaves ...
>
> François Rabelais, *Tiers livre*[1]

In his account of the origins of plant names, French writer François Rabelais (1483/94–1553) referred to the story in Ovid's *Metamorphoses* (1.545–52) in which the nymph Daphne is pursued by Apollo, who desires her. Fleeing him, Daphne pleads with her father, the River God Peneus, to "work some transformation, and destroy this beauty which makes me please all too well!" Her prayer is answered and "a deep languor took hold on her limbs, her soft breast was enclosed in thin bark, her hair grew into leaves, her arms into branches, and her feet that were lately so swift were held fast by sluggish roots, while her face became the treetop" (Innes 1955: 47). Before long, Daphne's metamorphosis into a laurel tree is complete. Ovid relates that Apollo, after embracing and kissing the transformed nymph, declares that laurel will henceforth be his tree, and that it will signify triumph.

From the fifteenth century onwards, artists began to depict the story. During the 1470s, for example, Florentine painter Piero del Pollaiuolo (*c.* 1443–96) depicted Daphne undergoing her metamorphosis (see Figure 6.1). In the painting, Apollo seizes her around the waist, but her arms have already become verdant laurel branches and her legs have taken root in the earth below. There is a relationship between Pollaiuolo's representation of Daphne's metamorphosis, which takes place within a Tuscan landscape, and Florentine statesman Lorenzo de' Medici's Ovidian poem "Ambra." In his own evocation of the

FIGURE 6.1 Piero del Pollaiuolo, *Apollo and Daphne* (1470s). Photo courtesy of National Gallery, London.

landscape around Florence, Lorenzo (1449–92) writes that "Among the leafless trees, the verdant laurel / Stands alongside the fragrant Cyprian myrtle ..." in a deliberate play on a form of his own name—Lauro (Thiem 1991: 127). The reference to laurel also draws attention to the source of his poem in Ovid's story of Apollo and Daphne. Lorenzo describes how the river god (and local river) Ombrone who, like Apollo, desires a nymph (Ambra), is thwarted when she turns into a rocky outcrop on the site of the future Villa Medici in Poggio a Caiano (the villa was named "Ambra"). Later in the poem, the shepherd Lauro, Lorenzo's alter ego, becomes jealous when the Ombrone river floods and "embraces" Ambra: "During this time of year, Ombrone, a lover / Swollen and proud encircles Ambra, she / Like a small island. Ambra, no less dear / To Lauro, jealous if his rival hugs her" (Thiem 1991: 132).

Pollaiuolo's decision to set the scene of Daphne's transformation in Tuscany may imply an allegorical reference to the triumph of Lorenzo's rule (Hartt and Wilkins 2013: 365). From Lorenzo onwards, the Medici family made use of laurel as a political symbol. In the Florentine *Giostra*, or tournament of 1469, Lorenzo's banner depicted a sun and rainbow above an image of a woman making a wreath from the leaves of a green, living branch of a laurel tree, which was otherwise dry (Ladner 1961: 315–16). In addition to the Apollonian imagery of victory (the wreath), the motif of the living branch symbolized regeneration, a meaning derived from the evergreen properties of laurel, which flourishes even after pruning. The same idea was exploited by subsequent members of the Medici, often in deliberate memory of Lorenzo, to symbolize the continuity and resilience of the family and its governance of Tuscany (Cox-Rearick 1984: 15–40).

The most famous image of laurel in Italian poetry appears, more than a century earlier than Lorenzo's "Ambra," in poet Francesco Petrarch's *Canzoniere*. Although the name of Petrarch's beloved Laura rarely, if ever, appears in his vernacular poems, there are multiple veiled allusions to her. The words "*l'aura*" (the wind) and "*lauro*" (laurel) both imply her presence.[2] The subsequent influence and familiarity of Petrarch's word-play can be detected, not only in later Italian-language poetry such as Lorenzo's, but also in Leonardo da Vinci's mocking remark that: "If Petrarch was so fond of laurel, it is because it has a good taste with sausages and roast thrush: I cannot set any store by their twaddle" (Kemp 1985: 203).

Leonardo's skeptical view of Petrarch's many references to Laura's name suggests that by the sixteenth century the *topos* had, in some quarters at least, become clichéd. Even so, it continued to exert an influence on poetry and art. In the Venetian artist Giorgione's painting known as "Laura," for instance, a dark-haired woman, wearing a fur-lined coat and exposing her right breast, is framed by the branches of a laurel tree. Some historians believe that the presence of the bare breast indicates that Giorgione (1477/8–1510) has portrayed not Petrarch's Laura but a Venetian courtesan, many of whom were accomplished poets—hence the laurel (Brown and Ferino-Pagden 2006: 208).[3] In other contexts, however, a bare breast can signify chastity and virtuousness, as can, in fact, laurel.

This additional symbolic meaning of laurel accounts for its presence in images of the Madonna. Ovid had referred to it as "the virgin laurel tree" (*Metamorphoses* 10.92) and, in the Christian era, it became a symbol of chastity due to the belief that it maintained its green foliage "uncorrupted" (D'Ancona 1977: 202). In Giovanni Agostino da Lodi's *Madonna and Child with Donors* (c. 1520, Museo Nazionale di Capodimonte, Naples), for example, a laurel tree forms a cross behind the Madonna and Child in a symbol of both chastity and triumph over death.

The symbolism of laurel also lent it to landscape design. During the sixteenth century, the artificial mounts that were constructed in the gardens of Tuscany and Lazio were associated with the idea of Mount Parnassus in central Greece, the dwelling place of Apollo and the Muses. In 1517, for example, the "Mons Vaticanus" (Vatican Mount) in the Vatican gardens was described as a "new Parnassus" and, later in the century, the Villa Lante in Bagnaia was characterized as the home of the Muses (Cellauro 2003: 42).

Garden mounts were often planted with laurel—the emblem of Apollo after Daphne's metamorphosis—and decorated with fountains of the winged horse Pegasus as well as statues of the Muses.[4] In the extensive gardens that Francesco I de' Medici laid out (from 1568) at the Villa Medici (now Demidoff) in Pratolino, for example, there was a mount depicting Parnassus, Pegasus, and the Muses, augmented with a concealed hydraulic organ and planted with laurel trees.

This brief and selective history of the symbolic uses of a single naturally occurring motif, ranging from the poetry of Ovid, Lorenzo de' Medici, and Petrarch to painting of the fifteenth century and on into the late Renaissance garden, is suggestive of the many uses of plants in a wide spectrum of cultural contexts. The mythological, poetic, religious, and political meanings of plants and trees, such as laurel, imply the existence of a de facto symbolic language. In Early Modern Europe, this language was, however, not a single, unified lexicon of motifs and associated meanings, but rather, a range of intersecting, sometimes complementary, and at other times contradictory, discourses. The first part of this chapter will explore some of the cultural meanings of plants in sixteenth-century Europe, chiefly Italy, through reference to three of these symbolic systems, as they might be called: first, plants as emblems; second, plants as "signatures"; and, third, plants as marvels.

VISUAL SYMBOLS: PLANTS AS EMBLEMS

Plants of various kinds play an important role in the moralizing emblem books and iconographical handbooks of the sixteenth and seventeenth centuries. For example, in scholar Andrea Alciato's *Emblematum liber* (Book of Emblems, 1531), laurel, to stay with the example discussed above, has specific symbolic meanings. "Emblem CCXI" of the illustrated edition of *Emblematum liber* published in Padua in 1621 depicts a laurel tree flanked by an altar on the left and a tripod on the right (Thuilius 1621: 881) (see Figure 6.2). Twigs of laurel are shown burning on the altar, sending up plumes of smoke. A dove appears between the legs of the tripod, which is surmounted by a dolphin, and a shield, decorated with a cameo-like profile head of a man crowned with a laurel wreath, leans against the tree.

Emblems are composites of image and text. In this case, the image is accompanied by two distichs (pairs of lines of verse), which read: "The laurel foreknowing the future, offers symbols of security: placed under the pillow it makes dreams come true," and "A laurel is due Charles for defeating the Phoenicians: such wreaths grace victorious heads." The first distich refers to ancient lore in which laurel served as an apotropaic defense against evil powers, as a source of prophetic wisdom when burnt, as a protection against lightning strikes when worn in a wreath, as an antidote to disease, headaches, and poison, and as a source of inspiration (Heckscher 1992: 209–10). The burning laurel of the emblem invokes these traditions, as does the tripod, which was associated with the prophetic utterances of the famed oracle of the god Apollo at Delphi, Greece (Heckscher 1992: 208n1). The second distich alludes to the Emperor Charles V—Holy Roman Emperor (1519–56), King of Spain (1516–56), and Archduke of Austria (1519–21)—and his military victory in North Africa in 1535, which resulted in twenty thousand Christian slaves being liberated from Tunis.[5] Here Alciato has exploited the established meaning of laurel, and especially of laurel wreaths, as a symbol of victory.

The emblematic meanings of plants, which were systematically established and codified by Alciato, iconographer Cesare Ripa (*c.* 1560, Perugia–*c.* 1622), and others, were exploited by Renaissance artists and architects. For the sake of an example, in art, acanthus could

FIGURE 6.2 Laurel: emblem 211 from Andrea Alciato's *Emblematum liber* as published by Joannes Thuilius (1621). Image courtesy of The National Gallery of Art, Washington, DC.

symbolize Christ's Passion, because of the similarity of its leaves to spearheads (D'Ancona 1977: 34). Thus, in painter Piero della Francesca's *Flagellation* (*c.* 1455–60), the bound figure of Christ appears in a loggia supported by Corinthian columns with acanthus leaf capitals in a probable allusion to the wound that he received on the Cross from the Roman soldier Longinus' spear thrust. In other contexts, however, acanthus could have different meanings. Roman author and architect Vitruvius (70/8–15 BCE) stated in his *Ten Books of Architecture* (4.1.8) that the proportions of the Corinthian columnar order were modeled on the figure of a young woman (Rowland 1999: 55). Here, the acanthus leaves of the column's capital signified ornamentation and adornment (in comparison, for example, with the pared-back and, according to Vitruvius, masculine Doric Order) and, in general terms, a gender distinction that was applied to building types.

RESEMBLANCE: PLANTS AS SIGNATURES

In the sixteenth-century writings of Swiss physician and philosopher Paracelsus (Philippus Aureolus Theophrastus Bombastus von Hohenheim) and Italian scholar Giovanni della Porta, among others, plants were conceived as "signatures." This doctrine was founded on the principle that plants and animals had been formed by God in such a way that their appearances signified their medicinal utility. Plants were, in the words of English statesman and humanist Sir Thomas More (1478–1535), "natural hieroglyphs" (Agamben 2009: 36). Paracelsus wrote, for example, that the Euphrasia flower (or "Eyebright") cures eye diseases and maladies, "because it has in itself the *anatomiam oculorum*; it has in itself the shape and image of the eyes, and hence it becomes entirely eye" (Agamben 2009: 37).[6] In his *Phytognomonica* (Vegetal Physiognomy), the Italian natural philosopher Giambattista della Porta illustrated several flowers that, through resemblance, could be used to cure the eyes (Porta 1588: 135) (see Figure 6.3).

Philosopher Michel Foucault (1926–84) has argued that the "doctrine of signatures" was one of the principal foundations of sixteenth-century epistemology (1966, 1970). He

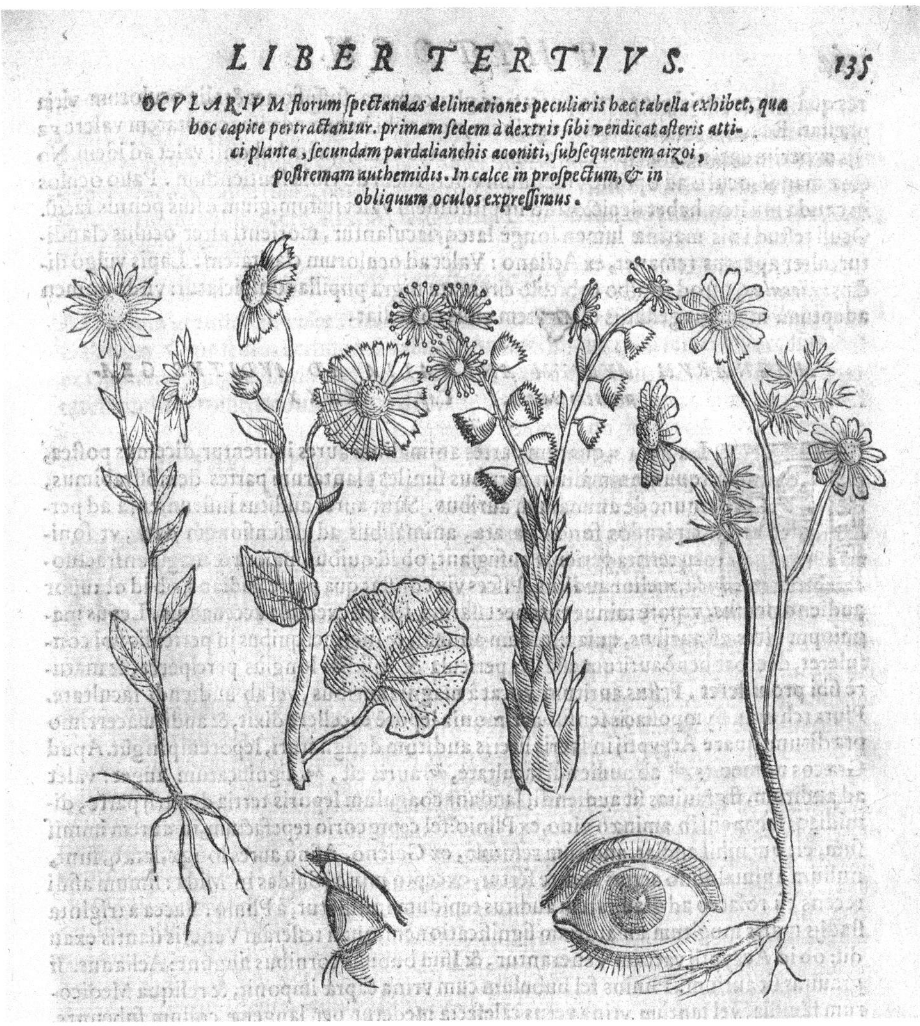

FIGURE 6.3 *Flores oculares*, to be used, in accordance with the doctrine of signatures, to cure eye diseases (Porta 1588: 135). Image courtesy of Williams Special Collections, Monash University, Australia.

claimed that the defining characteristic of knowledge in all its forms during this period was its dependence on "resemblance" as a constructive principle:

> ... it was resemblance that largely guided exegesis and the interpretation of texts; it was resemblance that organized the play of symbols, made possible knowledge of things visible and invisible, and controlled the art of representing them. The universe was folded in upon itself: the earth echoing the sky, faces seeing themselves reflected in the stars, *and plants holding within their stems the secrets that were of use to man.*
>
> (1970: 17, emphasis added)

Resemblance took four main forms. Foucault called the first of these *convenientia*, which refers to the notion that everything in the world is linked in a great chain of resemblance through physical adjacency (*convenientia* is literally "coming together"). One of the potential consequences of *convenientia* is the mingling of properties, as in the zoophytes (hybrid creatures believed to be half plant and half animal) discussed by Nicolò Serpetro (Findlen 1994: 65). The so-called Scythian Lamb or "borometz," for example, was thought to grow on a stalk and to sustain itself through eating the surrounding vegetation (Duret 1605: 330) (see Figure 6.4). As French poet and diplomat Salluste Du Bartas (1544–90) wrote, the borometz "dies that day / That they have brouz'd the neighbor grasse away" (1589: 66; Prest 1981: 51; Sylvester 1621: 181).

Foucault's second form of resemblance is *aemulatio*, which—like a mirror and its reflection, or a sound and its echo—enables things to resemble one another across distances (unlike *convenientia*, which relies upon physical propinquity). The German

FIGURE 6.4 The Scythian lamb or borometz, from Claude Duret's *Histoire admirable des plantes* (Marvelous History of Plants, 1605: 330). Image courtesy of Wellcome Collection, London.

alchemist Oswald Croll's *De signatura rerum* (On the Signatures of Things, 1609) provides an example of the relevance of the concept to the plant world:

> The stars are the matrix of all the plants and every star in the sky is only the spiritual prefiguration of a plant, such that it represents that plant, and just as each herb or plant is a terrestrial star looking up at the sky, so also each star is a celestial plant in spiritual form, which differs from the terrestrial plants in matter alone ... the celestial plants and herbs are turned towards the earth and look directly down upon the plants they have procreated, imbuing them with some particular virtue.
>
> (Foucault 1970: 20)

There is evidence that this astrological maxim had some influence on the layout and study of plants within the confines of the Early Modern Italian botanical garden: for example, in a letter dated June 30, 1626, Florentine herbalist Zanobi Bocchi stated that he had planted herbs in the botanic garden in Mantua by "applying the planets" (Tongiorgi Tomasi 1983: 21).

The third form—*analogy*—combines the powers of both *convenientia* and *aemulatio*, facilitating resemblance across space but simultaneously incorporating a notion of links or bonds. Leonardo da Vinci's writings on the analogy between the human body and the earth provide an example. At the beginning of his projected treatise on water, Leonardo wrote that the ancients called "Man" a "lesser world" (or a microcosm): both the human body and the world are composed of the same elements: earth, water, air and fire. Just as the body has bones as a framework for the flesh, so does the earth have stones for the same purpose. The equivalent of the lungs and "pool of blood" inside the body are the oceans which rise and fall "with the breathing of the world" (Leonardo was referring to tides). The veins, which carry blood in rivulets throughout the body, are analogous to the rivers and waterways emanating from the ocean. Leonardo went on to compare the body of the earth with a vegetative body or plant. In his explanation of the capacity of water to rise even to the mountain tops, he gave the example of "the water [which] rises from the lowest part of the vine to the branches that are cut." This is, he claimed, analogous to the process in which "from the lowest depth of the sea the water rises to the summits of mountains, where, finding the waves broken, it pours out and returns to the bottom of the sea" (Richter 1977: 46–7).

The fourth and final form of resemblance is *sympathy*. Foucault stated that: "Sympathy is an instance of the *Same* so strong and so insistent that it will not rest content to be merely one of the forms of likeness; it has the dangerous power of *assimilating*, of rendering things identical to one another, of mingling them, of causing their individuality to disappear ..." (Foucault 1970: 23). The sunflower—imported to Europe from the Americas in the sixteenth century—is a case in point. In German scholar Joachim Camerarius the Younger's plant-derived emblem book, *Symbolorum et emblematum ex re herbaria*, where it is a symbol of God's grace, the sunflower is depicted straining toward the sun (1590: 59; MacDougall 1991: 155). Sympathy is, however, mitigated by *antipathy*, which ultimately prevents the destructive corrosion of identities. Foucault cited Italian polymath Gerolamo Cardano's *De subtilitate* (On Subtlety) in explanation: "It is fairly widely known that the plants have hatreds between themselves ... it is said that the olive and vine hate the cabbage; the cucumber flies from the olive ..." (Cardanus 1550; Foucault 1970: 24).

Foucault's ascription of these concepts to sixteenth-century culture in general or to what he called the Renaissance "episteme" has been criticized by historians who argue that the doctrine of signatures was never as widely adopted as he claims. It has been shown, for example, that Foucault's theory of resemblance is drawn from Neoplatonism and occult traditions and that he does not acknowledge the greater influence of Aristotelian frameworks for understanding, in particular, the natural world (Maclean 1998). The doctrine of signatures, in other words, is not representative of the whole thought of the period, which was significantly more diverse than Foucault claims. Yet even so, there is evidence that the ostensibly esoteric ideas of Paracelsus, della Porta, and Croll were considered to have practical applications, as the example of the botanic garden at Mantua suggests.

ART AND NATURE: PLANTS AS MARVELS

The Early Modern period was fascinated by stories of transformation and hybridity such as, respectively, that of Daphne in Ovid's *Metamorphoses* (or Ambra in Lorenzo de' Medici's poem) and the legendary Scythian Lamb discussed earlier.[7] As Rabelais noted, some plants were named after the men or women who had been transformed into them. The idea of transformation is a leitmotif of the *grottesche* (grotesque decorations), which proliferated all over Italy during the sixteenth century. Plant forms played an important role in nearly all of these decorative schemes, both ancient and Early Modern. Painter and writer Giovanni Battista Armenini, for example, described the ancient *grottesche* of the vault of a grotto in which harpies were depicted, their breasts transforming into leaves (Barocchi 1971–7: 2701). In one of the earliest frescoes from the Renaissance for which an extant contract specifies the inclusion of *grottesche*—the Libreria Piccolomini in the Duomo at Siena—the sinuous tendrils of plants merge with human and animal figures to create new hybrid forms.[8]

The metamorphic imagery of *grottesche* has corollaries in natural history. French botanist and historiographer Claude Duret, for example, discussed the so-called Credulity Tree as well as the Scythian Lamb in his *Histoire admirable des plantes* (Marvelous History of Plants, 1605). As he wrote, it was believed that when the leaves of the Credulity Tree dropped from their branches they turned into either fish or birds depending on whether they fell onto water or land (Prest 1981: 50).

These legends and motifs are all concerned with plants and their transformations into animals and vice versa. For this reason, they recall Foucault's point about the epistemological importance of the concept of *convenientia*, which rests on the premise that all things are linked in a great chain of being. The metamorphoses that occur through *convenientia* are, at least in the examples discussed up to this point, the result of a natural process. In other cases from the period, however, transformation occurs through the artificial simulation or manipulation of plants. More perhaps than any other site, the garden provided a virtual laboratory for these experimental practices.

Dominican monk Francesco Colonna's romance *Hypnerotomachia Poliphili* (1499) provides two remarkable fictional examples of completely artificial plants and gardens. In the course of his quest for the nymph Polia, the protagonist Poliphilo encounters gardens made entirely of glass and silk. In the first, topiary box trees, cypresses, beds of simples, and flowers are all imitated in glass. A pleasant fragrance even "emanated from the flowers, which had been rubbed with an ointment and watered" (Godwin 1999: 124). Despite the fantastic character of Colonna's narrative, these artifices are not as far-fetched as they might first appear. In the earliest treatise exclusively dedicated to the uses of flowering plants in gardens—*De florum cultura*—Sienese professor Giovanni

Battista Ferrari offered advice about how to change a plant's color or scent through making cuts in its roots and pouring dye or perfume into them (Ferrarius 1633).

In Colonna's second garden, there were "box-trees and cypresses with golden stems and branches, appropriately seeded with gems, and the containers were filled with simples that Mother Nature would have envied, flowering and most desirable with every exquisite colour, and fragrant just like the glass flowers" with the difference that all the plants were made of silk instead of glass (Godwin 1999: 127). The reference to Nature's "envy" draws attention to a central theme of the Renaissance garden: the competition or collaboration between art and nature.

There are differing views about the priority of art or nature during the period. The Florentine sculptor Baccio Bandinelli believed that "the things one builds must be the guide and superior to those one plants," when he was asked to design a fountain for the Boboli garden in 1551 (Lazzaro 1990: 27). In contrast, other mid-sixteenth-century writers about gardens defined the relationship between art and nature as fluid and collaborative rather than oppositional in character. In a letter of 1541, Jacopo Bonfadio wrote that in the garden the interaction of art and nature produces more extraordinary effects than either could achieve on its own. According to him, this collaboration resulted in a "third nature" (Hunt 2000: 53–4). A few years later, in his treatise *La villa* (1559), Bartolomeo Taegio expanded on the point: "nature incorporated with art is made the creator and connatural of art, and from both is made a third nature, which I would not know how to name" (Lazzaro 1990: 9). Garden fountains and grottoes were frequently thought to exemplify this idea. In 1543, for example, the philologist Claudio Tolomei wrote an enthusiastic letter about fountains in which he claimed that it was difficult to distinguish between the respective contributions of art and nature (Lazzaro 1990: 61).

Plants also played a role in the collaboration of art and nature in the Renaissance garden. Although, so far as we know, there is no direct equivalent of Colonna's glass and silk gardens, many designers sought to emulate nature to a greater or lesser degree. The "rustic style" pioneered by the artist and architect Giulio Romano in Rome and Mantua and developed further by the architect Sebastiano Serlio in his treatises on architecture, for instance, incorporates columns that are carved to resemble roughly hewn and stripped tree trunks (perhaps in a reference to Vitruvius' account of the origins of architecture), such as those designed by French architect and author Philibert de l'Orme (1514–70) at Anet, France (Morgan 2007: 122). Another example is provided by the columns of the Fountain of the Owl in the garden of the Villa d'Este in Tivoli which are entwined with realistic-looking, polychromatic, but simulated, tendrils of ivy.

Other garden structures foregrounded the exchange between art and nature, such as the treehouses that were constructed at many sites. In draughtsman and painter Giovanni Guerra's drawing (1598, now in the Albertina, Vienna) of the treehouse in the garden of the Villa Medici in Pratolino, spiral stairs are depicted winding their way around the thick trunk of an oak tree. Guerra noted in an inscription that the tree had been "adapted to practical use" (Lazzaro 1990: 55). Treehouses of this kind were regarded as epitomizing the benefits of the collaboration between art and nature in the garden. In his description of another treehouse at the Villa Medici in Castello, painter, architect, and writer Giorgio Vasari commented that:

> In a meadow to the east of the garden [Nicolò] Tribolo planted a holm [oak tree] so thickly covered with ivy that it looked like a thicket, and it was approached by convenient wooden steps, at the top of which is a resting-place, with seats about

> it, with backs, all of living green, and in the middle a marble table with a vase of variegated marble, into which water is brought by a pipe which spouts in the air and is carried off by another pipe. The pipes for the water are so covered by the ivy that they cannot be seen, and the water is controlled by taps. It is impossible to describe how the water is carried along the branches of this tree, to sprinkle people and to make fearful hissing sounds.
>
> (Hind 1927: 3:174; Vasari 1568: pt 3, 2:407)

French philosopher and essayist Michel de Montaigne, who visited Castello in 1580, was struck by a "chamber among the branches" of the tree which was "so closed in by this verdure that there is no view out except through a few apertures that must be opened up by pushing aside the branches here and there" (Frame 1943: 931–2; Meusnier de Querlon 1774: 1:258). The walls of the chamber were, in other words, artfully composed of branches and leaves and the hydraulic technology employed to produce the *giochi d'acqua* ("water-tricks") carefully concealed to enhance both the element of surprise and the impression that the water has a natural source. The illusions of the suspended room and the water tricks are achieved through the collaboration of art, nature, and science.

The emulation of natural forms in the garden by art and architecture was motivated by the principle of "nature as model." This cultural concept joined other ideas about plants—as emblems and as signatures—from mythography to heraldry and from science to astrology, suggesting the multivalency and symbolic utility of plants during the Renaissance. Rabelais' excursus in *Le Tiers Livre* on the "wonderful qualities" of the plant that he called *pantagruelion* is representative, despite its fictional status (though *pantagruelion* resembles hemp and flax). We are told that the plant is so-called because Pantagruel first discovered it; the word "*pantagruelion*" means, literally, "Pantagruel's thing." Rabelais wrote that it was feared and loathed by thieves, who found it more inimical than the water-lily was to lecherous monks—a formulation that recalls both the emblem literature and the iconography of Renaissance art (lilies signify purity and, by extension, the Virgin Mary); the thieves' fear of *pantagruelion* resided its use to make hangman's nooses. Moreover, the antipathy of thieves to *pantagruelion* is not dissimilar to that of "cabbage to vine" or "onion to eyesight" in an echo of Cardano (and Foucault) on the antipathy of plants (1546: 339; Screech 2006: 605). Foucault's discussion of resemblance is again recalled in Rabelais' second explanation of the plant's name: "It is also called *pantagruelion* from similarity; for when Pantagruel was born into this world he was as tall as the plant in question ..." (Screech 2006: 606). The "doctrine of signatures" is further evoked by one of the medicinal uses of *pantagruelion*. According to Rabelais: "If you boil its roots in water it will relax tense sinews ..." (Screech 2006: 607). Finally, he continued, the myriad and marvelous qualities of *pantagruelion* are so extraordinary that the gods of Olympus themselves are terrified that Pantagruel's children will discover an even more powerful plant,

> by means of which human beings will be enabled to visit the sources of the hail, the sluice-gates of the rains, the smithy of the thunderbolts; they will be able to invade the regions on the Moon, penetrate the territories of the Signs of the Zodiac and settle there, some in the golden Eagle; some in the Ram; others in the Crown; others in the Harp, and others in the silver Lion, sitting down at table with us and taking our goddesses to wife, the only ways in which humans can be deified.
>
> (Rabelais 1546: 346–7; Screech 2006: 609)

PLANTS AS SIGNIFIERS OF POWER

Plants continued to function in a rich cultural tradition of symbolic systems into the seventeenth century—as emblems, signatures, and marvels—as visual symbols to a range of concepts and meanings. But by the end of the period under consideration here, by the middle to the end of the seventeenth century, the epistemological understanding of such systems had been transformed along with the understanding of the meaning of plants within those systems. In 1688, Parisian painter Charles de la Fosse executed a painting for the Trianon de marbre, the auxiliary palace at Versailles most closely associated with plants, and especially flowers, that filled the royal gardens of the Sun King, Louis XIV. The interior of the palace was decorated with floral still-life paintings together with a series of mythological works illustrating scenes from Ovid's *Metamorphoses* and *Fasti* (Days of the Year) including among them the stories of Apollo and Daphne, Zephyr and Flora, Iris and Jupiter, and Clytie and Apollo (see Figure 6.5). Ovid's telling of the tale of the nymph Clytie explained that after the Sun's seduction of Leucothoe, the jealous Clytie spread the word of Leucothoe's fall, leading her father to bury her alive for her transgression. The heartbroken Sun shunned Clytie and condemned her to watch and follow for eternity his daily journey across the sky. Explained Ovid,

> For nine days she tasted neither food nor drink, but fed her hunger only on dew and tears. She never stirred from the ground: all she did was to gaze on the face of the sun

FIGURE 6.5 *Clytie Transformed into a Sunflower, c.* 1688, Charles De La Fosse (1636–1716). Commissioned by Louis XIV of France for cabinet du couchant du grand Trianon, Versailles. Image by Trancrede, courtesy of Wikimedia Commons.

> god, as he journeyed on, and turn her own face to follow him. Her limbs, they say, became rooted to the earth, and a wan pallor spread over part of her complexion, as she changed into a bloodless plant: but in part her rosy flush remained, and a flower like a violet grew over her face. Though held fast by its roots, this flower still turns to the sun, and although Clytie's form is altered, her love remains.
>
> (Innes 1955: 109–10)

The flower into which Clytie metamorphosed, described as "like a violet" in Ovid's text, had traditionally been understood to be the heliotrope, a member of the borage family, which was long known in Europe for its distinctive "heliotropism"—the tendency for its flowers to follow the movement of the sun over the course of the day. And yet when Charles de la Fosse depicted the myth in 1688, his Clytie was transformed into a sunflower, a flower clearly *not* resembling a violet (Renaux 2008: 112). Unlike both heliotrope and the Ovidian myth, the sunflower, or *Helianthus annuus*, was new to western Europe, though it had been cultivated in North America for thousands of years (see the Introductory Chapter of this volume). By 1633, the flower had already been adopted as a symbol of fidelity and loyalty when it was deployed by Flemish painter Anthony Van Dyck (1599–1641) to communicate his faithful service to his patron, Charles I of England (Peacock 2006). By the late seventeenth century, as de la Fosse's painting shows, the spectacular flower had supplanted the violet-colored heliotrope in the ancient Ovidian myth. The sunflower demonstrates continued European fascination with the "plant as marvel," its marvelous tendencies having been appropriated for its use as a visual symbol expressing fidelity. But its use by Charles de la Fosse communicated so much more than that. For the presence of the flower, and the presence of the sunflower in particular, in his rendering of "fidelity" is indicative of a transformation of the culture of plants in European culture that had taken place in the sixteenth and seventeenth centuries and a transformation in their symbolic deployment such that the sunflower could convey intellectual, scientific, and cultural sophistication along with imperial connections and geopolitical consciousness. The expression of these qualities by a single flower—and therefore in a symbolically and visually economical manner—is the consequence of what Foucault attributed to an evolution in the relationship between the sign or symbol and that which it signified. In this evolution, occurring as Europe passed intellectually from the Renaissance into what he labeled the "Classical" age, the symbol, having acquired complexities of meaning and knowledges, became capable of communicating an idea without spelling out the symbolical relationship (Foucault 1970: 63). In other words, a single flower could signify not only the older symbolic reference to fidelity through its symbolically analogous heliotropism, but also newer meanings of plants. The sunflower has taken its place in the seventeenth-century French rendering of the Clytie myth not only because of its behavioral resemblance to the quality to be illustrated (loyalty or fidelity), but also and importantly because of its "natural history." To continue thinking with Foucault, "natural history," he argued, resides in the space between the signs or characters used to express the growing realm of knowledge and the great "tangle of nature" (Foucault 1970: 73). "Natural history finds its locus," he continued, "in the gap that is now opened up between things and words—a silent gap, pure of all verbal sedimentation, and yet articulated according to the elements of representation" (Foucault 1970: 129–30), representation that is animated by the "perpetual possibility of imaginative recall" (Foucault 1970: 59). The sunflower, therefore, was not only an effective botanical symbol of fidelity, but also a signifier of all that the imaginative recall of its natural history entailed: its exotic origins in South America, the imperial efforts required to rescue it from America and bring it to Europe, the curiosity that led to its collection, the practical knowledge requisite for its successful cultivation, and the intellectual endeavor underpinning the scientific inquiry into its heliotropism. In short, the sunflower simultaneously echoed heliotropically Clytie's eternal curse and animated the history of the Sun King's botanical and therefore political power.

The intellectual and botanical transition represented by the substitution of the sunflower for the heliotrope was in part the consequence of a changing culture of flowers and plants that occurred over the sixteenth and seventeenth centuries. And that changing culture transformed how they were incorporated into emblems and portraits and even propaganda. If, in the sixteenth century, Giorgione played with laurel, Laura, and Petrarch in his portrait of a Venetian courtesan, something very different was at work when French miniaturist and engraver Nicolas Robert (1614–85) incorporated flowers into the portrait of Gaston, duc d'Orléans, brother of Louis XIII, the second Bourbon king of France.[9] Depicted within a laurel wreath framed by a pink cartouche bearing the Bourbon *fleur de lys*, the medal of the Order of Malta, and the palm and laurel within the royal crown, the portrait identified Gaston as a member of the Bourbon dynasty. The attributes of the Bourbons were accompanied by military paraphernalia. The shield, drum, spears, and armor referred to Gaston's military service. But the other items in the portrait were just as significant to the identity of the duke. The armillary sphere, sextant, and books were signifiers of Gaston's intellectual life, for he fashioned himself a *curious* man. The Renaissance passion for collecting and studying the texts, medals, and sculptures of antiquity had, by the sixteenth century, expanded and transformed to include the collection and study of a wide range of "curiosities" from medals and antiquities to books, prints, fossils, shells, and animals. Gaston d'Orléans, though royalty, was part of the intellectual and social world of the *curieux*. He assembled an impressive cabinet of curiosities that he displayed at Blois (central France). And it is in this context that Nicolas Robert presumably decided to include the two floral pendants hanging from either side of the cartouche in the portrait. Plants, and especially flowers, occupied a prominent place in the world of the curious. So the flowers in the portrait of Gaston d'Orléans signified the duke as a fashionably intellectual man, they also represented the flowers that were themselves collected.

COLLECTING AND COMMODIFYING: PLANTS AS CULTURAL SIGNIFIERS

Just as important as the presence of flowers were the varieties of flowers represented: the floral pendants framing the duke's portrait included tulips, anemones, ranuncului, narcissi, carnations, pinks, and hyacinths. Collectively, they form a far different bouquet from roses, violets, and lilies, the blossoms traditionally incorporated into European literary and visual floral references to biblical and/or classical texts, figures, and virtues. The flowers chosen by Robert to enhance the portrait were "florists' flowers," the species of flowers mostly highly desired and collected by the "curious florists," among which Gaston d'Orléans was one of the most important. Beginning in the sixteenth century, a wide range of elite men, from dukes and counts to lawyers, doctors, monks, and pharmacists began to aggressively collect plants. Animated in part by efforts to gather specimens from the rapidly expanding post-Columbian world, some collected plants towards the goal of reassembling the long-dispersed "Garden of Eden," while others were intent on unlocking the potential medicinal powers of plants native and new to Europe. The resulting collections formed the first botanical gardens, gardens that would facilitate the newly important study of botany at universities and medical schools and academies such as those at Leiden and Padua and Paris.

Other collectors, however, were more interested in new species, among them a large number of flowering bulbs, mostly reaching Europe from eastern and eastern Mediterranean sources, that had aesthetic and exotic appeal for those who sought them and the gardens in which they were planted. Chief among these flowering bulbs were the anemones, ranuncului, hyacinths, and especially tulips that constituted the "florists' flowers" collected by the curious. Some built collections of all of the most fashionable flowers, while others worked at assembling collections around a particular species for which they sought to acquire every available variety, every available color and blossom variation. The desire and fashion for the new and exotic drove science: deliberate and aggressive attempts at selective breeding by collectors, resulted in many new color variations and double blooms of various species. These efforts were particularly effective in expanding the number of ranunculus, anemone, and tulip cultivars, representations of which fill the many florilegia created to document their existence and ownership. Gaston d'Orléans was a collector of each of these flowers. Indeed, Nicolas Robert came to paint his portrait as the Duc d'Orléans had specifically engaged Robert to paint "portraits" of the florists' flowers growing in his gardens at the chateau of Blois. By the time Gaston hired Robert, he had already proven himself to be a talented painter of flowers. Robert was responsible for painting the flowers in the exquisite *Guirlande de Julie*, a manuscript anthology of poems about flowers collected by Charles de Montausier (1610–90) from his literary friends in Paris and illustrated with Robert's paintings.[10] The poems and paintings of the *Guirlande* (which were bound together in perfumed red Moroccan leather) formed a metaphorical garland of flowers with which Montausier hoped to win the hand in marriage of Julie d'Angennes, daughter of Catherine Vivonne-Savelli, the Marquise de Rambouillet, who hosted the most fashionable salon in Paris.

While Robert's paintings for Montausier illustrated the floral poetry in the *Guirlande*, under Gaston d'Orléans' employ, Robert documented the flowers growing in the duke's garden. His beautifully rendered paintings on vellum amounted to a catalog of the duke's living collection. These flower paintings created the core of the "*vélins*" (vellums). After Gaston d'Orléans' death in 1660, the collection passed into the hands of his nephew, Louis XIV, who stepped into the role of patron, employing Robert to continue to add to the *vélins* until Robert himself died in 1685. Subsequent to his death, other royal flower painters would be employed to add to the collection. It continued to grow through the eighteenth century and would eventually result in the collection of over four hundred paintings now housed in the Múseum National d'Histoire Naturelle, Paris (Heurtel and Lenoir 2016).

While the *vélins* perhaps constitute the finest collection of flower portraits, it was by no means the only one. Rather, the Bourbon project was representative of the flower collectors' impulse to document the specimens in their collection. Some were painted. Others were engraved and printed for sale in florilegia such as French painter and illustrator Daniel Rabel's *Theatrum Florae* (Rabel 1627). Still more were documented not visually, but in printed catalogs and books like Pierre Morin's (Morin 1651). The documentation left the visual and printed record of what grew in their parterres.

Some of the catalogs were botanically driven, others more economically so, which importantly raises the question of value. The rarity of some plants, with many being imported from the Near East, Mediterranean, and eventually Asia and the Americas, meant that they could be difficult and expensive to acquire, attributes that only enhanced their desirability. Similarly, as collectors and gardeners became increasingly successful in breeding different colors, shapes, and endless variations among the florists' flowers, flowers became an object of fashion. Anemones and tulips were among the most "fashionable" flowers in the seventeenth century, with double and variegated anemones, and variegated tulips deemed the most desirable—and hence the most expensive. The desire for tulip bulbs was so great that speculation on the price of bulbs escalated, especially in The Netherlands. Speculation was greatest around the bulbs that were supposed to produce red and white striped flowers. In a story that has been told many times, the "bizarre" red and white tulips were actually caused by a virus that was undetectable in the bulbs. So speculators hoping to cash in on the craze purchased bulbs at ever-increasing prices, until the tulip bulb bubble burst in 1637. While it has been demonstrated that few Dutch people actually lost fortunes speculating on tulip bulbs, tulips came to symbolize irrational spending and speculating (Goldgar 2007). Flowers, by their very nature as the short-lived beautiful stage in the cycle of botanical life, had long been incorporated into expressions of beauty and critical reminders of the short-lived nature of human—and especially female—beauty. Flowers were reminders not only of beauty, but also of death and decay. As such, flowers were perfect components of still-life paintings. The genre blossomed in the seventeenth century, and not coincidentally in simultaneity with the rise of floriculture and flower painting. The expanding number of floral species, growing floricultural literacy among elites, fashionability of flowers, and sheer aesthetic beauty of flowers converged with a buying public. Any flower could communicate the reminder that life is fleeting, but in the wake of the "tulip mania," the tulip became the most common floral mode of expression in the *vanitas* genre. While some still-life paintings delivered their message subtly through a browning or fallen petal on an otherwise lovely blossom, others like French painter Philippe de Champaigne's stark still life, joining a striped tulip with a skull and hourglass, left no uncertainty as to its meaning: life, like beauty, was short. And then you die.

Flowers were also drafted into the communication of serious, if less macabre messages. The flower mania and collapse of the tulip bulb market in The Netherlands led many to associate flowers with the risk of financial speculation and the foolishness of being caught up by the whims of fashion. In 1637, the year the market collapsed, Dutch engraver and publisher Crispijn van de Passe II delivered a critique in the form of Flora reigning over a ship of fools, *Flora's Mallewagen* (Flora's Wagon of Fools, see Figure 6.6), while engraver and illustrator Pieter Nolpe depicted tulip transactions taking place inside a fool's cap (*Floraes Gecks-kap*, 1637).

The association of flowers with foolish and fashionable consumption persisted long after the tulip mania. In 1688, the French literary figure Jean de la Bruyère penned a literary portrait of a flower collector in his *Caractères* as a critique of fashion:

> The florist has a garden in the faubourg, where he goes shortly after the sun rises until he goes to bed; you would think that he planted himself there and had taken root among his tulips and in front of the *Solitaire* [a tulip variety]; he stares, he rubs his hands, he bends down, he looks at it more closely; he has never seen anything so beautiful; his heart is filled with joy; he leaves [the *Solitaire*] for the *Orientale*, then he

FIGURE 6.6 *Flora's Mallewagen* (Flora's Wagon of Fools), 1637. Attributed to Crispijn van de Passe II, 1637, northern Netherlands. Image courtesy of Rijksmuseum, Amsterdam.

> goes to the *Veuve*, he moves on to the *Drap d'or*, from there to the *Agathe*, from where he finally returns to the *Solitaire*, where he rests, sits down, loses himself, and forgets his dinner .... God and nature are in all that he does not admire, he does not go beyond the bulb of his tulip which he will not abandon for a thousand *écus*, although he will give it to you for nothing when tulips are neglected, and carnations have prevailed.
>
> (La Bruyère 1691: 483–4)

The "perpetual possibility of imaginative recall" (Foucault 1970: 69) allowed for the survival of the stereotype of the flower collector into the eighteenth century, when a version of La Bruyère's critique reappeared in eighteenth-century philosopher Denis Diderot's *Encyclopédie*. But more importantly for our purposes here, the uses of the tulip and flower collector as interrelated cautionary symbols of the brevity of life and the foolish spending of one's life demonstrate that over the course of the seventeenth century, the new culture of flowers was rewriting their symbolic meaning, meaning acquired in the new contexts of consumption. In Foucaultian terms, the "signifying element [the flowers] and the signified element [the cautionary message] are linked only in so far as they are (or have been or can be) represented, and in so far as the one actually represents the other" (Foucault 1970: 67). The mania for tulips made them a requisite feature in the *vanitas* painting, while the role of fashion in shaping flower collecting made them an effective (while attractive) means to critique modern patterns of consumption. Flowers were signifiers of their own natural history, their own consumption—but simultaneously signifiers of the broader concepts to which they drew comparison.

The consumption of flowers by collectors and investors alike, as we have seen, led to the inventorying of collections in the cabinet and in the form of the catalog. The catalog might be a painted record of the plants in one's collection, like Gaston d'Orléans' *vélins*, or German still-life painter Jacob Marrel's flower paintings used to advertise tulip varieties for sale, or they might be published catalogs of collections such as Pierre Morin's *Catalogues de quelques plantes à fleurs qui sont de présent au jardin de Pierre Morin le jeune* (Catalogs of What Flowering Plants are Present in the Garden of Pierre Morin the Younger; Morin 1651). Painted or printed, the catalog posed a problem—it situated the collection of objects in the intellectual gulf between object and language. For the flowers, listed or painted, demanded description and classification. Collectors needed to tout their botanical treasures, and sellers needed to advertise their wares. Naming and classifying, therefore, can be seen as a semiotic by-product of the collection of plants. The task of the collector and thinker alike was to use language to name and describe—to name and describe in a manner that communicated knowledge of the plant under consideration. While collector/sellers struggled with naming as a means to distinguish one object in their collection from another, the collector-cum-scientist struggled to name in a manner that created knowledge. In proposing his system of classification (that would eventually be replaced by the Linnaean), botanist Joseph Pitton de Tournefort (1656–1708), wrote that, "To know plants is to know with precision the names that have been given them in relation to the structure of some of their parts ... The idea of the character that essentially distinguished plants from one another ought invariably to be one with the name of each plant" (Tournefort 1694: 1; quoted in Foucault 1970: 139). While the plant continued to be a culturally powerful signifier, thinkers sought clarity in naming and classifying that moved beyond the symbolic.

The same rigor was demanded of representation. In 1676, Denis Dodart, on behalf of the Academie Royale des Sciences, undertook the publication of the *Mémoires pour servir à l'histoire des plantes*. Printed by the Imprimerie royale, it was the first of several volumes planned for telling the academy's "history of plants." Nicolas Robert, Abraham Bosse, and Louis-Claude de Chastillon were among the artists employed to illustrate the plants included in the history. Abraham Bosse was tasked with representing the *mandragora* (mandrake) (see Figure 6.7). Bosse chose to depict the plant in a manner that referenced the complex "history" of the plant. Not only was mandragora believed to be an aphrodisiac, but it was also believed to be used by witches for the purpose of inducing in their husbands a deep sleep to enable the witches to fly off to their Sabbath. It was, therefore, a dangerous plant. And a gendered plant. Believed to exist in "male" (single-rooted) and "female" (forked root) forms, harvesting the plant was even more dangerous: it was thought to emit a deadly shriek when pulled from the ground, the female form shrieking the loudest. Bosse, like others before him, chose to allude to mandragora's treacherous associations in his illustration of it. Above the waist, Bosse's "female" mandragora was an accurate representation of the leaves and flowers of the plant. Below, however, his mandragora evoked a female human body, complete with torso, thighs, and even genitalia. Though such visual gendering had become conventional, by 1676 it was no longer acceptable. The members of the academy rejected Bosse's mandragora, calling it a "ridiculous affectation" (quoted in Stroup 1990: 82). So Bosse eliminated the anthropomorphic elements of the image and it was included in the publication. But the episode is illustrative of the intellectual transformations that had taken place over the

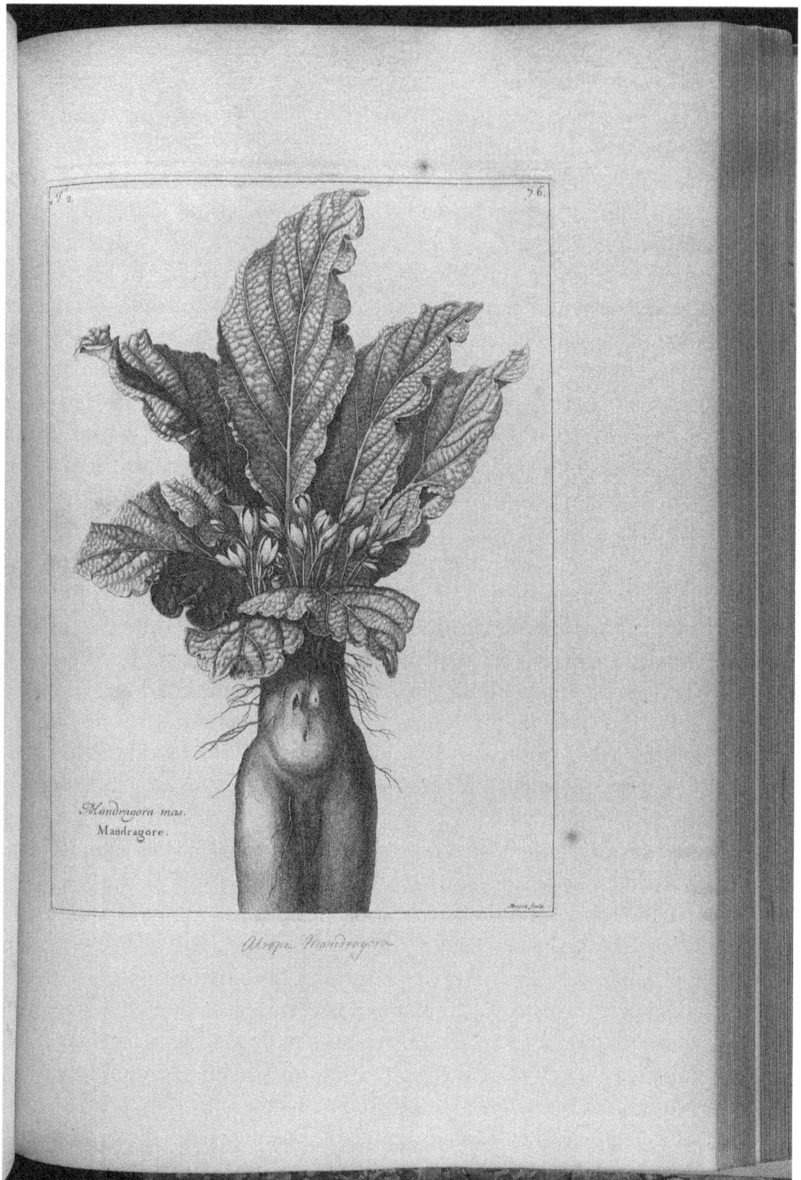

FIGURE 6.7 "*Mandragora mas*. Mandragore. Atropia Mandragora," Abraham Bosse (1602–76) Illustration of a Mandrake root depicting a humanoid root. From *Estampes pour servir à l'histoire des plantes* (Reflections Serving as a History of Plants), vol. 2. Plates by N. Robert, A. Bosse, and L. de Chastillon. Image courtesy of Wellcome Collection, London.

course of the seventeenth century: the plant was no longer a dangerous marvel, but rather an object to be studied.

The process of the de-mystification of the world of plants, however, did not mean that they lost their utility as symbols and signs. For while the frivolous consumption of fashionable flowers might serve as a cautionary symbol, the very act of (successful) cultivation of plants and flowers was increasingly used as an expression of plenty and prosperity, and with particular effectiveness by the Bourbon kings of France. Agrologist Olivier de Serres (1539–1619), for example, in his *Théâtre d'agriculture et mesnage des champs* (Exposition on Agriculture and the Tending of the Fields), suggested a link between plants and the state. First published in 1600, this was one of the most important early French books on estate management. The work understood the proper management of the estate to encompass all sorts of agricultural endeavors, from the rearing of livestock, the management of the vineyard, and the planting of grain and vegetable crops, to the creation of a pleasure garden. And the successful estate, the work suggested, was key to the successful management of the state. Olivier de Serres dedicated the volume to Henri IV, who was depicted on the title page enthroned in a garden parterre atop a triumphal arch bearing the inscription at the top of the arch: "SECVRITAS PVBLICA," or "Public Security." The message of the *Théâtre d'agriculture* was that the success of each estate, the success of each seigneur managing each estate was the success of the state. The ability of the estate to feed the family of the seigneur and the families of the peasants working the land for the seigneur, and the ability of the estate, through the cultivation of grapes for wine to drink, and flax and hemp to clothe them, together with the rearing of livestock for meat, milk, cheese, and wool, together with the pleasure garden, meant not only the well-being of the families of the estate, but the well-being and therefore the security of the state. Later in the seventeenth century, his grandson Louis XIV (1638–1715) would claim responsibility for the prosperity and security of the state through a range of plant-based political iconographical schemes. Historians have made much of Louis XIV's abandonment of the mythological mode of representation of his power when he ceased to have himself depicted as Jupiter or Apollo (Burke 1992). But he did not abandon symbols and signs altogether, favoring the use of plants, flowers, fruits, and vegetables to communicate the plenty and prosperity he sought to provide for his subjects. One of the last building projects by Louis XIV, the chateau of Marly, illustrates the transformation. An auxiliary palace in the park at Versailles, Marly was supposed to be decorated by painter Charles Le Brun in the allegorical style that he had used in the palace of Versailles, where Louis XIV as Apollo or as himself but supported by the gods and goddesses of Mount Olympus, staked a claim to his legacy. As work on Marly continued, the scheme was abandoned in favor of Flora, Aurora, and Pomona—deities of floriculture, agriculture, and horticulture. By the time the decoration of Marly was completed, however, in the later 1680s, the deities had been retired in favor of the flowers, plants, and fruits they represented (Néraudau 1986: 248). If anything, then, as flowers and plants had come to represent the ideas and meanings inherent in their natural attributes, cultivation, and consumption, they became even more potent symbols in Early Modern culture.

CHAPTER SEVEN

Plants as Natural Ornaments

JILL FRANCIS

Gardens are, by their very nature, ephemeral. They change according to the seasons, they reflect new fashions and trends, they vary according to the wealth and status of their owners, and they can be altered on a whim, according to taste. As statesman Sir Robert Sidney wrote to his estate manager at Penshurst in Kent in 1605, "the little garden, since it is so forewards, may goe for this year: if I do not like it, I can alter it the next." There are many elements that make up a garden, including its form, its structure, its layouts, and its borders, as well of course as the trees, shrubs, fruits, flowers, and vegetables that are cultivated within it. In gardens of the upper classes, man-made structures such as terraces, steps, and walks as well as balconies, seats, and mounts provided a variety of vantage points from which the garden could be appreciated and admired, and water features such as fountains, cascades, and grottoes all contributed to the creation of spectacular visual, auditory, and sensory experiences. But of all these aspects, it is the plants that are the least permanent and, for the garden historian therefore, the most elusive.

This chapter sets out to trace the place of plants as natural ornaments in the Early Modern era, with particular focus on the gardens of sixteenth- and seventeenth-century England and Wales. This is an especially exciting period in which to consider the place of ornamental plants in the garden, as it was during this time, and for the first time, that plants and flowers were being introduced into gardens for no other reason than their decorative value and their beauty (Francis 2018: 165–8; Zalum 2013: 81). Prior to this, and right across the social scale, the primary function of most gardens—and the plants within them—was to provide food, medicines, dyes, and scents for use in the household. Plants were included in the garden according to their uses and virtues. As John Gerard, the Elizabethan author of *The Herball* observed of plants for the garden, "the delight is great, but the use greater, and joined often with necessitie" (1597: Epistle Dedicatorie).

However, at the same time, the idea of plants and flowers as natural ornaments in the garden was becoming an increasingly important one. The Early Modern era was one of great change in European society, and as the seismic effects of the Renaissance, the Reformation, the invention of the printing press, and the "discovery" and exploration of the New World swept across Europe, they brought with them a corresponding shift in the way people viewed their world. New possibilities, new desires, and new aspirations were fueled as rare and exotic goods, accompanied by travelers' tales, both by word of mouth and in print, poured into the ports of Europe. And these, of course, included rare and exotic plants from Asia and the Americas which were increasingly being sought after by enthusiastic collectors, gardeners, and garden owners alike. Alongside this, a changing culture surrounding the consumption of luxury goods and new attitudes to the accumulation of wealth and the visible representation of status contributed to the increased significance of the ornamental garden as it developed as a place in which rare and costly plants could be displayed. This phenomenon was also reflected in the new genre of practical gardening literature that had been appearing in England since the middle of the sixteenth century, and culminating in what was arguably one of the most outstanding horticultural books of the period, John Parkinson's *Paradisi in sole Paradisus terrestris: A garden of all sorts of pleasant flowers*. Published in London in 1629, this book is an apparent response to the changing times, being the first gardening book in English to take the creation of a flower garden as its main subject. By placing plants and flowers firmly at both at the center of his book and at the center of the garden, Parkinson was affirming for the first time that "beautifull flower plants" might be included in the garden, not for their uses and virtues, but for no other reason than for their ornamental value.[1]

It is an analysis of these changes and their implications for the Early Modern gardener and garden owner that will form the main focus for this chapter, but first, the role that plants—despite their undoubted utilitarian function—have always played in the ornamentation of gardens needs to be considered.

"FOR MEATE OR MEDICINE, FOR USE OR FOR DELIGHT"

The fact is that ever since "God Almightie first planted a garden" (Bacon 1625: 266), it has been axiomatic that gardens contain plants, and whether trees, fruits, herbs, or flowers, they have added a vital aesthetic aspect to any garden, from the grandest and most flamboyant of upper-class gardens to the more modest kind of plot tended by the majority of people. However, at whatever level, the inclusion of ornamental flowers and plants does not necessarily appear to have been their most noteworthy feature.

Italian Renaissance gardens, for instance, were characterized by their structure, with emphasis on a symmetrical layout, shaped by architectural features such as terraces, steps, and balustrades, and punctuated with fountains, statues, and grottoes (Strong 1979: 15; see Figure 4.4 of this volume). Although it is known that these gardens did contain flowers (Duthie 1990: 82), contemporary observers rarely mention them and it is the overall impression of controlled proportion and beauty that has remained for future generations to admire. Having said this, however, plants were still an essential part of this ordered arrangement. Traditional evergreen varieties, of known hardiness and predictable habit, were used to create precisely cut hedges and borders to map out the permanent geometric designs, trees were set in corners or ordered rows, and pots containing plants and flowers were similarly carefully placed in order to highlight the structure, beauty, and harmony of the garden.

Upper-class English gardens, themselves influenced by the Italian and French *compartimenti* systems, were similarly laid out with walks, alleys, and bordered beds which gave permanent order and beauty to the garden (Jacques 1999: 32–3), without relying on the fleeting adornment of colorful flowers. English author, Gervase Markham (*c.* 1568–1637), provided precise instructions on how this might be achieved. He devoted two chapters of his book *The English Husbandman* to the laying out and "beautifying" of a modest garden plot (1613: 112–26). He described in detail how to form alleys and quarters and recommended privet, box, lavender, and rosemary as suitable plants for bordering them as these slow-growing shrubs could be clipped frequently and kept in good order. He also suggested the use of colored gravels and sands, not just for the alleys and walks but also for in-filling the geometrically patterned squares. In a final paragraph, he did concede that there is another way of beautifying gardens and that is the "quaint, rare" but "best eye-pleasing" practice of planting flowers to provide color. But even then, he advocated planting them in blocks of "one kinde and colour," the flowers thus contributing to the overall effect, rather than to show off the beauty of individual specimens. The problem with this method of ornamenting any garden, as recognized by many of his contemporaries, was that this kind of planting did not last for the whole year. Native English garden plants of this period tended to be summer-flowering varieties such as roses, gillyflowers, carnations, and columbines and, by their very nature, they bloomed, faded, and died within a few short months. For much of the year, therefore, there would be little in the way of flowering plants in the English garden, so the importance of using evergreen hedges, trees, and shrubs to give permanent shape and structure to the garden—using them as natural ornaments—became even more significant.

An extremely rare depiction of just this type of garden can be seen in the background of a portrait of statesman and philosopher Sir Thomas More and his family by Rowland Lockey, 1593–4 (see Figure 7.1). This late Elizabethan garden reveals a very simple layout, which comprises one hedged compartment enclosed by a high wall at the far side, which contains five squares, or quarters, and an L-shaped bed, each bordered with low hedging and surrounded by wide paths. There are just three trees, and although these are not planted symmetrically, the overall impression is of a geometric and ordered pattern within the garden. In common with the handful of other representations of gardens in

FIGURE 7.1 Rowland Lockey, after Holbein, *Thomas More and Family*, 1594 (detail) showing the layout of the garden in the background. Victorian and Albert Museum, London. Photo from V&A Images/Alamy Stock Photo.

paintings of the late sixteenth and early seventeenth century, there is little to suggest that ornamental flowers were a major feature of this garden, although interestingly, in this case, there are some colorful cut flowers prominently displayed inside the house where the family are seated. It is possible that these were grown in another area, where they could be cut without disturbing the overall design of the garden shown here.[2]

These examples demonstrate one way in which plants had traditionally been used—and would continue to be used—as natural ornament in the garden, but of course, there were other ways. Early practical garden manuals tended to focus on the cultivation of culinary and medicinal plants, such as cabbages, leeks, onions, parsley, dill, and mints for instance, as well as a range of "physick" herbs for "curing sundry griefs," although contemporary authors such as Thomas Hyll and William Harrison also mentioned the growing of more unusual varieties such as of melons, gourds, "pompions," and cucumbers (Harrison 1587: 208; Hyll 1577: 28–9).[3] However, although such plants were clearly grown as food—John Gerard and John Parkinson both recommended ways in which they could be prepared for the table—they also added an aesthetic element to the garden, being used, for instance, and as advocated by Thomas Hyll, to form decorative arbors (22). Grown over juniper or willow poles, these prolific climbing plants provided a pleasant shaded walk or place to sit (see Figures 7.2 and 7.3).

An example of this in practice is provided by Sir John Oglander, who gardened on his Isle of Wight estate at Nunwell in the early decades of the seventeenth century. In his "commonplace" book (compilation of knowledge), in which he described in detail the range of fruiting trees being cultivated in his garden, he noted that he had a vine and an "infinito" of raspberries growing over a frame in the bowling green (OG/AA/29).[4] It is not entirely clear whether or not the bowling green was still used for its intended purpose, but it is certainly possible that Sir John had created a shaded recreational area with these climbing fruits, similar to that described and illustrated in literature of his time.

Hyll also devoted a few paragraphs of his book to "delectable" and "pleasaunte floures" such as marigolds, columbine, lavender, gillyflowers, pinks, and carnations, but these were

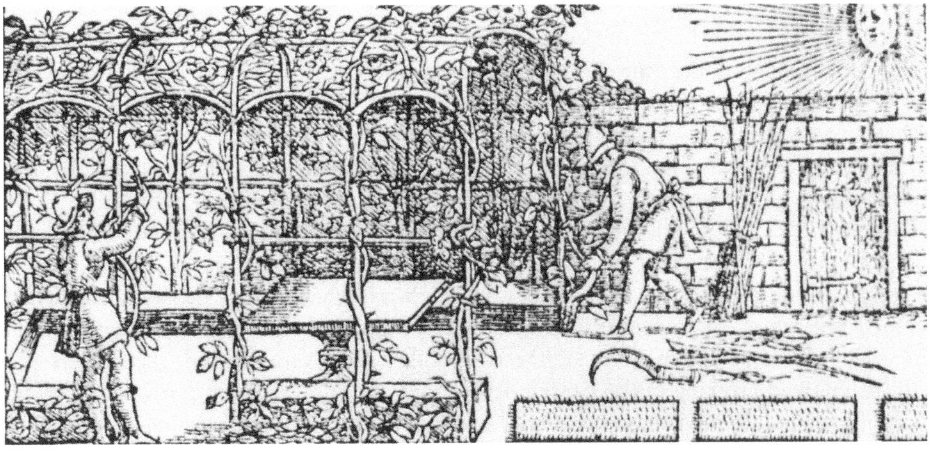

FIGURE 7.2 Woodcut illustration from a contemporary garden manual of gardeners training vines over an arbor "in arch manner." From *The Gardeners Labyrinth* [sic] by Thomas Hyll, using the pseudonym Didymus Mountain, with Henry Dethick and published in 1577. Photo from Universal History Archive/Universal Images Group/Getty Images.

FIGURE 7.3 Woodcut illustration from a contemporary garden manual of vines trained into “a square forme” (Hyll 1577). Photo by Culture Club/Getty Images.

all grown along with the other herbs already mentioned and, as he also noted, they could be dried and used to “ beautifie and refresh the house” (28). Indeed, in the Early Modern world, scented plants and flowers played an essential role in overcoming unpleasant and ever-present odors. In 1598, Paul Hentzner, a German visitor to Greenwich Palace, noted the floor strewn with rushes “after the English fashion” (Hentznerus 1612: 135; Walpole 1757: 47) and John Parkinson observed that germander, commonly used for bordering garden beds, was even “more used as a strewing herb for the house” (1629: 456). It should also be remembered that most plants, whether they be herbs, medicinal plants, vegetables, or fruit trees will inevitably produce flowers as part of their natural lifecycle: although this may not be the reason that they were planted, the flowers can still be enjoyed growing in the garden before they are harvested for drying or cooking, or being made into medicines, household dyes, scented waters, and cordials at the end of the summer. It is important to realize that the presence of flowers in the garden was not synonymous *only* with the ornamental pleasure garden.

And so it was with orchards. Although their main purpose was obviously to produce fruit, for the clergyman William Lawson the orchard provided delights to the senses in all seasons: the sight and fragrance of the blossoms in springtime, the green leaves and "mantle of the earth" in the summer, the colorful, scented, and tasty fruits in the autumn, and the trees standing in "comely order which so ever way you look" at any time of the year (1618: 53, cf. Thick 2003). Ralph Austen, writing a few years later in 1653, added the sounds of the birds, the rustling of the leaves, the feel of the cool, fresh air under the shady trees, and the smell of freshly dug earth to the list of sensuous delights provided by the trees in the orchard. Here, he wrote, "Profit and Pleasure meet and imbrace each other" (35–6).

Sir Thomas Tresham (1543–1605) planted two orchards in his garden at Lyveden New Bield in Northamptonshire in the late sixteenth century, and this rare instance of an extant and relatively undisturbed Elizabethan garden offers a perfect example of this juxtaposition of productivity and ornament. Reached via a series of terraces which rose up from the manor house to the impressive lodge at the top of the garden, the conventionally arranged "lower orchard" consisted of approximately three hundred trees, including many varieties of apple and pear as well as cherry, plum, and walnut, all set out in orderly rows. The Elizabethan statesman, Sir Robert Cecil observed that this was "one of the fairest orchards that is in England" (Eburne 2008: 121) and William Lawson would certainly have approved of the "comely order." The adjacent "moated orchard," however, was a much more elaborate affair. Tresham's extant notes reveal that this orchard was to be planted in a series of ten concentric circles—evidence of which can still be seen in the garden today. It would seem that the outer circles were to be planted with cherry and plum trees, while the remaining inner borders were to be planted with a combination of roses and raspberries (Eburne 2008: 123). The orchard was to be surrounded by a moat on four sides (although in fact the fourth canal was never built) with a terrace to one side, offering protection from cold northerly winds and, more importantly perhaps, a viewpoint for the visitor over this intricately arranged planting pattern. In addition, there were four spiral mounts, one at each corner which would have enhanced this view. It seems clear then, that as well as being a productive area, the moated orchard was also intended to provide an element of surprise and delight to anyone walking from the house, up through the grounds to Tresham's famous garden lodge.[5]

Most people of course were not in a position to plant such extensive orchards, but even the addition of a just a few trees would have provided not only food and drink for the household, but also natural ornament for the garden. Plants, then, whether "for meate or medicine, for use or for delight" (Parkinson 1629: 1) have always had a vital role in the ornamentation of any garden, but, as demonstrated so far, their aesthetic value was generally perceived as secondary to their uses and virtues. However, by the early seventeenth century, all this began to change, as the ephemeral beauty of rare plants and flowers became reason enough to include them in the garden.

"ALL ... SORTS OF OUTLANDISH FLOWERS"

There are a number of factors that need to be considered in attempting to account for this change, but the major one is simply that such plants were becoming increasingly available. The Renaissance came somewhat late to English shores, reaching its zenith during the second half of the sixteenth century with the literary flowering of playwrights such as Shakespeare and his contemporaries, but as the long reign of Elizabeth came to a close, it was inevitable, perhaps, that the new century and the reign of a new monarch would bring with it new aspirations and possibilities for the future. The arrival of exotic plants and flowers from the New World has been well recorded, and rare plants, which had been arriving in mainland Europe throughout the previous century, were now making their way to England from all over the world. In his *Herball* of 1597, John Gerard described many plants that he had obtained from "forreine places" (Epistle Dedicatorie), but just three decades later, in 1629, John Parkinson was writing of Gerard that "since his dates we have had many more varieties, then he ... ever heard of" (Epistle to the Reader). The choice of plants available to the gardener or garden owner was now far greater than it had been at the end of the previous century.

It was not just plants of course—as London developed as a center of global trade, the availability of rare and exotic goods from all over the world engendered a new spirit of conspicuous consumption that could be displayed in gardens as well as anywhere else. As the traditional moralistic differentiation between needs and wants—which favored the first, but cast a suspicious eye on the second—gradually became eroded during this period, so too the accumulation of luxury goods began to lose its moral implications. Formerly associated with decadence, idleness, and corruption, the acquisition and display of luxury goods now found a new acceptance, as these commodities were relabeled as rare, exotic, refined, expensive, and imported (Peck 2005: 6–8).[6] This gave rise, among other things, to a new culture of collecting, demonstrated through, for instance, the Dutch community of *leifhebbers* (fanciers) who acquired precious bulbs from their own or their friends' travels, or from ships coming into port, and then exchanged both the plants and their horticultural knowledge within a circle of like-minded connoisseurs (Goldgar 2007: 23). Cabinets of curiosities were established by collectors, such as the one visited by John Evelyn in April 1644 at the famous garden of Pierre Morin in Paris, where he viewed a "rare collection of Shells, Flowers and Insects" (De Beer 1955: 133), or that "collection of rarities preserved at South Lambeth neer London"—created by John Tradescant the Elder and augmented (its content's published) by his son John—which contained a huge variety of animal relics, tribal artefacts, weapons, etc., as well as hundreds of plants (Leith-Ross 1984: App. III).

Perhaps the best-known example of a rare and exotic plant being imported and incorporated into English gardens is that of the tulip. Now seen as quintessentially Dutch, in the minds of Early Modern gardeners the tulip was "a strange and outlandish plant" (quoted in Goldgar 2007: 38). It was new, never having been seen before in this part of Europe; it was exotic, being brought to Europe from Constantinople, but before that originating in the Middle East; it was enigmatic and mysterious, apparently changing color from one year to the next, and perhaps most importantly, it appeared in an infinite variety of colors, shapes, and forms. It could be any shade of red, purple, white, or yellow; it could be a single color, striped, or variegated; it could have rounded, pointed, or feathered petals. And because of its status as exotic, curious, rare, and beautiful, it was also expensive, making it a collectable object of desire amongst wealthy connoisseurs.

Although one of the best-known facts about the seventeenth century Dutch tulip trade is probably the short-lived madness of "tulipmania," a time when prized bulbs reputedly changed hands for the price of a house, the fascination with tulips began much earlier and continued long after the mid-1630s when this temporary crisis occurred.[7] For the reasons noted above, tulip bulbs had always been costly items, desirable because of their rarity and beauty as flowers for the garden. According to John Gerard, his friend James Garet, the apothecary, originally from Flanders, who lived and worked in London, had been growing and experimenting with tulips in his garden for at least twenty years (Gerard 1597: 117). It is likely that these had been introduced to him by his fellow-countryman, Carolus Clusius (1526–1609), the most famous botanist of his age and a central figure in horticultural activity in Europe. Clusius had a huge network of correspondents with whom he exchanged botanical information, seeds, and bulbs, and this network was responsible, among other things, for much of the distribution of the tulip throughout Europe. This correspondence demonstrates that Clusius possessed tulips as early as 1570, and it is known that he visited his friend and fellow-botanist James Garet at his home in London in 1571 (Goldgar 2007: 22, 34; Harkness 2007: 26; see also Chapter 5 of this volume). It seems a reasonable assumption, therefore, that between them, Clusius and Garet can be credited with the introduction of the tulip into England during the latter decades of the sixteenth century.

Gerard refered to the tulip as "a strange and forrein flower," describing just fourteen varieties in his *Herball.* Always anxious to attribute plants with their uses and virtues, he wrote that:

> There hath not been anything set down of the ancient or later writers, as touching the nature or vertues of the *Tulipaes*, but are esteemed especially for the beautie of their flowers.
>
> (1597: 120)

And this is the essential point—tulips had no medicinal or culinary uses: they were grown for no other reason than their beauty and their ornamental value.

Thirty years later, John Parkinson described well over a hundred varieties of tulip, occupying twenty-four folio pages of his book, with four full-page woodcuts illustrating thirty different kinds. He no longer described them as strange and foreign, but did still comment on their rarity and curiosity, noting that there were none who were "not delighted with these flowers" (1629: 9). Another thirty years on and Sir Thomas Hanmer, one of the leading plantsmen of this age and author in 1659 of *The Garden Book*, continued to extol the virtues of tulips as "the Queen of bulbous plants, whose flower is beautiful in its figure, and most rich and admirable in colours and wonderfull in variety of markings." Interestingly, he also observed that the plants were "so well knowne in England" that they needed no description (Elstob 1933: 18). Nevertheless, it is clear from both his *Garden Book* and from his extant manuscript notes that Hanmer was a great lover of tulips and that not only did he spend significant sums of money on buying a wide variety of bulbs both at home and abroad, but he also exchanged bulbs with an eclectic group of fellow tulip-fanciers including the Shropshire nurseryman, John Rae, fellow Welsh MP, Sir John Trevor, and the parliamentarian General John Lambert (B1663).[8] Even as they became more popular and less exclusive, tulips were clearly still prized as a beautiful plant to be given pride of place in the ornamental garden.

The tulip, then, encapsulated the fundamental change in the way in which ornamental flowers were now regarded, epitomizing a shift in the way Englishmen came to view their gardens. No longer were gardens simply useful and productive; they became fitting places in which to display the ephemeral beauty of fine flowers.

But the tulip was just one example of a flowering plant, so rare and exotic at the beginning of the century, that was to eventually find its way into the gardens of Early Modern England. There were many others and they were coming into the country, and particularly into London, at such a rate during the early decades of the seventeenth century that, as Parkinson acutely observed, there were now so many beautiful new plants available that people "have not known either what to chose or what to desire." It is this that he set out to remedy in his book—and, therein, "play the Gardiner." Rather than simply listing all plants according to their uses and virtues, as had been the case with earlier herbals such as Gerard's, Parkinson selected plants and flowers in order to "set forth a Garden of all the chiefest for choyce, and the fairest for shew" (1629: Epistle to the Reader). And, as has been demonstrated, he now had plenty to choose from.

As well as being an author, Parkinson was also a practical gardener himself—he made his living as an apothecary, and plants were the tools of his trade. He established and maintained an extensive garden at Long Acre near Covent Garden just outside the City of London and of the 484 plants recorded as growing there, over two thirds were exotic new varieties. It is highly likely that in fact there were far more plant introductions in Parkinson's garden than this, but regrettably they remained undocumented. Other gardens from around the same period that have been more rigorously cataloged such as William Coys' garden at Stubbers in Essex in 1620, or John Tradescant's garden in Lambeth in 1629–33 also list hundreds of plants acquired from around the world (Riddell 1986: 113). But these were exceptional gardeners with exceptional gardens. What then, of other gardeners and their gardens? What evidence is there of an increased interest in the flower garden in England at this time? The remainder of this chapter will consider this question from two points of view. First it will examine the networks of exchange that developed in response to the increased interest in and demand for exotic plants, before exploring some of the extant evidence available for the planting of seventeenth-century flower gardens in England and Wales.

"A FRIEND OF MINE BROUGHT IT UNTO ME": INFORMAL NETWORKS OF EXCHANGE

Throughout the seventeenth century, new plant introductions continued to arrive on English shores via multifarious routes. Plantsmen were sent to buy plants from markets on the Continent for their wealthy clients—John Tradescant the Elder, for instance, traveled extensively throughout the Low Countries and France buying plants on behalf of Sir Robert Cecil for the gardens at the newly acquired and rebuilt Hatfield House (Leith-Ross 1984: 29). Travelers and adventurers journeyed across the globe, bringing back strange plants that were then distributed among friends, acquaintances, or simply those who were interested in them—or indeed, might know what to do with them! The Dutch network of horticulturalists has already been mentioned in this context, and Gerard's and Parkinson's books abound with examples of plants and seeds received from "beyond the seas" as gifts. Plant hunters were commissioned to seek out plants to bring back to England, while others—perhaps most famously the John Tradescants, elder and younger—left these shores to collect plants for themselves.[9] Later in the century, men such as John Evelyn and Sir Thomas Hanmer, living in exile in France during the Civil Wars, eventually returned to England with new ideas, plants, and contacts.

Once they arrived, by whatever means, how were these plants then distributed to the gardeners of Early Modern England? The informal networks of friends, relations, and other interested parties have already been mentioned and of course this method of exchange was already well established as a means of obtaining plants for the garden. The memorandum books and correspondence of Sir Thomas Temple, for instance, detail many examples of fruit trees, cuttings, and grafts simply being obtained from other people's gardens. In 1630, he recorded receiving walnut and quince trees "of the best sort," as well as apple trees from a gentleman acquaintance, James Prescott, and more from his brother. He accepted grafts of pear trees from his son-in-law, who had in turn received these from a Mr. Symonds and which he had "praised much." In another letter, written in 1631, he instructed his estate steward Harry Rose to ride to nearby Stratford-upon-Avon to gather some of the finest buds from the vines growing in Mr. John Hall's garden there. News of Mr. Hall's grapevine had traveled to Sir Thomas at Burton Dassett via his sister-in-law, and although it appears that Mr. Hall and Sir Thomas were not acquainted, nevertheless, Hall was apparently happy for Temple's gardener to take the cuttings (ST38, STT2285, STT2287).[10] Other examples are provided by Sir Robert Sidney, who in 1610 sent quince trees to his friends (Historical Manuscripts Commission 1942: 244) or Lady Margaret Hoby who gave herbs from her garden to a neighbor (Moody 1998: 145), and the frequency of such anecdotal references would indicate that this free exchange of plants was normal practice and, indeed, had no doubt been so for centuries.

Evidence that such informal networks continued to thrive well into the 1650s is found in the notebooks and correspondence of English politician Sir Thomas Hanmer (2nd Baronet, 1612–78)—but by now, rather than mundane native plants and fruit trees, the focus was on the acquisition and distribution of new ornamental flowers and plants. Hanmer noted many gifts of bulbs received and given: not surprisingly, he lists many cultivars of tulip, but also narcissi, iris, and colchicum, as well as other plants such as a "yellow jasmine" and a "double striped pomegranate" that he had particularly admired in the garden of John Lambert. Many of Hanmer's fellow garden enthusiasts were named in his notes and clearly, once one person had obtained a particular specimen, however they might have done that, then bulbs, off-sets, cutting, and seeds were passed around the circle of interested friends and acquaintances for cultivating in their own gardens (B1663).

However, apart from these informal, non-commercial exchanges, and providing further evidence for the increased interest in and demand for exotic flowers and plants for the garden, the seventeenth century began to witness the burgeoning of an organized commercial nursery trade, centered, inevitably, on London. While household account books from the period indicate that native plants, trees, and hedging were clearly being purchased at a local level via a network of gardeners either working on estates or independently, and although there is evidence of the existence of a few provincial nurseries, the main business of buying and selling plants was concentrated in the capital (Francis 2018: 345–52). People came to London to buy plants and trees, which were then transported, sometimes over great distances and at great expense, to grace the gardens of their rural country estates. In March 1623, Sir Thomas Aubrey took delivery of twenty-seven trees sent from London to his South Wales estate (Bowen 2006: 55, 60); in October of the previous year, Sir Arthur Ingram arranged for the transport of trees and roses by ship from London to Hull to be planted in his gardens in Yorkshire (WYL178/4).[11] So there is plenty of evidence to suggest that, for those with the appropriate means, it was possible to obtain whatever plants they wanted for their gardens and orchards. But thus far, there are still few clues as to where these purchases were actually being made.

"THEM THAT MAKE THEIR LIVING OF IT": THE RISE OF THE NURSERY TRADE

It has been suggested that, prior to 1660, there was little in the way of an organized commercial plant trade in England, and that the Restoration represented, as in so many other areas of life, a turning point in the establishment of such a trade (Harvey 1972: 7).[12] However, while this latter point may well be true, the fact is that there is evidence, such as that cited above, to indicate the development of a thriving trade in plants and flowers in the early years of the century. A variety of sources show, for instance, that in London there were commercial nurseries supplying fruit trees and other native plants to gardeners. John Parkinson mentioned a John Millen of Olde Street as a reliable source of good fruit: "He hath stored himself with the best only and he can sufficiently furnish any," he wrote in 1629 (575). Parkinson also noted that John Tradescant had bought fruit trees from Millen, and Thomas Johnson, who published an updated version of Gerard's *Herball* just a few years later, also agreed that anyone who wanted gooseberry bushes, apricot, peach, pear, plum, apple, or cherry trees for their garden, that they were "to be had with Mr John Millen in Old Street" (Johnson 1633: 1456). This particular nurseryman clearly had a good reputation among renowned gardeners. Earlier, John Gerard had recommended Henry Banbury of Toothill Street near Westminster, Richard Pointer of Twickenham, and Master Warner from Horsey-Down as suppliers of fruit trees (1269). John Parkinson commented elsewhere in his book that gentlemen who do not intend to keep a nursery themselves, as was the case on many country estates, must instead "buy their trees ready grafted from them that make their living of it" (1629: 538)—clearly implying the prevalence of such practitioners. One other nurseryman—or "florist"—named by Parkinson, Johnson, and Tradescant is Master Ralph Tuggie, who specialized in cultivating English flowering plants such as carnations, roses, and auriculas.[13]

From these few examples, and there are others, it can be seen that there was some kind of established trade in fruit trees and native plants and flowers, but what has yet to be discussed is any evidence of nurseries trading in the new range of plants that had been entering the country since the turn of the century. Some kind of trade must have existed because, although there is little in the way of firm evidence of selling—such as catalogs of plants and seeds for sale (Harvey 1972: 7)—people were buying plants, and they must have been buying them somewhere! Early references are scant, but for instance, in 1609, the court commentator John Chamberlain reported in a letter from Ware Park, the home of Sir Henry Fanshaw that "we now have four or five flowers from Sir Rafe Winwood that cost twelve pound" (McClure 1939: 290). In the early 1630s, Sir John Oglander noted in his memorandum book that he had purchased "French flowers" which cost him 10*s* a root, as well as all sorts of tulips (OG/AA/29).[14] In 1649, a carrier was paid 3*s* 9*d* for bringing a box of flowers from London to Tavistock, the Devonshire home of the Earl and Countess of Bath (Gray 1996: 79). The extremely large sums being paid for these ornamental plants would indicate their rarity and value—10*s* a root or more than £2 each, compared to, say, the 1*s* 6*d* "a peece" paid by Sir Thomas Aubrey for his fruit trees sent from London or the 2*s* 6*d* paid by him for a thousand hedging plants for the estate (Bowen 2006: 60, 124). Although anecdotal evidence such as this reveals that exotic and expensive plants were clearly being purchased, it still offers no information as to where these plants were being obtained.

Given this paucity of evidence, therefore, it is fortunate that there is one contemporary documentary source that does shed light on just such a trade in ornamental plants; it is the wealth of information contained in the aforementioned manuscript papers of Sir Thomas Hanmer of Bettisfield. Among these, and of particular interest here, is one of two small notebooks kept by Hanmer between 1654 and 1657, which contains not only details of the planting of gardens at Bettisfield, Haulton, and Lewisham, but also a unique record of plants purchased during this time from various nurseryman in London and abroad (B1663). He listed plants purchased in England, France, and Holland, often with a note of the prices paid. Throughout the notebook Hanmer specifically named a number of individuals, of whom at least twelve were clearly in the business of selling plants, because in every case, Hanmer indicated how much he had paid for plants from these people. A further three were also likely to be nurserymen, as addresses were supplied that would indicate a business premises, although there is no more detail. And although two, located in Worcestershire and Shropshire respectively, were much nearer to Hanmer's home in North Wales, and two others were located in Paris, the remainder were, as far as can be ascertained, all based in London.

A Mr. Moulart, for instance, was mentioned on three occasions: twice as a supplier of flowers admired by Hanmer in other people's gardens, and once as a supplier to Hanmer himself. According to a list made by Hanmer in 1654 of plants "bought of Moulart" it would seem that this was a specialist supplier of recently introduced bulbs including colchicums, jonquils, fritillaries, and iris. He paid 5*s* for two purple and white irises, which he first saw in the garden of Mrs. Seely in Shooe Lane.

Hanmer was particularly fond of "bears ears" or auriculas, and it appears that his main supplier of these was a "Humphries of Woollstable in Westminster." Hanmer bought several varieties from there in 1654 and took some of them to his garden at Haulton. Humphries' reputation as a specialist in auriculas was also noted by John Rea in his *Flora* where he mentioned particular varieties of this pretty ornamental flower named after "those that raised them," and he includes Humphries in this list (Rea 1665: 153) (see Figure 7.4).

Other sources also indicate that Westminster was an area that attracted a number of nurseries—Ralph Tuggie and Henry Bunbury both ran their businesses from there, and "Walker of St James," mentioned by Hanmer in his notebook as having many Virginian plants, may well have been located in this area as well. On the other side of town, Hanmer visited Smyth of Greenwich, where he bought gillyflowers (or "gilliflowers," *Dianthus*) for 18*d* each and pots to put them in for 2*d* each. He also bought a range of other plants to set in the borders at his garden in Lewisham, as well as "3 pannes" of bears ears (auricula) seeds for 14*s*. On another occasion he bought bear's ears plants from Smyth at 1*s* a root, again for the garden at Lewisham, which would have only been just a mile or so from this Greenwich nursery. Hanmer also frequented another nursery which was actually on the doorstep in Lewisham, where he was supplied with a wide variety of flowering plants, including hyacinths, narcissi, iris, anemones, ranunculi, and columbines. Again, as in the case of Moulart, this nursery clearly specialized in the new range of flowering bulbs and plants.

From this brief overview of nurserymen known to Hanmer, it is clear that there were a significant number of established flower nurseries in London in the decades leading up to and including the 1650s and beyond. That they were reasonably prevalent can be inferred from the fact that Hanmer was apparently able to choose suppliers located either close to the garden at Lewisham, well out to the east of London and south of the river, or otherwise located on the western outskirts of the city, more convenient for transporting plants back to his gardens at Bettisfield and Haulton in North Wales. The fact that Hanmer, like Sir Thomas Aubrey and Sir Arthur Ingram, felt it necessary to deal with nurseries in London for the supply of these kinds of plants would seem to indicate that they were not available more locally and that trade was centered very much on the capital. From the albeit scant information provided here, it would appear that some nurseries were quite specialized, dealing mainly in bulbs, as in the case of Moulart, or auriculas in the case of Humphries. The Lewisham nurseries seem to have been slightly larger concerns, selling a wider variety of plants and also garden equipment such as pots and spades. What is absolutely clear, however, is that all these nurseries were catering for

FIGURE 7.4 Auriculas from John Parkinson, *Paradisi in sole Paradisus terrestris*, 1629. Image courtesy of Research Library, The Getty Research Institute/archive.org.

a new and growing market for ornamental flowering plants in a way that simply had not been seen before.

There is plenty of evidence, then, to suggest that people were, by one means or another, obtaining new and exciting plants for their gardens. So how did they go about displaying them in their gardens—using these plants as natural ornaments?

"A GARDEN OF PLEASANT FLOWERS"

Extant evidence for the planting of early-seventeenth-century gardens is scant and disparate, offering only occasional specific references to flowers, but nevertheless, such references do offer further insight into the growing interest in ornamental flowers for the garden. There are, for instance, a number of references to the gardens at the home of Sir Henry Fanshaw at Ware Park in Hertfordshire. In his *Elements of Architecture*, the much-traveled diplomat and writer Sir Henry Wotton (1568–1639) observed that despite the many beautiful gardens he had seen in Italy and France, the garden at Ware Park was "surely without parallel among foreigne Nations." He went on to describe how the flowers were set precisely, according to their colors and their season, so that they resembled "a piece not of Nature, but of Arte" (1624: 110; Hunt and Ellis 1975: 48–9).

John Chamberlain, another regular visitor to Ware Park, agreed that here you could "see as fresh and flourishing a garden (I thincke) as England affoordes" (McClure 1939: 1:227), and Lady Ann Fanshawe noted in her memoirs of her father-in-law Sir Henry's garden that there was "none excelling it in flowers" (1907: 9). Although it is obviously significant that all these commentators remarked particularly on the flowers, unfortunately, these observations actually reveal very little about the garden beyond that fact that it was much admired—and exceptional.

More useful, perhaps, are extant descriptions of two contemporaneous Yorkshire gardens: one belonging to the Reverend Walter Stonehouse, "A Modell of my Garden at Darfield, 1640" (Gunther 1922: 348–51) and the other, a more recently discovered plant list contained in the Garden Notebook of Sir John Reresby of Thrybergh, covering the years 1633–44 (Woudstra and O'Hallaran 2008). Stonehouse's list consists of a catalog of 450 plants that was written up in 1640, to which a further 416 were added over the next four years, together with a detailed plan of his "best garden," which was made up of five geometrically patterned beds, marked with numbers corresponding to the plant lists (see Figure 7.5).

How exactly plants were displayed in Early Modern gardens is frustratingly difficult to ascertain, but contemporary illustrations indicate sparse planting, either in beds or individual pots—presumably in order to show these expensive items off to their best advantage (see Figures 7.6 and 7.7).[15] And despite the detail of Stonehouse's plan, it is still not really possible to tell whether the plants were to be planted in blocks of color, or singly, but clearly the display of individual specimens was important.

Although, in common with other gentry gardeners of this period, much of his interest was in fruit growing, Sir John Reresby also included in his list an extensive number of tulip varieties as well as over four hundred herbaceous plants that were growing in his garden. Reresby was a near neighbor of the Reverend Stonehouse and had inherited the estate at Thrybergh Hall in Yorkshire on the death of his father in 1628. Evidence for his enthusiasm for gardening began with the enclosing of orchards and gardens around the Hall with dry stone wall and the commencement of the keeping of his garden notebook in 1633 (Woudstra and O'Hallaran 2008: 137). The list reveals his passion for "outlandish" plants: flowers such as anemones, colchicum, fritillaria, and lychnis, to name but a few, as well as the many varieties of tulips already mentioned. The list is comparable to John Tradescant's plant list of 1634, Stonehouse's list referred to above, and those recorded as growing in Parkinson's garden at Long Acre—indeed, many of the plants listed are cross-referenced to pages in *Paradisus terrestris*—and it leaves little doubt that Sir John Reresby was an enthusiastic flower gardener and that it is likely that his garden was, again, somewhat exceptional. Further evidence substantiates this view. Some twenty or so years after his father's death, Reresby's son was to write in his memoirs:

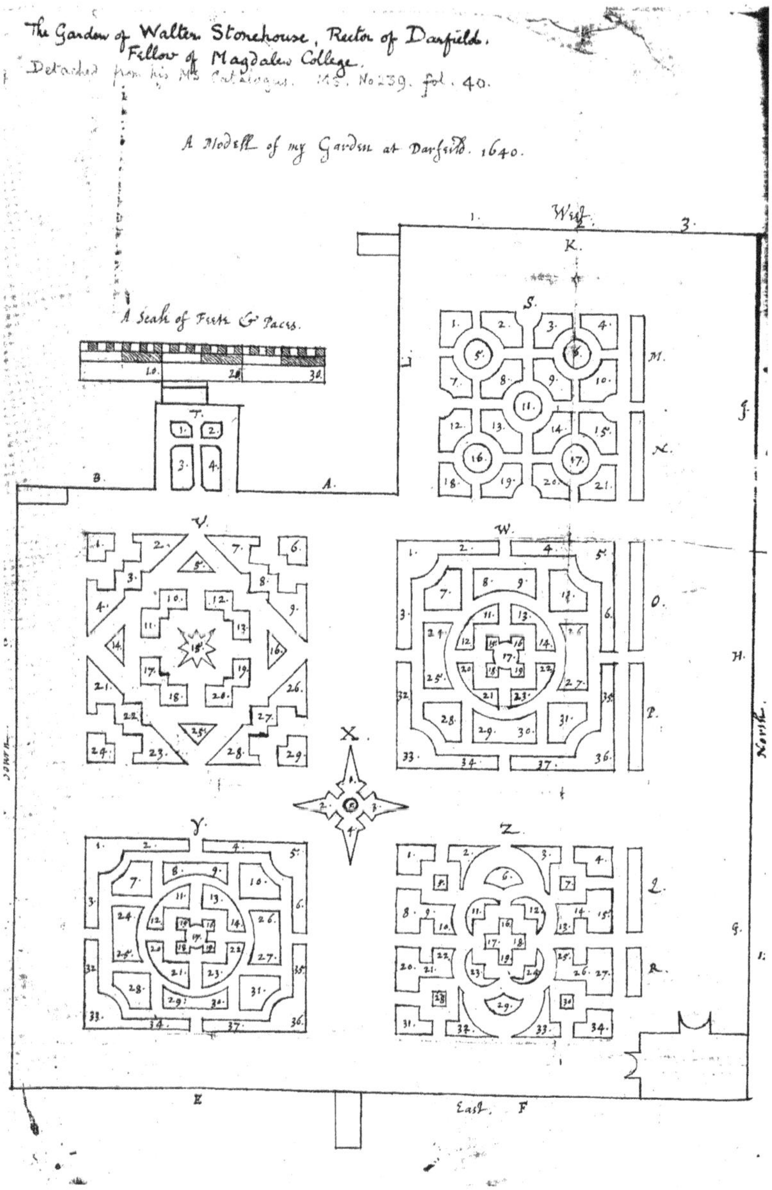

FIGURE 7.5 Plan of the Reverend Stonehouse's garden at Darfield in Yorkshire, 1640. Magdalen College Library, MS No. 239, f. 40. By kind permission of the President and Fellows of Magdalen College, Oxford.

> My father was exactly curious in his garden, and was of the first that acquainted that part of England (so far north) with the exactness and nicety of those things—not only as to the form or contrivance of the ground, but as to excellency and variety of fruits, flowers, greens, in which he was rather extravagant than curious, for he placed his pleasure not only innocently but pleasantly in it.
>
> (Cartwright 1875: 78)

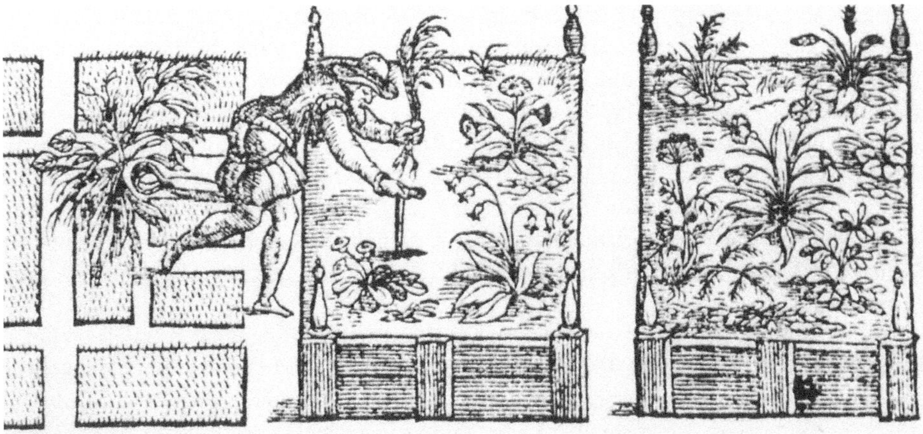

FIGURE 7.6 Woodcut illustration of gardener planting flowers singly in beds (Hyll 1577). Photo from Universal History Archive/Universal Images Group/Getty Images.

FIGURE 7.7 Cornelius Johnson, *Arthur, Ist Baron Capel and his family*, c. 1641, detail. National Portrait Gallery, London. Photo by Robert Alexander/Getty Images.

Reresby's son would only have been about ten years old at the time of his father's death, so it seems unlikely that this came from personal recollection, but indicates instead that his father's garden had a reputation for introducing new and innovative ideas to this part of England—"so far north." He did comment on the layout of the garden, but emphasized more his father's wide range of plants and extravagant choices. Not only was Reresby both knowledgeable and skillful ("exactly curious"), but he also took great pleasure in his garden. There is an implication that while he exercised a careful control over the wider estate, when it came to the ornamental flower garden, he was prepared to be more lavish. We have no way of quantifying his extravagance, but the fact that it was noteworthy must be significant.

Another flower garden about which exists some detail is one created by Sir Thomas Temple at Burton Dassett in 1631. As was the case with most country estates, much of the garden area was given over to orchards, areas for vegetable growing, and to hops, but there were plans for a small ornamental garden which would "yeild to me some sweette ffloweers." In his written instructions to Richard the Gardiner, who was to be employed to build this new garden, Sir Thomas gave a great deal of information about how he wished it to be laid out. The garden was to be enclosed, first with an upright hedge and then "paled" with a wooden fence, although the wall of the house and some other outbuildings also appear to form part of the enclosure. He further instructed that the garden should be divided into three parts, one part to be an "alley for passage," the other two parts to each be divided into five beds containing traditional English summer-flowering varieties such as roses, gillyflowers, violets, and primroses—there is no sign here of anything more exotic, not even tulips which by this time had been gracing some English gardens for nigh on thirty years (ST38, STT2347). It is probably significant that by this time Sir Thomas was an old man, and here he was simply displaying a desire for a pleasant place, filled with familiar fragrant flowers and plants in which to while away the remainder of his days. Despite new fashions and trends, the place of personal taste is always an important consideration.

Further evidence of the increased enthusiasm for growing ornamental flowers in the garden is provided in the aforementioned commonplace books of Sir John Oglander. Unlike his northern compatriots, Sir John did not provide much horticultural information about flower varieties and so on—from the extant manuscripts it seems that he did not keep detailed plant lists—but he does include an interesting description of what was obviously his ornamental flower garden:

> An[d] in the Parlour garden, in one knott all sortes of Gilliflowers, in the other knott all sortes of ffrench fflowers, and Tulippes of all sortes.
>
> (OG/AA/29)

According to this description, the garden was divided into two areas: one for gillyflowers—traditional English flowers—and the other reserved for more exotic species—specifically all sorts of French flowers and tulips. As shown above, these plants were costly items, but to spend large sums of money on flowers was clearly not unheard of. Sir John Reresby's extravagance has already been mentioned and Oglander, who was equally careful with his money in other matters, wrote of himself that he had been "so foolish as to bestowe more moneyes then a wise man would have in fflowers for the Garden" (OG/AA/29).[16]

For one last example, it is instructive to turn once more to Sir Thomas Hanmer who described a wide variety of ornamental flowers and plants in his *Garden Book*. The intensely personalized and particular nature of these descriptions would suggest that they reflected very closely the flowers cultivated in his own garden at Bettisfield in North Wales. As already noted, these included infinite cultivars of tulip—including one first grown at Bettisfield, *Tulipa* 'Agate Hanmer'—as well as other flowering bulbs such as narcissus, crown imperial, anemones, ranunculi, martagon lilies (*Lilium martagon*), and cyclamen. He also described more traditional gillyflowers and roses and his other special favorites, "bear's ears" or auriculas.

It is significant that Hanmer was clearly happy to include all these flowers in his garden, whether new, fashionable, or otherwise, and it would seem that plants that had been considered rare and exotic in the early years of the century had now become commonplace. Indeed, Sir Thomas represents an example of a gardener who has taken onboard the somewhat radical notions being advocated just thirty years earlier by John Parkinson, epitomizing a legitimate and exclusive interest in the ornamental flower garden. Hanmer noted himself that this was a recent phenomenon, observing that "now ... some spare no charge amongst other things in procuring the rarest flowers and plants." He also observed that although the Italians and Germans and then the French and Dutch had been "diligent enquirers and collectors of [...] rarityes" for some years past, it was only recently that this practice had been taken up in England. He reiterated the novelty of these ideas a number of times: "this way of beautifying ... comes apace into fashion" he wrote; gardeners who were unfamiliar with these plants had not caught up with the new ways, "being for the most part inexpert and dull." Elsewhere, he noted that planting was now different to how it was in his father's time, when beds were "hedg'd about with privet, rosemary or other tall hearbs which hide the prospect of the worke" as opposed to the present time when "all is now commonly neere the house layd open and exposed to the view of the chambers" (B1667). The concern to display the flowers in such a way that they could be seen and admired both in the garden and from the house was now paramount.

It would seem that the concept of the ornamental garden embellished with choice flowers, that Parkinson was introducing to his readers some thirty years earlier as something new and different, had now become accepted practice.

CONCLUSION: "THE ORDERING OF THE GARDEN OF PLEASURE"

Having established that the seventeenth century saw many changes in English gardens, it is worth pausing finally to consider how these changes manifested themselves in a practical sense in the garden and how they affected the work of the gardener.

New plant introductions meant that new ways of gardening had to be learned in order to cultivate them successfully. Many of the new plants described in detail by John Parkinson were spring-flowering bulbs such as crocuses, tulips, iris, anemones, and so on. He admired these fine flowers for many reasons, but for him, their outstanding quality was that they "shew forth their beauty and colours so early in the year, that they seeme to make a Garden of delight even in the Winter time [...] the more to entice us to their delight" (1629: 8). It has been noted that traditional English plants tended to be summer-flowering varieties, but now, for the first time, there was the possibility of providing color and interest in the garden throughout the year. And what these exotic new introductions were providing was not just new varieties of flowers and plants, but also a whole new way of furnishing a garden. Sir Francis Bacon, in his famous essay, *Of Gardens*, penned in 1625, also advocated year-round planting, listing in glorious detail a cornucopia of horticultural profusion, including many recent plant introductions.

Parkinson, typically, took a more practical approach. As well as providing appropriate guidance on choice, he also offered to take upon himself the role of a "new Gardiner, to give instructions to those that will take pleasure in them," because in his opinion (and as Hanmer was still complaining some thirty years later) most gardeners were "utterly ignorant in the ordering of these Out-landish flowers." In other words, traditional gardeners employing traditional methods did not know how to deal with these new plants—and indeed, why would they? Parkinson himself had to build up his own extensive knowledge of the nature and habit of previously unknown plants, which he did essentially through trial and error. By planting new roots, seeds, and plants in his garden, he experimented, observed, and recorded how they grew. Only then did he consider himself to have the authority to pass on what he had learned: "This I do affirm upon good knowledge and certaine experience," he wrote. He began with the basics: which way up to plant the bulb for instance (14), but, more fundamental than this, it was now necessary to rethink the traditional cycle of planting. Bulbs, if they were to thrive, had to be planted in the autumn and left in the ground over winter in order for them to establish and be ready to produce their spring display of colour.

However, according to Parkinson, the "usual custom" was to plant gardens in the spring "which is the best time that may bee chosen for all English flowers" and then to clear the beds at the end of the summer, once the plants had finished flowering. So to plant the bulbs in the autumn not only required a complete change in thinking—and one can well imagine it being challenged by traditional gardeners who were set in their ways—but it also caused practical problems because the removal of the summer flowering plants at the end of the season would disturb the newly planted bulbs. The obvious solution was to create two separate beds, one for spring-flowering varieties and another for summer-flowering plants in order that "when one may be removed, the other may not be stirred" (13). As earlier noted, Sir John Oglander had one garden bed for gillyflowers and another for tulips and French flowers and this type of planting of course offered glorious new opportunities for display. Unlike traditional plants such as roses and gillyflowers, which have a somewhat rambling and wayward habit, flowers such as the "stately" crown imperial, narcissus, or tulip can be set in neat and orderly rows and will stay that way. In a manner reminiscent of Sir Henry Wotton's hyperbolic praise of the gardens at Ware Park, Parkinson recommended, "Tulipas may be so matched, one colour answering and setting of another, that the place where they stand may resemble a peece of curious needle-work, or a peece of painting" (14).

The foregoing has offered some glimpses of the changing role of plants and flowers as natural ornaments in the early- to mid-seventeenth-century garden. Although one should be cautious of trying to form a general picture from such sparse and specific examples, and of course it must always be borne in mind that ornamental gardens such as this were still very much the preserve of the upper classes, it is nonetheless clear that, in tune with the changing times, there was a discernible shift in attitudes towards the acquiring and cultivation of an ever-widening range of flowers for the garden. No longer did plants have to be attributed with uses and virtues in order to earn their place in the garden: for them to simply be delightful, pleasurable, and beautiful was now reason enough.

CHAPTER EIGHT

The Representation of Plants

More than Just a Pretty Face?

GILLIAN RILEY

"*Al mio gusto è più presto bello che buono*" (To my taste, better to look at than to eat).

To doctor and naturalist Costanzo Felici (1525–85) the alien tomato plant was just a pretty face, a decorative pot plant, not a serious fruit, as he described it in his *Letter on Salads*, an essay dedicated to the naturalist Ulisse Aldrovandi (1522–1605), which circulated in manuscript among fellow enthusiasts (Arbizzone 1986: 90). That was in the 1580s, quite some time after the arrival in Europe of plants from the New World. Meanwhile botanical gardens were springing up in Europe, largely in Italy, under the enthusiastic direction of physicians, pharmacists, botanists, and collectors. Plants as pharmaceutical ingredients, or for consumption as food or seasoning, or as exotic rarities, were being collected, grown, studied, and recorded in various ways. Artists and scientists shared the tasks with enjoyment and delight. The focus of this chapter is developments in Europe, with an emphasis on Italy.

NATURALISTS AND BOTANICAL ILLUSTRATION

It was not easy for Aldrovandi and his colleagues to break away from the authority of the written word of classical writers like Dioscorides, most of whose surviving manuscripts have rather crude illustrations. The verbal description was paramount; Pietro Andrea Mattioli's translation of the original Greek text into Italian is clear, and his own commentaries, discussing possible interpretations of it, often criticizing his contemporaries, are full of insights into the cultivation of plants, and their properties, and information about new discoveries from all over the world (1544 and later editions to 1577). But he and his fellow botanists saw that accurate illustrations were now essential. Black and white images printed from engraved wood blocks (also called "xylographic matrices") were made from accurately colored paintings in gouache and watercolor which, as well as serving as works of reference, became a passion and delight for collectors. The skills of the artist were an essential tool for the botanist, who felt handicapped without them, hence Felici's frustration at the lack of good draftsmen in his remote hilltop village. His correspondence with Aldrovandi, with its precise verbal descriptions of plants, gives some idea of what was expected of the artists. Pioneering Flemish botanist Carolus Clusius (1526–1609) had the advantage of being a gifted artist as well as botanist, and the Venetian nobleman

Pietro Antonio Michiel (1510–76) made some of the illustrations for the five volumes of his manuscript herbal showing the plants in his private botanical garden in Venice. The rest, over a thousand, were probably by Domenico dalle Greche, and are now in the Biblioteca Nazionale Marciana in Venice.[1] Eventually, a plant on its own, fruit, flower, or vegetable, would become the subject of a work of art in its own right.

Gherardo Cibo, who died in 1600 not long before Aldrovandi's death in 1605, was another nobleman of independent means, who gave up the idea of a career in ecclesiastical or diplomatic circles in Rome, and returned to the family home in Rocca Contrada, today Arcevia, in the Italian Marche, where he collected and preserved botanical specimens, and had the rare skills to paint them in their natural habitats. Many of his works show a plant filling the page, with a background of the countryside in which it grew, together with little thumbnail sketches of the botanist and artist at work, as they compare the living plant with images or descriptions in a book (see Figure 8.1). These could be interpreted as affectionate vignettes of Aldrovandi and his contemporaries. In many ways, his paintings, a collection of which is in the British Library, are a visual accompaniment to pharmacist Francesco Calzolari's description of a botanizing trip to Monte Baldo in 1554, emphasizing the importance of recording plants in their surroundings as well as the enjoyment Aldrovandi and his friends got out of doing this (Calzolari 1566; Sandrini 2007).[2]

Aldrovandi began by studying law, then botany as an aspect of his medical studies, and then embarked with great enthusiasm on collecting information about everything there was to know about the natural world. Felici was a humble physician who shared his enthusiasm. In his correspondence with Aldrovandi in Bologna, Felici lamented the difficulty of finding good artists in the remote hilltop town, Piobico, in the Italian Marches, where his parents lived, and even in Rimini, on the Adriatic coast, where he practiced as a doctor. He was exchanging specimens of animals, fish, and plants with the great naturalist, but despaired of getting accurate representations of them. "*Se vi havesse un depintore comodo vi daria molte figure, ma non so como me l'havere su quelli monti*" (If there were an artist here to hand, I could let you have many images, but I don't know how to get hold of one in these mountains) (Nonni 1982). This gets right to the heart of the matter: how to get accurate images to circulate alongside accurate written descriptions of the natural world? Not a philosophical or aesthetic dilemma, a purely practical and extremely frustrating situation. It has been shown in earlier chapters how the study of plants was being codified, not yet a scientific system, but with an awareness of the need for one, and how botanists and scientists were arguing, more or less amicably, about the definition and depiction of plants for medical or agricultural use. Dried plants and seeds are all very useful, but the dew on a fresh leaf or the bloom on a ripe plum, the glistening eye of a just-caught fish or the quivering scales of an enraged serpent, needed somehow to be recorded, if preserving them as specimens was not possible. Felici used to converse with fishermen as they brought in their catch at Rimini, noting the appearance and local names of the fish, always on the lookout for "monsters" or freaks—as Aldrovandi had during his early visits to in Rome, where he wandered around the fish market, and got to know the great French naturalist Rondelet and possibly overheard the renowned cook Bartolomeo Scappi bargaining for the best that could be had for the papal table. Twenty-first-century consumers are so used to the uniform appearance of fruit and vegetables, unblemished tomatoes or apples, or fish on the fishmonger's slab, identical farmed sea bass, or look-alike herrings, that the excitement of discovering variants or the minor differences in the same species, or the rare appearance of an unfamiliar creature is an uncommon feeling.

FIGURE 8.1 Botanical painting of *Galanthus nivalis* (snowdrop) on the right and a blue-flowering bulb, probably *Scilla bifolia*, on the left, with a botanist and a young man gathering plants on a mountain top. Colored drawings of plants, copied from nature in the Roman States, by Gherardo Cibo, *c*. 1564–84. Image from Album/British Library/Alamy Stock Photo.

Felici would bring specimens home to write a detailed description, laying them on the kitchen table, where as often as not the cat would get them, or leaving them to dry out in the ashes, a risky business. Sending a fresh fish all the way from Rimini to Bologna, over 100 kilometers (about 60 miles), several days' journey, was not a possibility: they needed to be dried, or reduced to skeletons. Freshly gathered plants fared a bit better, and dried ones quite well, so long as there was also an accurate account of what they were like when living, for blossoms faded and leaves withered, and seeds failed to sprout.

The printed book was the ideal means of circulating images and words to give physicians and naturalists a common body of information, and the means of comparing and discussing specimens. German theologian-botanist Otto Brunfels in *Herbarum vivae eicones* (Living Plant Images, 1530–6), German physician-botanist Leonhart Fuchs in *De historia stirpium* (On the History of Plants, 1542), and Sienese physician-botanist Andrea Mattioli in *Discorsi* (Commentaries, 1577) did this in their massive publications. Charles de l'Écluse, known as Carolus Clusius (mentioned earlier), was head of the great botanical garden at Leiden, after a distinguished period, since 1574, as plant collector and creator of a medical garden at the court of the Emperor Maximilian II, emperor of the Holy Roman Empire and ruler of the Habsburg Monarchy (1564–76). In Vienna he was able to call in rare specimens. His passion for accurate description was an inspiration to his contemporaries, with whom he had an extensive correspondence, and also to court artists like Giuseppe Arcimboldo, who made many studies of plants from life. Clusius created several albums of plant studies in color, the *Liber picturati*, now in Krakow, with annotations and some images by him. Clusius' books, published by Christophe Plantin in Antwerp, include *Rariorum aliquot stirpium ... historia* (History of Some of the Rare Plants, 1576, 1583a), and *Rariorum plantarum historia* (History of Rare Plants, 1601), and he also made translations of the works of Flemish physician and botanist Rembert Dodoens (1517–85), Spanish physician and botanist Nicolás Monardes (1493–1588), and Portuguese physician and herbalist Garcia de Orta (1501?–68)—the first into French and the last two into Latin—accompanied by clarifying notes.

With these books to hand, physicians and botanists could discuss the identification of plants and how to use them, often in Latin when uncertain of each other's vernacular. Aldrovandi wrote to Clusius in Latin, with easygoing fluency, both determined in their pursuit of accurate depictions and descriptions, exchanging expressions of great regard and small packets of seeds or bundles of bulbs. Clusius at Leiden became a sort of hub, a repository and starting point for plants and their enthusiastic collectors. His correspondence in the University Library at Leiden is now digitized, making it possible to appreciate his role as the most important catalyst for plant collection in northern Europe.

BOTANY AND ILLUSTRATION FOR COLLECTORS

This ferment of enthusiasm for new and exotic plants—some of which came into Europe via Spain from the Americas, while Dutch merchants exploited political and commercial contacts with the Far East and Africa and India—stimulated artists as well as scientists, so that comfortably-off collectors could have on their walls paintings of the treasures they were avidly gathering from far and wide to grow in their gardens. This shift from medicine and science to "botany for collectors" was a healthy hotbed for the fine arts, with some artists flourishing in both areas, a cross-fertilization that enriched all concerned.

Fuchs included in *Historia stirpium* a portrait of his team: Albrecht Meyer, the artist who painted the plant from life, Heinrich Fullmaurer, the draftsman who drew a black-and white version on a block of pear wood, and the engraver, Rudolph Speckle, who cut it in relief, ready for printing (Fuchs 1542: end pages) (see Figure 8.2). Clusius wrote of how he needed to sit alongside his artists, making sure they got every small detail of a plant exactly right. Aldrovandi fully appreciated this, and during his lifetime kept several artists

FIGURE 8.2 Leonhart Fuchs, *De historia stirpium commentarii insignes* (Notable Comments on the History of Plants), 1542. His three craftsmen—Albrecht Meyer, Heinrich Füllmaurer, Rudolph Speckle—who between them painted the subject from life, drew it in black ink on a wooden block, and carved this as a woodblock for relief printing. Image courtesy of Wellcome Collection, London.

and engravers hard at work making stunning paintings and from them wood blocks, to give black-and-white images of every item in his huge and ever-expanding collection (see Figure 8.3). Thousands of paintings and wood blocks have survived. They are on view in the Aldrovandi Museum in Bologna and can all be seen online.[3]

FIGURE 8.3 Ulisse Aldrovandi (1522–1605), Italian botanist, naturalist, and physician, late sixteenth century. Photo by Ann Ronan Pictures/Print Collector/Getty Images.

Much of this huge body of work was not published until after Aldrovandi's death, but the enthusiasm and zest with which he pursued knowledge, and the uses he made of it were an inspiration to his contemporaries. Without the resources of the Medici in Florence or the Hapsburgs in Vienna and Prague, Aldrovandi financed his studies and collections with lecturing and teaching, and gained universal respect. Visitors flocked to see his collections, his library, and the various botanical gardens he helped to set up in Pisa, Padua, Florence, and Bologna. Like Clusius he had a vast correspondence with the great and the good, but also with modest local enthusiasts like Costanzo Felici. But he published little of his collection in his lifetime, in Bologna and Frankfurt, while Clusius was well served by the Plantin Moretus printing and publishing house in Antwerp (Duroselle-Mellish 2016).

As well as being a resource for botanists and physicians, the botanical garden became the equivalent of a *cabinet de curiosités* or *Kunstkammer* for rich collectors, who shared the enthusiasm and excitement of discovering and exchanging exotic plants. Prince Bishop Johann Konrad von Gemmingen at Eichstätt in Bavaria commissioned a luxurious publication, the sumptuous *Hortus Eystettensis* (Eichstätt Garden) published by Basilius Besler in 1613, with magnificent colored copperplate engravings of the rare and decorative plants in his botanical garden. They include chili peppers and tomatoes among many rare plants from the New World.

A GROWING TASTE FOR REALISM: FROM BOTANICAL ILLUSTRATION TO FINE ART

The enthusiasm for realism was not confined to scientists; much earlier, artists and their clients had enjoyed infiltrating scenes from daily life into religious paintings—soft furnishings, fruit and flowers, things to eat. In the previous century Books of Hours were vehicles for charming details of everyday life; the *Hours of Katherina van Kleef* (Catherine of Cleves) were commissioned in 1440 at the time of her wedding to Arnold van Egmont, and completed seven years later, by which time the marriage had become eroded with bitterness and hatred, making the placid domestic scene of Mary and Joseph and the infant Christ, sitting comfortably at home in a typically modest Low Countries dwelling all the more poignant. Every detail of the hearth, cooking implements, and furniture is there. The typical late medieval International Gothic decorations in the margins subtly blend in with the more naturalistic violets, columbines, and carnations that have crept in; the border around a nativity scene includes life-sized pea pods, with their dried and already yellowing peas within, symbols of fertility, but also reminding us of the *erwtensoep* (pea soup) of the Low Countries.[4]

Earlier still, around the 1380s, several lavishly illustrated versions of the *Tacuinum sanitatis* (Maintenance of Health) were produced in Milan in the studio of architect, sculptor, painter, and illuminator Giovannino de' Grassi, where a serious ninth-century Arabic health handbook received a "decorative" treatment. These were luxury manuscript picture books for the rich, with illustrations of everything edible including fruit and plants, and a brief note on their health-giving and nutritional properties, along with enchanting glimpses of everyday life. There are peasant women bundling leeks up for the market, lords and ladies picking ripe apples, and peasants harvesting the newly fashionable asparagus crop. One of the versions of the *Tacuinum* belonged to Archbishop George of Lichtenstein, bibliophile and patron of the arts, who commissioned the Bohemian artist, Master Wenceslas, to decorate the walls of one of the towers on the fortifications of the town of Trento, in northern Italy, with a cycle of frescoes of the Twelve Months, transforming a defensive citadel into a pleasure bower. The same comforting mixture of happy peasants working the land, and posh nobles disporting themselves, throwing snowballs or hunting and fishing, also has details of plants and agricultural life, and fruit and vegetables in shops and markets.[5]

The accurate realism of this depiction of everyday life was new, as is clear when one compares the illustrations to *Le livre de la chasse* (Book on Hunting) by Gaston Phébus (1331–91), Count of Foix and Béarn, in a manuscript produced in Paris about the same time (Strubel 2011). The image of a group of dedicated veterinarians tending hunting dogs with various health problems is moving in its tenderness and compassion, but compared with Giovannino de' Grassi's portrayal of Gian Galeazzo Visconti and his dogs in a page from the *Visconti Hours* it seems naive and crude. Grassi has shown the owner of the animals gazing fondly upon them, while they gaze back in besotted adoration, with a psychological and graphic realism that is entirely modern. Grassi's work evidences the fresh breezes of the early Italian Renaissance wafting away the last dying gasp of International Gothic. A similar effect can be seen in the paintings of Ventian painter Antonio Crivelli a generation later (*c.* 1430–95), where the rich ornate textiles of the saints have all the intricacy and complexity of the by then backwards-looking International Gothic, while the depiction of plants is fresh and simple. Many of Crivelli's paintings have swags of fruit, flowers in vases, and birds, usually with multiple significance, but of interest here as realistic rendering of plants at that time. A life-sized cucumber and apple, the cucumber symbolizing the purity of Christ and the apple the fecundity of his mother, are placed prominently in the foreground of Crivelli's *Annunciation with Saint Egidius* of 1486, in the Franciscan church of the Annunciation in Ascoli Piceno in the Italian Marches. This complex work was commissioned by the city fathers of this prosperous little town to celebrate a state of partial political independence granted by the Papacy. The realism and modernity of Crivelli's treatment of the Annunciation, with the Bible story in an up-to-date setting, with a fashionable young woman in a comfortable modern home, surrounded by all the attributes of prosperous urban life, many of them fraught with symbolism, are notable. The apple or quince could mean good things and bad things, the Fruit of the Tree of Knowledge (good), or the means by which Eve tempted Adam, and so brought about the Fall of Man (bad); or it can signify fruitfulness in general, or, in images of the infant Christ, a symbol of redemption. The ridged and knobbed skin of the cucumber (also a symbol of the purity of Christ), and the pink streaks of the apple are evident. Crivelli placed these fruit, along with quinces, pears, and plums, in most of his paintings, still-life elements waiting to burst forth as works of art in their own right, precursors of Caravaggio's *Basket of Fruit* in the Ambrosiana, Milan, almost a century later.

Swags of fruit, classical sculpture, as well as Christian theology, can also be found in works by Crivelli's contemporary Andrea Mantegna (*c.* 1431–1506), of which the most astonishing is his *Madonna della Vittoria*, celebrating a more or less successful battle conducted by Francesco Gonzaga, husband of Isabella d'Este, who commissioned the painting in 1496. The Virgin and Child sit in a bower of greenery and citrus fruit, along with apples and gourds and various birds, all with symbolic attributes. The beauty of the fruit, glowing amongst dark green fronds, might have been inspired by an uncharacteristically cheerful event thirty years earlier, when the young Mantegna set off with a group of antiquarian friends on a day trip to Lake Garda, searching for classical inscriptions, taking full advantage of lakeside hospitality to celebrate their discoveries, and take their leisure among the fragrant lemon groves in the crisp October sunshine.[6]

Citrus fruit had been cultivated in the benign microclimate of Lake Garda since the thirteenth century, providing good income for local entrepreneurs who had a more rapid access to markets in northern Europe than producers in the south of Italy. The citron was a luxury fruit, essential to the Jewish Festival of the Tabernacles, Sukkot, where the ridged and grooved *etrog* was carefully selected and handled in this celebration of harvest and fertility. An angel is holding a citrus fruit in the same ritual way in a painting by Paolo Morando, about 1514, of the Virgin and Child and St John, and another work in the National Gallery, London, about the same time, by Girolamo dai Libri, *The Virgin and Child with St Anne*, shows a flourishing lemon tree, with blossoms and fruit, alluding to the Jewish belief that biting on the pitom or tip of a citrus fruit could ease the pains of childbirth, appropriate for St. Anne, the patron saint of women in childbirth.

A few years later in Rome, between 1517 and 1518, the rich banker Agostino Chigi commissioned architect and painter Raphael (Raffaello Sanzio da Urbino, 1483–1520) to decorate the loggia, an open ground-floor terrace, of his villa with the story of Cupid and Psyche, an ancient pagan celebration of fruitfulness and fecundity, at the same time reflecting a modern connoisseur's delight in his plant collection. The villa, now known as the Farnesina, was an escape from the crowds and chaos of the banking district of Rome, a rural retreat across the Tiber, in what is now Trastevere, surrounded by market gardens and orchards, the ideal site for his own splendid pleasure gardens, based on the idea of a Roman *vivarium*, where everyday agriculture was juxtaposed with statues, fountains, and fragments of classical architecture, artfully displayed among vistas of plants, trees, and exotic specimens. The loggia of the villa was open to the elements along its length, linking the indoor space to the voluptuous gardens on either side. There is little documentary information about Chigi's creation and maintenance of these gardens. There is, however, stunning visual evidence of their contents in the work of Raphael's assistant, Giovanni da Udine, who decorated the festoons dividing the spandrels of the vaulted ceiling of the loggia with swags of vegetables, fruit, and flowers, over 170 kinds—stunningly beautiful images of familiar plants grown nearby, along with rare specimens from the recently discovered Americas, treasured by collectors and botanists. It is as collector rather than botanist that Chigi sought out and displayed these horticultural prizes; he was growing maize in 1510 long before it got into printed books, as were the Medici in Florence, who cultivated it as a serious crop (Caneva 1992; Markey 2016: 21, 82). His gardens were just some of the proudly displayed evidence of his *magnificenza*, the glory and status that was seen as so much more than the spoils of a rich banker.[7] Maize from the New World appears in the frescoes along with the large yellow pumpkin, and little round zucchini and beans, as well as more familiar fruit and vegetables, many cultivars of apples, a quince, cherries, onions, shallots, leeks, and a breathtakingly beautiful silvery cabbage (see Figure 4.5 of this volume). These entrancing portraits of fruit and vegetables survived on the frescoes of his villa after Chigi's early death in 1520 and were unharmed during the Sack of Rome in 1527. Chigi's collections of Roman sculpture, coins, jewels, artifacts, architectural fragments, and modern works of art were renowned in his time, but after his death were dispersed amidst acrimony and squabbling among his heirs and creditors. It is not known what became of the gardens, or whether visiting botanists and tourists were able to see the frescoes, which marked a breakthrough in the depiction of plants. Chigi's contemporaries seem to have appreciated Giovanni da Udine's portraits of fruit and vegetables more than the last generation of European art historians, whose disparagement of what they saw as lowly subjects skewed their vision; in fact the loggia's festoons were admired by contemporary painters and botanists, and similar swags of fruit and vegetables can be seen in later work by Giovanni da Udine in the Vatican, and in frescoes by Francesco Salviati in the Palazzo Vecchio in Florence, commissioned by Duke Cosimo in 1543–5, with maize and tomatoes grown in the Medici gardens and presumably eaten by them in the 1540s.

Illustrations in printed books seem clumsy in comparison with these works of art, but they gave botanists an overview of a plant, showing all the parts essential for classification, at different points in its growth, whereas the artists were free to picture a vegetable or fruit at its maturity, combining them irrespective of the constraints of seasonality. Which is what Vincenzo Campi's genre scenes did. His *Fruit Vendor* of the 1580s, in the Brera Gallery Milan, shows a lush display of fruit and vegetables across the seasons, with the cherries of early summer alongside the mulberries, quinces, and pumpkins of autumn, with the newly fashionable asparagus and artichokes isolated in the bottom left-hand corner, and the humble cabbage bottom right.

When the need for accurate botanical illustration collided with this delight in representation of the beauty of the natural world, painters and their clients, as well as scientists and physicians, commissioned and enjoyed works of art like this, with the religious element, often the title of the painting tucked away in the background or omitted altogether. Campi's painting was commissioned by the Affaitati family, rich merchants of Cremona, who had trading and cultural connections with Antwerp, where the artists Pieter Aertsen and Joachim Beuckelaer were producing market and kitchen scenes, often replete with symbolism, and usually including a bible story, with beautifully accurate fruit and vegetables. This was in the 1580s, quite some time after after Chigi's early death in 1525. Did his commercial links with northern Europe stimulate artists and their patrons in Italy? Certainly, the contacts between Antwerp and the northern provinces of the Low Countries encouraged a flourishing demand for the genre and market scenes that were dragging still-life subjects into focus, and giving us examples of meticulously rendered fruit and vegetables. These were appreciated in Italy, where cultural and trade factors influenced the tastes of patrons, and in the Hapsburg courts of Vienna and Prague, where naturalists, botanists, and painters shared the mounting enthusiasm for the accurate depiction of nature (see Figures 8.4 and 8.5).

In 1562 Giuseppe Arcimboldo had left his native Milan for a glittering career in the service of Maximilian II, later to become Holy Roman Emperor, and was a favorite at the courts of Maximilian's son Rudolph II in Vienna and later in Prague. Rudolph had the resources to employ twenty-four painters as well as sculptors, goldsmiths, and instrument makers. Arcimboldo was appreciated for his skill in organizing festivities and cultural

FIGURE 8.4 *The Four Elements: Earth*, Joachim Beuckelaer (1533–74), Flemish. Photo by Universal History Archive/Universal Images Group/Getty Images.

FIGURE 8.5 *Market Scene*, 1569, Pieter Aertsen (1508–75). Photo by Fine Art Images/Heritage Images/Getty Images.

events and engineering projects, and his botanical and animal studies, as well as strange and intriguing composite portraits, where heads in profile or full frontal, composed of fruit, flowers, vegetables, or even small animals, books, dead wood, and fish, could be read as likenesses of people, while his upside-down, invertible portraits resembled an inanimate object like a bowl of vegetables one way up, and, startlingly, revealed a human face when turned the other way. These works have been variously interpreted over the ages, dismissed by some as in bad taste and embraced by others, as by the Surrealists, for their daring. Today, his composite heads, described as "serious jokes" by serious historians, can also be seen, as they were by Archimboldo's contemporaries, as joyful botany, using the skills of painters of natural history in a pleasurable way, gently mocking the exuberance of High Baroque and the pretensions of official portraiture. His composite picture of Rudolph as Vertumnus, the Roman god of fertility, growth, and regeneration, was sent to the emperor as a token of friendship after Arcimboldo's return to his native city Milan in 1587 (see Figure 8.6). It uses fruit and vegetables with an affectionate sense of humor; as official portraitist for many years, Arcimboldo was familiar with the details of his subject's face, and knew how to give it the treatment that pleased the sitter. This version would have met with delighted recognition, with two rosy apples for the imperial cheeks, two azaroles for the lobes of the fleshy lower lip, the shine on a black cherry giving a twinkle to the regal eye, while wild raspberries imply teeth, and ripe cherries the jewels on his ornate bonnet, and hirsute fronds of wheat and millet for the ginger beard, contrasting with the bold black grapes that might be his well-groomed hair. The inherent symbolism, demonstrating the peace and productivity of the Hapsburg empire, and the benign flowering of talent under the rule of Rudolph, was indeed *serio ludere* (serious play) with complex intellectual content, as well as a simple pleasure in the witty use of fruit and flowers, and Arcimboldo's amazing skill in representing them. Indeed, the Imperial court was effervescing at this time with the excitement of collecting rare and exotic plants and animals, with botanic gardens and menageries to house them when alive, and museums for preserved specimens of them when dead, along with libraries of printed books and albums of paintings and drawings to record them in detail (DaCosta Kaufmann 2010: 11). Rudolph's Cabinet of Curiosities or *Kunstkammer* was far more than a collection of strange rarities, more an essential work of reference for scientists and artists, as can be seen from correspondence with Ulisse Aldrovandi, who had commissioned life studies by Arcimboldo for his own collection in Bologna. The sad fate of the Persian lily, a rare plant from Hungary, on the point of flowering, is told with some exasperation in a letter from Franciscus de Paduanis to Aldrovandi (Kaufmann 2010: 123); it seems that before painting it, Arcimboldo left his painterly task to attend a wedding and by the time he returned the lily had bloomed, faded, and disintegrated.

Another skillful artist employed by the Holy Roman Emperor was Joris Hoefnagel (1542–1601), a Flemish painter from a well-off mercantile family in Antwerp. His meticulous studies of plants and animals were only one aspect of a wide range of skills; he drew maps and made topographical studies, and contributed several albums of drawings and watercolors of plants and animals. He is perhaps best known today for his *Wedding in Bermondsey,* a large topographical painting, from which food historian Ivan Day has recreated the traditional heavy oval Bride Cake, borne in a sturdy linen sling by a strong woman in the wedding procession; the minute attention to detail in this very small item in the large painting made it possible to analyze the fruit and spices in the recipe, breaking through the crust of the massive cake, a neat example of Hoefnagel's meticulous observation. The same skills were applied to manuscript illumination in the 1590s. Hungarian calligrapher Georg Bocskay's *Mira calligraphiae monumenta* (Model Book of Calligraphy), a wildly innovative version of the *cancellaresca corsiva* (chancery cursive style), produced in the 1560s, was completed thirty years later with Hoefnagel's images. The calligrapher had left space for illustrations, and here Hoefnagel's delicate, inventive groupings of fruit, flowers, and insects accompany the exuberant Baroque penmanship in what was one of the last examples of private manuscript production, already superseded by the proliferation of printed books which met the increasing demand for shared, not exclusive, knowledge. Depicted here are some plump ripe tomatoes, not to be found in Arcimboldo or other contemporaries, as well as everyday fruit and vegetables, from broad beans to cherries.[8] Hoefnagel illustrated other manuscripts, but also contributed albums of plant studies to the Imperial collection.

It is not known if painter and illustrator Giovanna Garzoni (1600–70) had seen or heard of this manuscript when she produced one of her own, during her studies as a young woman in Venice in the 1620s.[9] Her calligraphy is more flowing and relaxed than Bocskay's, but the striking thing about the manuscript is Garzoni's already assured touch in her miniature portraits of fruit and flowers, which can also be seen in her work, discussed below, for the Medici in Florence. Another notable illustrator was the Flemish physician Anselm Boethius de Boodt (1550–1632), whose enthusiasm for depicting the natural world produced twelve volumes of albums, containing 728 images of animals, birds, and plants, many of them by himself, and carefully captioned with the accuracy one might expect from the emperor's doctor and director of the botanical garden in Prague. His albums are little known, having gone to Bruges with De Boodt when he retired on the death of Rudolph in 1612 and remained in private hands, but have recently passed to a new private owner, who has generously put them on loan to the Rijksmuseum in Amsterdam.[10]

The Emperor Rudolph was the link between many artists who probably knew each other. Hoefnagel's work was known to German painter Georg Flegel (1566–1638), who produced many plant studies, as well as still lifes with food and drink (Wettengl 1993). His collaboration with Lucas van Valckenborch the Elder (*c.* 1535–97), another painter who worked for the Emperor, produced scenes of figures in landscapes together with close-ups of domestic life. An elegant banquet from the early 1590s with eight servants languidly placing a lavish meal before an aristocratic young couple and an older man, includes a goose, a duck, a chicken, a rabbit or hare, arranged on a white tablecloth strewn with fragrant fresh flowers and herbs—rosemary, lavender, carnations, and violets. Written descriptions of banquets often refer to the strewn cloth, but images are rare, and here Flegel's wonderfully accurate flowers are even more fascinating than his careful presentation of the food.

Another collaboration by the two artists made in 1595 is a genre scene, *Allegory of Summer*, with a formal garden in the background, and two young women, one of them in the garb of a noblewoman sitting amongst an array of fruit and vegetables, with an exquisite vase of flowers on a shelf to one side, while the other young woman, probably a servant, has a neat armful of plucked flowers and a basket of identifiable leaves and herbs (see Figure 8.7). The contrast between the artifice of the flower arrangement and

FIGURE 8.6 *Vertumnus—Rudolf II*, c. 1590, Giuseppe Arcimboldo. Rudolph II (1552–1612), Holy Roman Emperor from 1576, as Vertumnus, ancient Roman god of seasons who presided over gardens and orchards. From the Stoklosters Slutt, Balsta, Sweden. Photo by Art Media/Print Collector/Getty Images.

the random tangle of leaves for the kitchen is clear enough. This is a rare sighting of the herbs that were replacing exotic spices in the kitchens of the rich. One can find images of them rendered in careful detail in botanical manuscripts and printed herbals, but rarely, as here, as ingredients. It is interesting to note that the young woman on the right is shelling broad beans, not arranging flowers. Collectors of rare and unknown flowers and plants were happy to see them in such genre scenes where the humble activities and accoutrements of kitchen and market place were as congenial as aristocratic flower pieces.

Herbs and flowers are strewn over the linen bathsheet of an aristocratic lady in a portrait by French painter François Clouet of 1571. She sits serenely in her flawless, naked beauty in a tub of walnut wood, while a more robust, but cheerful, wet-nurse feeds a swaddled baby, a small child grabs at grapes, and a servant brings more hot water from the kitchen. In the bottom left foreground there are still-life elements—a vase of flowers, a metal bowl of fruit, and herbs and blossoms to perfume the bath—laid out on the cloth like the specimens in a botanical album. This was when French artist Jacques le Moine de Morgues (*c.* 1533–88), having avoided death in Florida when the Spaniards attacked and slaughtered the French colony there, escaped with his life but no drawings, and went on to illustrate the geography of the New World and its products, but also albums of beautifully observed botanical studies. He later, as a persecuted Huguenot, fled France to live and work in England, where among his work is a full-length watercolor, *A Young Daughter of the Picts,* a well-built young woman, but tattooed all over with accurately drawn flowers, including some plants from the Americas.

STILL LIFE

The exquisite vase of flowers that figures in many genre scenes has muddied the waters of art history for decades. When investigating the origins of the still life the flower arrangement was held to be key in determining *the* "moment" when a single still or inanimate subject broke free from its surroundings and got to have a frame all of its own. But why focus on a vase of flowers when small still lifes of raw or cooked ingredients featured in large genre scenes? A cluster of root vegetables in Beuckelaer's *Earth,* one of his series *The Four Seasons*, a market scene of 1569, or the artichoke and asparagus, exotic luxuries, in the bottom left-hand corner of Campi's *Fruitseller*, 1578–81, were there waiting to be isolated and framed.

It is possible that vegetables, kitchen stuff, were much too vulgar and earthy for the refined sensibilities of modern art critics, while flowers had not only their rarity and beauty but a heavy load of symbolism. So the vase of flowers might have skewed our vision of what people wanted to hang on their walls, and why. A vase of flowers in a niche by Dutch painter Roelant Savery in about 1600 has been claimed to be the first independent still life. Roelant Savery was employed by Rudolph, and was renowned for his studies of plants, animals, and birds, including rare creatures like the dodo. An endearing detail in one of his visions of Paradise shows two playful leopard cubs surrounded by beautifully painted tulips and irises, and in all of his animal works there is a meticulous attention to plant life. There are flower paintings by the German artist Ludger tom Ring in the *Vienna Codex*, an album of nature studies in the Emperor's collection, dating from the early 1560s, and a sketch of a basket of flowers by him done as early as 1562.[11]

There may have been other paintings that were not flower arrangements, but groupings of vegetables and fruit on their own, most importantly in Arcimboldo's upside-down or

FIGURE 8.7 *Allegory of Summer*, 1593, Georg Flegel (1566–1638). Photo by Fine Art Images/Heritage Images/Getty Images.

"composite" works. *The Gardener* of 1590 appears to be a bowl of casually arranged vegetables, carefully depicted, until turned the other way up, when the tawny onion becomes a rosy cheek, two carrots the fleshy lips, and a bundle of radishes the shirt collar of the rustic gardener. Ultimately, it is advisable to avoid a "who got there first" approach to the origins of still-life painting (DaCosta Kaufmann 2010: 120). There could well have been many others, like the *Basket of Fruit* by Caravaggio (Michelangelo Merisi da Caravaggio) in the Ambrosiana, 1597–8, often said to be the very first example of an independent still life. Baskets of fruit were painted and written about in the ancient world, and were collected by Renaissance scholars and antiquaries and so accrued a great many erudite associations. But at the time of painting this, and later, the *Supper at Emmaus*, 1601, where an almost identical basket is on the table, it is probable that in every tavern between Milan and Rome, *un po' di frutta* (some fruit) at the end of a meal would be apples and grapes served in a basket like this one, a familiar everyday object. It is also true that Caravaggio and his patrons were responding to the call of the Counter Reformation to bring the Catholic faith to the everyday world of ordinary folk. The theme and the details of the painting were referring to both the here and now, and at the same time appealing to the erudition of his sponsors, Cardinal Francesco Maria del Monte and Cardinal Federico Borromeo, by including the heavy load of symbolism they enjoyed. The meal was what one might expect in a decent *hosteria* (inn) in the everyday world: bread, wine, a chicken or guinea fowl, and a basket of fruit, served on the conventional tablecloth over a Turkey carpet by a waiter with his long linen napkin. The blemishes on the fruit were normal, but could be interpreted as metaphors for the corruption lurking inside images of perfection, the rot within, as we find later in so many Dutch genre and still-life paintings, and the fish-shaped shadow of fruit, indicating fish as a symbol for Christ.

The depiction of plants in Caravaggio's circle in Rome became an art form in its own right, distinct from botanical illustration, but deploying the same observational skills. When Caravaggio was arrested for mayhem, as recorded often enough in his police record, the lethal weapons in his possession on one occasion turned out to be a pair of compasses and a magnifying glass, essential for the stunning flower and fruit paintings he, and a mysterious colleague, now called the Master of Hartford, were producing in Rome during the late sixteenth and early seventeenth centuries. Flowers in a circular glass flask, with all the magical tricks of reflection and distortion, along with luxury vegetables, such as artichokes and asparagus spears, as well as other fruit and vegetables, became an end in themselves, not just parts of a bigger picture with a story or people attached.

Meanwhile the Medici rulers of Florence were stimulating the economic and cultural growth of Tuscany in the seventeenth century, and the depiction of plants had become both a scientific necessity as well as an aesthetic pleasure. Artists were employed by the Medici to record the rare specimens in their collections, and enhance their country villas with depictions of plants and flowers in the gardens and orchards that surrounded them, culminating in the late seventeenth century with Bartolomeo Bimbi's group portraits of different fruit, all neatly numbered and captioned. Bimbi's work takes us into the early eighteenth century, but well before that Grand Duke Francesco had invited Jacopo Ligozzi to come and work for him in Florence in 1577 where he produced superb gouache paintings of plants in the ducal collection, provoking the openly expressed envy of Ulisse Aldrovandi, who could only occasionally, and with much diplomacy, "borrow" Ligozzi to paint items in his own collection in Bologna. Ligozzi's plants are vibrant with life; one can almost see the leaves fluttering, as little birds attack a branch of figs, bursting with juice, or watch the stems of the angelica plant turning from pale green to pale pink, anticipating their appeal when candied and chewy. His agave plant is shown with its dramatic sturdy leaves, a complete contrast to the ethereal Cypress vine (*Ipomoea quamoclit*); both came to Europe from Mexico in the mid-sixteenth century, and were cherished as rare ornamental plants, by collectors knowing little of their uses (Tongiorgi Tomasi and Hirschauer 2002: 42). Sadly, the Florentines did not become acquainted with the syrup drawn from the agave's leaves, a useful source of sweetness, as well as a brewed version of this, pulque, or the distilled version, mescal. The cypressvine morning glory or Cypress vine (also called cardinal vine, not the plant now known as water spinach, *Ipomoea aquatica*) was painted as a delicate climbing plant, its feathery leaves fluttering and trembling. Ligozzi's skills in reproducing the almost metallic toughness of the agave and the fragility of the vine were greatly appreciated, and he spent the rest of his life in the service of the Medici in Florence.

A much younger painter, Giovanna Garzoni (mentioned earlier), would have been familiar with Ligozzi's work when she was employed in Florence by the young Ferdinando II from 1642–51, whose interest in scientific developments and botanical studies was the ideal stimulus for her skills. Her delicate still-life portraits of bowls of fruit and vegetables were executed with meticulous touches of gouache on vellum, small-scale "miniatures," like those in her calligraphy manuscript, that her clients wanted for their personal enjoyment, where botany gave place to interior decoration (see Figure 8.8). A bowl of cherries has the stalks waving restlessly, while at its foot lie two broad bean pods and some already dry beans. A monumental pomegranate, bursting open to reveal its juicy red seeds, sits in a white maiolica dish in a strangely surreal rocky landscape, surrounded by yellowing leaves, fluttering in an unexplained breeze, on which hovers a pale-yellow dragonfly, almost indistinguishable from them, while a snail and two chestnuts sit in the foreground. Flies and insects may have been Garzoni's acknowledgment of the heavy burden of symbolism in still-life and genre paintings of the Low Countries, but here she seems to be saying that this is just the way it is in our kitchens and homes, a fly on the fruit, a snail with the beans, etc. A bowl of three different kinds of artichoke reminds one of the Medici's encouragement of local agriculture as well as botanical rarities, as can be seen in the vegetable markets of Florence and Rome today, with the tiny artichokes of early spring, eaten whole and raw, *a pinzimino* (with a little salt and olive oil), and the later cultivars cooked in many different ways—fried or stewed with parsley or *mentuccia,* a kind of mint, and garlic, or deep fried as *carciofi alla giudea* (Jewish fried artichokes), while today *carciofi alla romana* (Roman-style artichokes), the trimmed hearts and leaf bases, stuffed with garlic and herbs, are stewed slowly in olive oil and water, stalks uppermost, to give a melting reduction of the plant to a soft mash of which nothing is wasted. Garzoni's dish of freshly picked asparagus, wreathed in exquisite pink carnations, is redolent of its luxury status, even though by then asparagus was grown as a lucrative cash crop in the Veneto, coming in barge-loads to the markets in Venice, as Castelvetro recorded earlier in the century.

Don Lorenzo de' Medici died in 1648 before Garzoni had finished his commission, *The Old Man of Artimino*. This complex work, described by Garzoni in her accounts book as *di molta faticha* ("done with a lot of hard work"), seems to be a celebration of the products of the Tuscan countryside, something the Medici rulers were keen to promote. Along with a nice cured ham and some salami, cheeses, eggs, and two hens, clutched by the old man, Bencino Brugnolaio, there are several edible plants—an artichoke, beans,

FIGURE 8.8 *Cherries and Carnations*, Giovanna Garzoni (1600–70), watercolor on parchment. Galleria Palatina & Appartamenti Reali di Palazzo Pitti, Florence. Photo © Raffaello Bencini/ Bridgeman Images.

celery, and some cherries. This combination of scientific studies, botanical enthusiasm, and fine art was characteristic of the depiction of plants in Florence in the seventeenth century. Bartolomeo Bimbi's vast group-portraits of fruit for the Medici are a contrast to the intimate studies of Garzoni. They are today displayed at Poggio a Caiano, one of the Medici villas in the countryside around Florence, where the different kinds of citrus fruit, apples, pears, plums, cherries, and figs were scientifically bred as experiments in horticulture rather than as part of an aristocratic pleasure garden. Bimbi depicted these fruit in bowers of greenery or tumbling over each other in a picturesque landscape, each one neatly numbered and listed in a cartouche at the bottom of the painting, a breathtakingly beautiful combination of scientific accuracy and aesthetic pleasure. His patron Duke Cosimo III was responsible for some remarkable portraits of monsters and rarities in the vegetable world, in which Bimbi depicted these giant growths with dramatic lighting effects, in surroundings that emphasized their remarkable size. A pumpkin, weighing about 53 kg (160 pounds), was brought to Florence all the way from Pisa in a donkey cart on a hot sultry day in 1711, welcomed *con molto strepito,* by a very noisy crowd, cheering it on its way to the painter's studio, where he set to work in the dying light, to immortalize the already unhealthily mottled monster on the verge of bloated liquefaction, and cut a wedge out of the pumpkin dramatically to light up the composition with its golden flesh (see Figure 8.9).

Meanwhile back in Bologna Aldrovandi was supplying Philip II of Spain with paintings of plants and animals drawn from life (maybe copies of existing works), and in return receiving samples and information about rare and unknown specimens from the Americas.

FIGURE 8.9 *The Pumpkin*, Bartolomeo Bimbi, second half of the seventeenth century. Image from The Picture Art Collection/Alamy Stock Photo.

The Hapsburg empire also provided contacts with Spain, and it was in Seville that Diego Velasquez (1599–1660), soon to become court artist, was painting scenes of low life, later known as *bodegones* (from *bodega* "tavern") showing lowly people preparing and eating ordinary food. There are not many plants to be seen in his work, but *Christ in the house of Martha and Mary* in the National Gallery, about 1618, shows, among the sauce for some fried sea bream, garlic and a fresh red chili pepper, rarely mentioned in cookery texts of the time, but probably chosen here as a popular seasoning in common use. These appear again in 1646 in Bartolomé Esteban Murillo's *The Kitchen of the Angels* where, immobilized in a trance, a Franciscan friar neglects his kitchen duties and is helped by a crew of angels, whose ingredients include chili peppers and tomatoes.

For food historians, Spain is so a rich source of images of chilies and tomatoes that evidence from cookery texts and primary sources can be overlooked. Other food plants from the New World like vanilla, maize, haricot beans, some varieties of squash, were certainly also cultivated and eaten, but it is not possible to track their uses solely in paintings. Spanish fine art does not tell us a great deal about plants, although flower arrangements abound, and is not an accurate guide to what was happening in kitchens.

Juan Sánchez Cotán (1560–1627) painted vegetables with a mesmerizing intensity, suspended from lengths of string to form a parabola in the empty dark recess of a *cantarero*, an improvised open-air larder on a cool shady windowsill. The geometry of the composition, from about 1600, might imply deep metaphysical interpretations, but the reality is a simple reminder of the painter's austere vegetarian diet, which he adopted after leaving Toledo for a monastic life in a Carthusian order in Granada. The cabbage, melon, quince, cucumber were typical, but before that he had filled similar compositions with fish and fowl and other good things to eat. Cardoons were a favourite with Sánchez Cotán, with their significant parabolic curve, and the pinkish, off-white tones of spiny leaf base and tasty stems, less watery, more pungent than celery, with the bitter-sweet meatiness of artichokes.

CONCLUSION

Representations of plants can tell us a great deal about kitchens and pharmacies as well as the museums of collectors and the galleries of connoisseurs, and it is in these sometimes pungent and always fascinating locations that their place in the history of art can often be appreciated. Representations of plants from the period under consideration here, importantly, also accurately portrayed plants in nature—that is, as part of a living, natural ecosystem. Weiditz (discussed above), who provided woodcut illustrations for Brunfel's *Vivae herbarum eicones* (Living Images of Plants), was the student of the German artist Albrecht Dürer (1471–1528). A keen observer of nature, Dürer, who painted his subjects in minute, accurate detail, has been credited with being the father of ecological illustration, even if his studies may have been motivated by a desire to create realistic settings for larger, more complex figured scenes (Egerton 2012). Of particular importance in this context is his 1503 watercolor and gouache composition entitled *Das große Rasenstück* (The Large Piece of Turf; see Figure 1.8 of this volume). This study of what appears to be an unordered group of wild grasses and other plants measures only 40.8 × 31.5 cm (16 × 12.4 inches), the plants depicted here being shown almost at life size and as viewed from ground-level. Here appear smooth meadow-grass (*Poa pratensis*), cock's-foot grass (*Dactylis glomerata*), creeping bentgrass (*Agrostis stolonifera*), along with English daisy (*Bellis perennis*), dandelion (*Taraxacum officinale*), birds-eye speedwell

(*Veronica chamaedrys*), greater plantain (*Plantago major*), hound's-tongue (*Cynoglossum officinale*), and yarrow (*Achillea millefolium*), all of them humble plants, not collectible exotica, some today considered weeds. In its realism, *The Large Turf* rivals contemporary botanical illustration, but where it differs from these is that it does not focus on sample specimens, portraying instead:

> individual variant members of these species, with each blade and stem and leaf and flower given a differentiated identity ... unlike the images in an ancient herbal or a modern textbook, *The Large Turf* doesn't visually isolate or distinguish its various plants. It presents them in a state of natural disarray, confused, interleaved, entangled .... It is a slice of living, chaotic undergrowth.
>
> (Lubbock 2008)

NOTES

Introduction

1 This observation is adapted from Michel Baudier, *Histoire générale du Serrail* (1662). Concluding his chapter on the kitchens of the Topkapi Palace he remarks: "*Si le lecteur trouve de l'ennui au récit de cette matière de cuisine, qu'il considère que, sans ce chapitre, les autres qui composent l'Histoire ne seraient point*" (The reader who finds this exposition of culinary matters tiresome may reflect that, without this chapter, the rest of History would not exist).

2 Belon (1558: ff. 78v–80v), translated in Clusius (1589: 85–7); Vasari (1568: pt 3, 2:402–9), translated in Hinds (1927: 3:168–75); Montaigne in Meusnier de Querlon (1774: 1:254–9). See in general Bellorini (2016).

3 Other early sources include a long footnote to a printed Latin psalter of 1516, inserted there for no good reason except that the editor, like Columbus himself, was Genoese. Attached to the text "*Their line is gone out through all the earth, and their words to the end of the world* [Psalms 19:4]," this not uninformative sketch begins: "In our times, owing to the marvellous daring of the Genoese Christopher Columbus, a nearly new world has been found and joined to the fellowship of Christians. Since Columbus himself often proclaimed that he had been chosen by God to fulfil this prophecy, it is not irrelevant to recount his life here. It was, then, Christopher surnamed Columbus, Genoese in origin, born of poor parents, who in our age by his own effort explored within a few months more lands and open sea than nearly all other mortals in all past centuries ..." (Justiniani 1516). Peter Martyr ab Angleria's title *De orbe novo decades* has a classical ring. It alludes to the *Decades* (series of ten books) into which Livy's history of ancient Rome was divided.

4 Monardes (1574: f. 98r); compare Clusius (1605: 75, 323); Amador de los Rios (1851–5: 1:348) "corbana." The Ecuador "cinnamon" that led Gonzalo Pizarro on his Amazonian wild goose chase was another species again, see Chapter 2.

5 Rauwolf's notes on plants, compiled to accompany his "herbarium vivum" or collection of pressed specimens, were published in Latin almost two centuries later (Gronovius 1755).

6 Both are now easily accessible. The *Libellus*, the Codex Badiano-De La Cruz, has appeared in several twentieth-century editions and translations (*Libellus* 1964 is the most recent). Sahagún's *Historia general*, the Florentine Codex, a vast work, can be read in English translation (Anderson and Dibble 2002); better still, the original, carefully written and extensively illustrated manuscript in Nahuatl and Spanish is available online. All twelve online volumes can be found by following links from this page: www.wdl.org/en/item/10612/.

7 The copy, with Clusius' ownership inscription and notes, is now at Cambridge University Library, where, as an undergraduate, I (Andew Dalby) consulted it fifty years ago. The inscription reads "1564 / Caroli Clusii A[trebatis] / Ulyssipone / vi. Calend. Januarius." This date, as written by Clusius in classical Latin style, equates to December 27, 1564, the sixth day, counting backwards inclusively, from January 1. The date has been variously interpreted, but C. R. Boxer (1963: 21, 25) shows that the book could not have reached Lisbon before September 15, 1564, and observes that Clusius was in Lisbon between September 1564 and January 1565.

8 See for example Oscar Tintori's catalog at www.oscartintori.it/prodotto/limone-ponzino-amalfitano/ (accessed April 1, 2021).

9 Engravings accompany the 1590 edition, which was published as the second volume of Theodore de Bry's *America*. Sunflowers, growing in kitchen gardens, appear on two of them (Hariot 1590: pls. 19–20). There are kitchen gardens but no sunflowers in the original watercolors by John White, which survive in the British Museum (Hulton and Quinn 1964). Given that Clusius had edited the illustrated sunflower description in Dodoens' 1583 herbal, and that the sunflowers in the 1590 engravings resemble that 1583 illustration, their insertion in these engravings may be his responsibility. The watercolors and engravings, arranged side by side, are available online: virtualjamestown.org/images/white_debry_html/jamestown.html (accessed April 1, 2021).

10 Here and above we have has translated from the manuscript copy by Nardo Antonio Recchi (f. 133r: available online: archive.org/details/demateriamedican00hern [accessed April 1, 2021]). At this point it differs little from the 1651 edition by Federico Cesi and other Lynceans, which, being easier to consult, is cited in the text.

11 Printed texts alone were searched for this case study, and others may well await finding.

12 Green grapes were picked for verjuice; full ripening for wine would take longer. Sugar cane, though not listed in this passage, was already being planted. Thus the groundwork was being laid for the West Indian monoculture of the sixteenth century and the consequent economic collapse of the seventeenth.

13 Peter Martyr distinguishes between *labruscae*, wild vines of Mexico, and *vineae*, plantations of the European cultivated species. He probably did not know that the "various kinds of flour" preferred in Mexico included maize mixed with chia (*Salvia hispanica*) and alegria (*Amaranthus cruentus*: cf. Berdan and Anawalt 1997: 33 and passim), the latter a close relative of the "blite" that Columbus had identified in Cuba. Both mixtures were better nutritionally than maize alone.

14 Andrew Dalby's thanks to Ray Sokolov for allowing him to use his translation, to which he has made a couple of minor changes. *Aloja* and *clarea* were both spiced wines, the former probably white and including wormwood (as does modern vermouth), the latter probably red and corresponding to sixteenth-century English clary. The banquet was more carnal than appears here: two long lists of meat dishes have been omitted.

Chapter 1

1 The acorns are an afterthought, added in the second edition (Harrison 1587: 168). The texts can be compared at english.nsms.ox.ac.uk/holinshed/texts.php?text1=1577_0078&text2=1587_0085#p991 (accessed April 3, 2021).

2 A corrody was a "pension" agreement made between an elderly person and a younger relative or a monastery for food, fuel, and sometimes shelter in exchange for the surrender of land.

Chapter 2

1 Belon is alluding to Plutarch, *Life of Artoxerxes* 3.2. It is stated by several Greek authors (see Dalby 2003: xi) that terebinth fruits and cress were the staple diet of young noble Persians, varied by whatever meat they could get by hunting.

2 Clusius' discussion of Sichuan pepper is arranged as a footnote to his revised Latin version of Garcia de Orta's *Colóquios*, which forms a separate section of *Exoticorum libri decem* (1605). Clusius cites Avicenna, *Canon* 2.2.266 "Fagara" (Alpagus 1527: f. 96v). The Latin translation of Avicenna's *Canon* had long been familiar as a medical textbook, book 2 of it

in particular as a medicinal herbal. The *Canon* was republished several times between the editio princeps (1473 or before) and the easily accessible edition by Andrea Alpago which I cite.

3 Orta is cited by chapter numbers of the 1563 edition because its page numbering is awry. These chapter numbers are equally valid for the translation by Markham (1913), but Orta is translated afresh in my quotations because Markham's translation often fails to get the meaning across.

4 Earlier Chinese voyages had culminated in the impressive expeditions led by Zheng He between 1405 and 1433, regularly visiting Java, Malacca, Ceylon, and Calicut, reaching Ormuz and beyond in 1414. The Galle trilingual inscription commemorates the Chinese landing in Ceylon in 1409 (Mills 1970; Pelliot 1933).

5 Mills' translation of the Chinese text, which literally reproduces its staccato style and was indispensable to me, has been rephrased slightly here to allow the underlying logic to shine through.

6 The last sentence is decisive. *Piper nigrum* is in fact now grown in Sumatra, but the pepper with slightly larger berries, usually hollow, is the native *P. cubeba*: "The stalked berries are a little bit larger than pepper corns, having a furrowed surface. Most berries are hollow" (gernot-katzers-spice-pages.com/engl/Pipe_cub.html [accessed April 3, 2021]).

7 For the fruit of the water plant *Euryale ferox*, utterly obscure in the Early Modern period but containing what is probably one of the most ancient luxury foods appreciated by genus *Homo*, see Dalby in Volume 1. It and the durian are both characterized by a spiny exterior enclosing several edible sections: that is the only resemblance between them.

8 The Brazilian species *Schinus terebinthifolius*, discovered later, has been naturalized elsewhere in the tropics and is the source of "pink peppercorns."

9 Acosta's text is, I think, the earliest reference to chocolate as a female aphrodisiac. The English translation misses this, but memorably contains the first appearance of the word chocolate in English: "The Spaniards both men and women, that are accustomed to the countrey, are very greedy of this *chocholaté*" (Grimeston 1604: 271).

10 Orta is possibly referring to Jacques de Vitry, not a Franciscan but bishop of Acre from 1216, author of *Historia Orientalis et Occidentalis*, in which he wrote that the banana was called "tree of paradise" and its fruit "Adam's apple" (Jacobus de Vitriaco 1597: 170).

11 See a fearsomely spiny orange branch in *Grandes heures d'Anne de Bretagne* f. 168, available online: gallica.bnf.fr/ark:/12148/btv1b52500984v/f344.image (accessed April 3, 2021) and three juicy oranges in the Loggia of Psyche, species no. 78 in the list by Anna Whipkey and Jules Janick at the illustrated online database hort.purdue.edu/newcrop/udine/ (accessed April 3, 2021).

12 While aware of its New World origin, Clusius rightly observes that the cashew is related to *Semecarpus anacardium*, the "marking nut" of India, on which he, like Orta before him, cites Avicenna, *Canon* 2.2.41 (Alpagus 1527: f. 77r; Orta 1563: f. 16v; Markham 1913: 32). C. Acosta makes no connection between the two fruits: evidence, possibly, that he knew only Orta's Portuguese text, not Clusius' 1574 translation.

13 Malmsey [*malvoisie*] in Léry's time was the sweet wine of Crete and southern Greece.

14 The early history of tomatoes in Europe is discussed by Rudolf Grewe (1987) and David Gentilcore (2010: 1–26). While noting the slightly earlier reference by Mattioli, Gentilcore identifies and cites the first documentary record of ripe tomatoes in an Italian basket in the household of Cosimo de' Medici in 1548. Gentilcore argues that the golden fruit of some of the earliest references is not tomato but tomatillo (*Physalis philadelphica*), and he interprets this recipe by Hernández as a tomatillo and chili sauce.

15 Girolamo Cardano, against whom Scaliger was reacting, had described the chili in 1550, adding correctly that it reached Europe from Hispaniola ("an island of the other world") in 1493, but had perhaps not tasted it. Scaliger certainly knew it in food.

16 It is not an easy task to assign the cultivated chilies seen by successive explorers and early botanists to the several species of *Capsicum*. The relevant species are *C. annuum, C. baccatum, C. chinense, C. frutescens*.

17 Gentilcore (2017: 192, 202) discusses the introduction of the tomato to Italy with special attention to the category problem that it posed. And still poses. The ever-changing introductory sentences of the Wikipedia article "Tomato," which usually describe it as "the fruit of the plant *Solanum lycopersicum*," are now and then anonymously altered to read "the vegetable of the plant *Solanum lycopersicum*."

Chapter 3

1 Further discussion may be found in the following: Davies (1956: 26–9); Morison (1978: 73–4); Quinn (1974: 98–100); Skelton (1962). For the Cabot Project, see www.bristol.ac.uk/history/research/cabot/publications/ (accessed April 3, 2021).

2 The letter (quoted in the Introduction of this volume) was the first of a series eventually gathered under the title *De orbe novo* (Petrus Martyr 1530). See also Janick and Caneva (2005).

3 All quotations from Oviedo are taken from the translations by Nina M. Scott (Myers 2007).

4 For more detailed discussion, see Cuttler (1989); Snyder (1976).

5 For more on maize, see Finan (1948).

6 For more on this see, Egmond (2010, especially chs. 11 and 12); Harkness (2007).

Chapter 4

1 Anthony Ashley Cooper, Fourth Earl of Shaftesbury, letter to James Harris, May 4, 1738, transcribed in Burrows and Dunhill (2002: 49).

2 Herodotus 7.31. The exact location of Kallatebos, the site of the plane tree, has not been identified, but it was probably near present-day Uşak. It cannot have been as distant as Ine Göl, as proposed by How and Wells (1912: 2:139).

3 Pliny the Elder, *Naturalis historia* 16.238. George Seferis [Giorgos Stylianou Seferiades], *Μυθιστόρημα* (1935), poem 15, in Keeley and Sherrard (1967: 36–7); Seferis (1950). See Stubbings (1946).

4 Ambrosoli (1997: 7), referring to Munich, Bayerische Staatsbibliothek, MS Lat. 337. On Mattioli's editions of Dioscorides, see Michael North, "Research Reborn: Dioscorides and Mattioli." Available online: circulatingnow.nlm.nih.gov/2016/01/06/research-reborn-dioscorides-and-mattioli/ (accessed April 3, 2021).

5 About 3 acres (Ash 1941: 13). "Quintius Cincinnatus obsessi consulis et exercitus liberator, ab aratro vocatus ad dictaturam venerit, ac rursus fascibus depositis, quos festinantius victor reddiderat, quam sumpserat imperator, ad eosdem iuvencos et quattuor iugerum avitum herediolum redierit" (Columella, *De re rustica* 1.praef.). For Washington's emulation of Cincinnatus, see Wills (1994).

6 *Mezzadria* was finally abolished in Italy after the Second World War, by laws passed in 1964 and 1982. The current prosperity of the Tuscan countryside affords cogent proof of how damaging the old feudal system was to human welfare and agricultural innovation. Discussion of the topic still carries an immense political charge in Italy. A good introduction is Snowden (1989: 8–70).

7 The classic work on climate history and the Little Ice Age is Lamb (1982) (often reprinted). See also e.g. Free and Robock (1999).

8 An early report on the discovery of the mosaics by Maria Antonietta Tomei of the Archaeological Superintendency, Rome (now reshuffled as the Soprintendenza Speciale per i Beni Archeologici di Roma, or SSBAR, spawn of the dread MIBACT, Ministero dei Beni e delle Attività Culturali e del Turismo), news.bbc.co.uk/2/hi/europe/6664941.stm (accessed April 3, 2021).

9 My translation. First published by Angelo Ridolfi (1810: 61–6), this is letter 216, dated December 10, 1513, in the one-volume Machiavelli edition of Ciliberto and Accendere (2018: 2873–7).

10 The menus recorded by Rossetti (1584) mention *tortelli di zucca* frequently: 71, 100, 115, 177, 215, 241, 243, 318, 335, 350, 357, 404, 415, 421, 449, 510, tortelli stuffed with sturgeon and squash, 487 (storione e zucche), and a dish with pieces of sturgeon and Zucca marina di Chioggia (*Cucurbita maxima*) "*pezzetti di storione impilotati con zucca marina*."

11 See Alex Revelli Sorini and Susanna Cutini, "Tortini di riso pel Moro," with recipe. Available online: www.taccuinistorici.it/ita/news/moderna/personaggi/Ludovico-il-Moro-e-il-dolceriso.html (accessed April 3, 2021).

12 He also insisted on having a black-skinned valet, see Benzoni (2006).

13 Scythes had been known since antiquity, but were not used in Europe until the twelfth–thirteenth centuries; see Bonicalzi et al. (1978: 20).

14 For "Spanish grass," see Sider (2005). The word "alfalfa" has been known in English since 1764 (see *Oxford English Dictionary*, www.oed.com).

15 Bay (2019) is the primary source for this summary and provides a detailed analysis of this garden.

16 As reported by Fabio Chigi, the future Pope Alexander VII, who was Agostino's younger brother Sigismondo's great-grandson, in *Chisiae familiae commentarii*, Biblioteca Apostolica Vaticana, MS Chigi a.I.1, ff. 33v–34r (cf. Cugnoni 1879: 65): "Celsus quidem Cittadinus vir eruditissimus narravit mihi malorum medicorum limoniorumque extitisse copiam immensum, audisseque se dictitatum a suis maioribus Augustinum postulasse a patritio quodam Neapolitano L. viviradices, habuisseque perhumaniter supra mille, omnesque inibi sevisse."

17 In a long poem called *Suburbanum Augustini Chisii* composed early in 1512, the humanist Biagio Pallai, who wrote, as we have seen above, as Blosio Palladio, called Chigi's *Viridario* the "true home of Venus" ("haec Veneris sit vera domus; cui tota rubentis Pomonae et viridis cesserunt Numina Florae"), and reassures Chigi that he is "*rex animo*" (Rowland 2005: 63).

18 Leonicenus 1492; Maria Conforti, "Testi antichi e nuovi saperi: botanica e materia medica." Available online: https://www-oed-com.udel.idm.oclc.org/view/Entry/4896?redirectedFrom=alfalfa#eid (accessed April 23, 2021); Paolo Pellegrini, Niccolò da Lonigo, Dizionario Biografico degli Italiani 78 (2013), http://www.treccani.it/enciclopedia/niccolo-da-lonigo_(Dizionario-Biografico)/ (accessed March 3, 2021).

19 Cynthia Pyle has traced the direct relationship between humanist literary scholarship and empirical study of nature. See e.g. Pyle (1996, 2010).

20 The UNESCO citation is available online: whc.unesco.org/en/list/824 (accessed March 3, 2021).

Chapter 5

1 Rauwolf's herbarium is in the collection of Naturalis Biodiversity Center, University of Leiden.

2 Cf. Blunt and Raphael (1979); Givens et al. (2006). Research since 2000 is modifying the total numbers but not the trend.

3 The *Carrara Herbal* (*c.* 1390–1405; Kyle 2017) and the *Codex Bellunensis* are in the British Library. Both are available online: www.bl.uk/manuscripts/FullDisplay.aspx?ref=Egerton_MS_2020 and at www.bl.uk/manuscripts/FullDisplay.aspx?ref=Add_MS_41623 (accessed April 3, 2021). The mid-fifteenth-century *Codex Roccabonella* (also *Codex Benedetto Rinio*) is in the Biblioteca Marciana in Venice. The text that accompanies the images of the *Carrara Herbal* is an Italian translation from medieval Latin. The Latin text of this originally Arabic work remained influential to the early sixteenth century. It was printed in 1473 as *Liber Serapionis agregatus in simplicibus medicinis* and in 1497 as *Liber Serapionis de simplici medicina*.

4 *Codex Bellunensis* ff. 20r and 28v (tamarind), and 95v (acacia); see Mariani Canova et al. (2006: 1:52–3).

5 This album is in the Wellcome Library, London, MS 336. Available online: wellcomelibrary.org/item/b18763571# (accessed April 3, 2021).

6 Cibo's albums are in the British Library, 2 vols, MS Add. 22332–3. Both volumes are available online: www.bl.uk/manuscripts/FullDisplay.aspx?ref=Add_MS_22332 and www.bl.uk/manuscripts/FullDisplay.aspx?ref=Add_MS_22333 (accessed April 3, 2021). The text accompanying the images in the Cibo albums is from Pietro Andrea Mattioli's Italian version of Dioscorides (Mattioli 1544 and later editions).

7 The Cibo herbarium is in the Biblioteca Angelica, Rome; *En Tibi* in Naturalis, Leiden; the Merini herbarium in the Museo di Storia Naturale in Florence; the Hurtado de Mendoza herbarium in the Escorial Library, Madrid. The pages of *En Tibi* are available online: http://bioportal.naturalis.nl/result?theme=en_tibi (accessed April 3, 2021). For surveys of herbariums in Europe, see Hurka and Neuffer (2011); Stafleu (1987); and Thijsse (2016).

8 An example is MS 346, Herbarius, Latinus (*c.* 1490) with an inscription by Johann Reuchlin, Wellcome Library, London.

9 The Gessner plant albums (*c.* 1540–65) are in the University Library of Erlangen-Nuremberg (Germany), *Historia plantarum*, 2 vols, *c.* 490 folios; and in the University Library of Tartu (Estonia), more than 150 folios. The volumes at Erlangen are available online: gateway-bayern.de/BV039778087 and gateway-bayern.de/BV040149451 (accessed April 3, 2021). The Fuchs drawings (*c.* 1535/40–64) are in the Oesterreichische Nationalbibliothek, Vienna, 9 vols. The Michiel albums (*c.* 1545–75) are in the Biblioteca Marciana, Venice, 5 vols (730 species), see also Chapter 8 note 1 of this volume. These particular collections originated in preparation for future publications.

10 The Oellinger herbal and the *Camerarius Florilegium* are in the University Library of Erlangen-Nuremberg (Germany). The Oellinger herbal is available online: gateway-bayern.de/BV040687699 (accessed April 3, 2021).

11 Cesalpino's work inspired Linnaeus and he has often, somewhat anachronistically, been regarded as the latter's forerunner.

12 The Garets are mentioned more often as sources than anyone else in Clusius' *Exoticorum libri decem* (1605); James Jr. forty times and Pieter thirty-two times, while James Jr. also occurs fourteen times in Clusius' *Rariorum plantarum historia* (1601).

13 Manuscript letters (9) from James Garet (Jr.) in London to Clusius (1583–1601, in French); manuscript letters (7) from Pieter Garet in Amsterdam to Clusius (1601–5, in Dutch), all in Leiden University Library, VUL 101. Cf. Egmond (2010: 175–207, esp. 202–3).

14 On its botanical results, see especially Pardo Tomás and López Piñero (1996); on its publication history, Andretta and Brevaglieri (2013: 70–4) and Mason (2009: 149–72). For more general Iberian colonial science and plants, see Cañizares-Esguerra (2006) and Bleichmar et al. (2008).

15 On Europeans bioprospecting in a colonial setting, see Schiebinger (2007).

16 Its first English translation appeared as early as 1598.

17 Four volumes of the Ratzenberger herbarium are in the Herzogliche Bibliothek Gotha, three in the Naturkundemuseum im Ottoneum, Kassel (Germany).

18 The vicissitudes of the Early Modern intercontinental drugs trade lie beyond the scope of this article, but see in particular Bartels (2003); Cook (2007); Fontes da Costa (2015); Guerra (1966); Roberts (1965); Walker (2008, 2013).

19 Naturally, Europeans were by no means the only intermediaries involved in intercontinental trade. See for such exchanges in the Portuguese empire especially Walker (2016).

20 Manuscript letter from James Garet to (September 9, 1589, in French); and manuscript letter from Charles de Tassis to Clusius (January 25, 1589 in French), referring to Garet, the potatoes and tomatoes; both Leiden University Library, VUL 101. Cf. Egmond (2010: 202–3).

21 Interestingly, Clusius did not incorporate Garet's suggestion, and does *not* connect the potato with the solanum family (1601: lxxix–lxxxi).

22 The text of the Spanish report of 1579–80 by Pedro Sarmiento de Gamboa is translated and discussed in Mason (2015: 2). Modern edition of the original manuscript by Batista González (2000).

Chapter 6

1 Rabelais (1546: 333–8); translation by M. A. Screech (2006). Lazzaro (1990: 24) discusses this passage and its implications for botanical nomenclature.

2 See Peter Hainsworth's introduction to Petrarch (2010: xxiii), for discussion of the poet's play on Laura's name.

3 Note Jaynie Anderson's comments on the lack of resemblance of Giorgione's sitter and Petrarch's evocation of his beloved (1997: 475–6): "All Renaissance readers knew that Petrarch's Laura, as portrayed by Simone Martini, was blonde, that she had blue eyes, that she was chaste, that she was an ideal beauty, and that she was unlike Giorgione's Laura, who is dark-haired, dark-haired, whose beauty is unidealised, and whose chastity is questionable. The reference to Petrarch in both portraits is not a simple one, and suggests an ironic *paragone*, both painters surpassing the poet's description."

4 By striking his hoof on Mount Parnassus, Pegasus created the spring Hippocrene ("horse spring"), haunt of the Muses.

5 Ancient Carthage, whose site lies close to modern Tunis, was a Phoenician colony: hence "defeating the Phoenicians."

6 See also Paracelsus in Waite's translation (1894: 189): "So the euphrasia or *herba ocularis* is thus called because it cures ailing eyes. The sanguinary herb is thus named because it is better than all others to stop bleeding. The scrofulary (*chelidonium minus*) is so called because it cures the piles better than any other herb. And so with many other herbs, of which I could cite a vast number, all of which were named on account of their virtue and faculty …"

7 For a study of the centrality of metamorphosis in sixteenth-century culture, see Jeanneret (2001).

8 As early as 1502, Pinturicchio was commissioned to decorate the Piccolomini Library in Siena Cathedral "with such fantastic forms, colours and arrangements as are now called grotesques (… *che oggi chiamano grottesche*)." See Kayser (1963: 20).

9 Nicolas Robert, Portrait de Gaston d'Orléans (1608–60). Original: Vélins, portefeuille 75 f. 2. Reproduction tirée de Archives du Muséum National d'Histoire Naturelle, série 6 vol. 1, 1926.

10 *La Guirlande de Julie* is in the Bibliothèque Nationale de France, Fonds français N.A.F. 19142, and is available online: https://gallica.bnf.fr/ark:/12148/btv1b8451620k (accessed April 4, 2021).

Chapter 7

1 Although this may have represented revolutionary new thinking in England, it was not such a new idea on the Continent. In 1576, Matthias de Lobel, a Flemish physician and botanist of international reputation, published a book, *Stirpium observationes* or "Observations on the history of plants" and, like Parkinson (who owned an extant copy of this book, heavily annotated in his own hand), he included flowers and plants purely for their ornamental value.

2 For more on depictions of gardens in Early Modern paintings, see Strong (2000).

3 These may seem somewhat exotic varieties to be growing in ordinary gardens, but it would appear that their cultivation was commonplace: cucumbers had been grown in England since medieval times: purchase of seeds appear in the accounts of the Archbishop of Canterbury's gardener at Lambeth for 1322 (Harvey 1974: 94) and notes on its cultivation appear in an instruction manual on planting a kitchen garden written "for the helpe and comfort of poore people" by Richard Gardiner of Shrewsbury (1599); in his *Herball* of 1597 John Gerard describes and illustrates seven kinds of cucumber, four kinds of melon, seven kinds of pompions, and two types of gourd (762–78). They are referred to by contemporary authors almost interchangeably, Gerard for instance noting that "doubtless the Muske Melon is a kinde of Cucumber" (770). It has recently been brought to my attention that "cucumbers," as usually translated from early medieval European sources, were not *Cucumis sativus* but *Cucumis melo,* or Chate melon (Paris et al. 2012).

4 The Oglander Manuscripts in the Isle of Wight Records Office are abbreviated to OG in citations.

5 The outline of the garden as laid out by Sir Thomas Tresham, together with the famous garden lodge, can still be seen at Lyveden today. The concentric circles which comprised Tresham's "moated orchard" were first identified from an aerial photograph taken in 1940 by a Luftwaffe pilot: Eburne (2008: 123). For more on the gardens at Lyveden New Bield, see Eburne (2008: 114–34) and Dix (2011: 170–1).

6 For more on the rise of conspicuous consumption in the early seventeenth century, see Peck (2005) and Thomas (2009: 110–46).

7 For a full discussion of this phenomenon, see Goldgar (2007), who convincingly argues that the extremes of tulipmania have been vastly exaggerated, the known "facts" being based on contemporary moralizing propaganda. For a brief period between the summer of 1636 and the spring of 1637 tulips were remarkably expensive. Because tulip bulbs spend most of the year out of sight beneath the ground, sales took the form of contracts not just for unseen goods, but also unknown goods—the changes that could occur in the bulbs from one year to the next were not always for the better. Contracts were passed on for higher and higher sums, the potential for double-dealing was huge and when the time came, buyers could not pay and sellers could not deliver. Not surprisingly, the bottom dropped out of this market very quickly. That tulipmania happened is not disputed, but it did not have the seismic effects on personal fortunes and the wider economy that we have been led to believe.

8 Sir Thomas Hanmer's manuscripts in the Bettisfield Estate Records, National Library of Wales, Aberystwyth, are abbreviated to B in citations.

9 For more on the lives and plant hunting exploits of the John Tradescants, see Leith-Ross (1984).

10 Sir Thomas Temple's manuscripts in the Temple Family Papers, Huntington Library, California, are abbreviated to ST in citations.

11 Sir Arthur Ingram's papers among the Temple Newsam Manuscripts, West Yorkshire Archive Service, Leeds, are abbreviated to WYL in citations.

12 John Harvey based this assertion on his study of early nurserymen and early plant catalogs, where he found little evidence of commercial trade catalogs of seeds and plants for sale until after 1660 (Harvey 1972, 1974). However, looking at this question from the point of view of those buying the plants as opposed to those selling them suggests a different conclusion.

13 Parkinson describes and illustrates two carnations named after Tuggie in *Paradisus terrestris* (1629: 313); Thomas Johnson mentions Tuggie in his revised edition of Gerard's *Herball*, coupling his name with both John Parkinson and John Tradescant (Johnson 1633: 589, 785); while Tradescant himself notes in the back of his copy of Parkinson's *Paradisus terrestris* that he had "4 more Roses whereof Mr Tuggy Hathe two" (Leith-Ross 1984: 199).

14 Although it is impossible to know exactly what Oglander meant by "French flowers," comparing this with other contemporary evidence of flower enthusiasts who also obtained plants and bulbs from France, often at high prices, it seems that there was some kind of implied prestige attached to buying plants from the Continent. By the early seventeenth century, the Continent, and particularly France, was seen as a primary source of everything that was luxurious, stylish, and fashionable (see e.g. Francis 2018: 217; Peck 2005: 13). Although Oglander gives no more details about the plants he has purchased, he clearly thinks it is noteworthy that they are French.

15 See for instance, Strong (2000: fig. 44, Artist Unknown, "View of a complete garden in the manner of de Vries' *Hortorum* Formae," *c.* 1620, and fig. 7.9).

16 As indicated in Sir John Oglander's "Observations in husbandrie" OG/AA/28.

Chapter 8

1 The catalog *Di sana pianta: Erbari e taccuini di sanità* (1988) reproduces twenty-one of these illustrations, probably by Domenico Dalle Grechi, from Biblioteca Nazionale Marciana, Venice, MSS It, cl. II, 26–30. Michiel's text has been published, but accompanied by only a small selection of the illustrations reproduced in monochrome (De Toni 1940).

2 British Library MS Add. 22332 and 22333.

3 His albums are available online: http:137.204.21.141 (accessed April 7, 2021). The Bologna collection of his printed works (some volumes hand-coloured) is also available online: amshistorica.unibo.it/ulissealdrovandi-opereastampa (accessed April 4, 2021).

4 The *Hours of Catherine of Cleves*, at the Pierpont Morgan Library in New York (M 917/945), is available online: www.themorgan.org/collection/Hours-of-Catherine-of-Cleves/thumbs (accessed April 4, 2021).

5 In general, see Arano (1976). Manuscript versions of the *Tacuinum Sanitatis* can be found at several online sites. The manuscript discussed here, now in the Austrian National Library, has been published in facsimile, most recently in 2004 (Unterkircher), and is available online: data.onb.ac.at/rep/10020524 (accessed April 4, 2021).

6 For more details see Riley (2007: 305–6).

7 There are two sources available online for these frescoes (neither of them very easy to find). The older one indexes the plants: hort.purdue.edu/newcrop/udine/default.html (accessed April 4, 2021). The newer one is a tour of the frescoes: vcg.isti.cnr.it/farnesina/ (accessed April 4, 2021).

8 This manuscript, now at the Getty Center, has been reproduced in facsimile (Hendrix and Vignau-Wilberg 1992) and the facsimile is available online: https://www.getty.edu/art/collection/objects/1487/joris-hoefnagel-and-georg-bocskay-mira-calligraphiae-monumenta-flemish-and-hungarian-fols-1-129-written-1561-1562-illumination-added-about-1591-1-596/ (accessed April 7, 2021).

9 See letter G in Casale (1996: 36–7); also Casale (1991: 120).

10 The album is available online: www.rijksmuseum.nl/en/collection/RP-T-BR-2017-1-1 (accessed April 4, 2021). See also www.rijksmuseum.nl/en/press/press-releases/tefafs-top-masterpiece-to-the-rijksmuseum (accessed April 4, 2021).

11 The album, still at the Austrian National Library, is available online: https://onb.digital/result/10009912 (accessed April 4, 2021).

BIBLIOGRAPHY

Acosta, Cristóbal. 1578. *Tractado delas drogas y medicinas de las Indias Orientales*. Burgos: Martin de Victoria.

Acosta, José de. 1590. *Historia natural y moral de las Indias*. Seville: Juan de León.

Agamben, Giorgio. 2009. *The Signature of All Things: On Method*. New York: Zone Books.

A'lam, Hūšang. 1989. "Botanical Studies on Iran, i. The Greco-Islamic Tradition." *Encyclopedia Iranica*, IV(4): 390–401.

Alciatus, Andreas [Andrea Alciato]. 1531. *Emblematum liber*. Augsburg: Heinrich Steyner.

Alpagus Bellunensis, Andreas (ed.). 1527. *Avicenne Liber canonis medicine*. Venice: Junta.

Amador de los Rios, José (ed.). 1851–5. *Historia general y natural de las Indias, islas y tierra-firme del mar Océano, por el capitán Gonzalo Fernandez de Oviedo y Valdés*. 5 vols. Madrid: Real Academia de la Historia.

Ambrosoli, Mauro. 1997. *The Wild and the Sown: Botany and Agriculture in Western Europe, 1350–1850*. Translated by Mary McCann Salvatorelli. Cambridge: Cambridge University Press.

Amigues, Suzanne. 2004. "Le silphium: état de la question." *Journal des savants*, 2(1): 191–226.

Anagnostou, Sabine. 2000. *Jesuiten in Spanisch-Amerika als Uebermittler von heilkundlichem Wissen*. Stuttgart: Wissenschaftliche Verlagsgesellschaft.

Anagnostou, Sabine. 2005. "Pharmazie auf internationaler Ebene: die Apotheke des Collegio Romano vom 16. bis 18. Jahrhundert." *Geschichte der Pharmazie*, 57: 57–63.

Anagnostou, Sabine. 2007. "The International Transfer of Medicinal Drugs by the Society of Jesus (Sixteenth to Eighteenth Centuries) and Connections with the Network of Carolus Clusius." In Florike Egmond, Paul Hoftijzer, and Rob Visser (eds.), *Carolus Clusius. Towards a Cultural History of a Renaissance Naturalist*, 293–312. Amsterdam: Edita.

Anderson, Arthur J. O. and Charles E. Dibble (trans.). 2002. *Bernardino de Sahagún: The Florentine Codex; General History of the Things of New Spain*. 12 vols. Salt Lake City, UT: University of Utah Press.

Anderson, Jaynie. 1997. *Giorgione: The Painter of Poetic Brevity*. New York: Flammarion.

Andretta, Elisa, and Sabina Brevaglieri. 2013. "Storie naturali a Roma fra antichi e nuovi mondi. Il *Dioscoride* di Andrés Laguna (1555) e gli *Animalia Mexicana* di Johannes Faber (1628)." *Quaderni Storici*, 142(1): 1–46.

Andrews, Jean M. 1992. "The Peripatetic Chili Pepper: Diffusion of the Domesticated Capsicums since Columbus." In Nelson Foster and Linda S. Cordell, *Chilies to Chocolate: Food the Americas Gave the World*, 81–93. Tucson: University of Arizona Press, 1992.

Arano, Luisa Cogliati. 1976. *The Medieval Health Handbook, Tacuinum Sanitatis*. New York: George Braziller.

Arber, Agnes. 1986. *Herbals, their Origin and Evolution: A Chapter in the History of Botany 1470–1670*. Edited by William T. Stearn. Cambridge: Cambridge University Press.

Arbizzoni, Guido (ed.). 1986. *Costanzo Felici: Del'insalata e piante che in qualunque modo vengono per cibo del'homo*. Urbino: QuattroVenti.

Arrizabalaga, Jon, John Henderson, and Roger French. 1997. *The Great Pox: The French Disease in Renaissance Europe*. New Haven, CT: Yale University Press.

Ash, Harrison Boyd (trans.). 1941. *Lucius Junius Moderatus Columella On Agriculture*, vol. 1. Cambridge, MA: Harvard University Press.

Ashley, Sir William. 1921. "The Place of Rye in the History of English Food." *Economic Journal*, 31(123): 285–308.

Ashley, Sir William. 1928. *The Bread of our Forefathers*. Oxford: Oxford University Press.

Astill, G., and J. Langdon (eds). 1999. *Medieval Farming and Technology*. Leiden: Brill.

Austen, Ralph. 1653. *A Treatise of Fruit Trees*. Oxford: Tho. Robinson.

Bacon, Francis. 1625. "Of Gardens." In Francis Bacon, *Essayes or Counsels*. 2nd edn., 266–79. London: John Haviland for Hanna Barret.

Bammer, Anton. 1990. "A Peripteros of the Geometric Period in the Artemision of Ephesus." *Anatolian Studies*, 40: 137–60.

Barbieri, Costanza. 2014. *Le "Magnificenze" di Agostino Chigi. Collezioni e passioni antiquarie nella Villa Farnesina*. Rome: Accademia Nazionale dei Lincei.

Barocchi, Paola (ed.). 1971–7. *Scritti d'arte del cinquecento*, vol. 3. Milan and Naples: Riccardo Ricciardi Editore.

Barrera Osorio, Antonio. 2006. *Experiencing Nature: The Spanish American Empire and the Early Scientific Revolution*. Austin: University of Texas Press.

Bartels, Karlheinz. 2003. "Exotica et Transmarina: Arzneidrogen-handel und Apotheke in Mittelalter und früher Neuzeit." In Christoph Friedrich and Sabine Bernschneider-Reif (eds.), *Rosarium Litterarum. Beiträge zur Pharmazie- und Wissenschaftsgeschichte. Festschrift für Peter Dilg zum 65.* Geburtstag, 33–48. Eschborn: Govi-Verlag.

Bath, Michael. 2008. *Emblems for a Queen: The Needlework of Mary Queen of Scots*. London: Archetype Publications.

Batista González, Juan (ed.). 2000. *Pedro Sarmiento de Gamboa: Viajes al Estrecho de Magallanes*. Madrid: Dastin.

Bauhinus, Casparus (ed.). 1598. *Petri Andreae Matthioli ... Opera quae extant omnia, hoc est, Commentarii in VI libros Pedacii Dioscoridis Anazarbei de medica materia ....* Frankfurt am Main: N. Bassaeus.

Bauhinus, Casparus. 1623. *Pinax theatri botanici*. Basel: L. König.

Baumann, Brigitte, and Helmut Baumann. 2010. *Die Mainzer Kräuterbuch-Inkunabeln: "Herbarius Moguntinus" (1484), "Gart der Gesundheit" (1485), "Hortus sanitatis" (1491); wissenschaftshistorische Untersuchung der drei Prototypen botanisch-medizinischer Literatur des Spätmittelalters*. Stuttgart: Hiersemann.

Baumann, Felix Andreas. 1974. *Das Erbario Carrarese und die Bildtradition des Tractatus de herbis. Ein Beitrag zur Geschichte der Pflanzendarstellung im Übergang von Spätmittelalter zu Frührenaissance*. Bern: Benteli.

Bay, Miriam. 2019. "Cultivating Myth and Composing Landscape at the Villa d'Este, Tivoli". PhD thesis, University of Birmingham, UK.

Beecher, Donald. 2006. "The Legacy of John Frampton: Elizabethan Trader and Translator." *Renaissance Studies*, 20(3): 320–39.

Bell, G. D. H. 1987. "The History of Wheat Cultivation." In F. G. H. Lupton (ed.), *Wheat Breeding: Its Scientific Basis*, 31–49. London: Chapman.

Bellorini, Cristina. 2016. *The World of Plants in Renaissance Tuscany: Medicine and Botany*. London: Routledge.

Belon, Pierre. 1553. *Les observations de plusieurs singularitez et choses mémorables trouvées en Grèce, Asie, Judée, Égypte, Arabie et autres pays estranges*. Paris: Gilles Corrozet.

Belon, Pierre. 1558. *Les Remonstrances sur le default du labour et culture des plantes et de la cognoissance d'icelles*. Paris: Gilles Corrozet.

Beneš, Mirka. 2013. "Gardens and the Larger Landscape." In Elizabeth Hyde (ed.), *A Cultural History of Gardens in the Renaissance*, 187–216. London: Bloomsbury.

Benocci, Carla. 2005. "Il nuovo modello dei giardini di famiglie e la rinascita dei Chigi alla fine del Cinquecento: La Villa delle Volte a Siena rinnovata da Agostino dello Spedale della Scala." In Carla Benocci (ed.), *I giardini Chigi tra Siena e Roma: dal Cinquecento agli inizi* dell'Ottocento, 73–116. Siena: Protagon Editrice per la Fondazione Monte dei Paschi.

Benocci, Carla. 2010. *Villa Lante a Bagnaia tra Cinquecento e Seicento: la chiesa in forma di villa*. Vetralla: Davide Ghaleb Editore.

Bentivoglio, Enzo, and Simonetta Valtieri. 1976. *Santa Maria del Popolo a Roma*. Rome: Bardi Edizioni.

Benzoni, Gino. 2006. "Ludovico Sforza, detto il Moro, duca di Milano." In *Dizionario Biografico degli Italiani*, vol. 66. Rome: Treccani. Available online: www.treccani.it/enciclopedia/ludovico-sforza-detto-il-moro-duca-di-milano_%28Dizionario-Biografico%29/ (accessed April 4, 2021).

Berdan, Frances, and Patricia Anawalt. 1997. *The Essential Codex Mendoza*. Berkeley: University of California Press.

Besler, Basilius. 1613. *Hortus Eystettensis*. Nuremberg: Besler.

Bilimoff, Michèle. 2015. *Promenade dans les jardins disparus: les plantes au moyen âge d'après Les grandes heures d'Anne de Bretagne*. Rennes: Editions Ouest-France.

Birkestad, Helewyse de. n.d. "A Time for Change: New World Foods in Old World Menus." Available online: www.medievalcookery.com/helewyse/files/newworld.pdf (accessed April 6, 2021).

Black, Christopher, 2001. *Early Modern Italy: A Social History*. London: Routledge.

Bleichmar, Daniela, Paula De Vos, Kristin Huffine, and Kevin Sheehan (eds.). 2008. *Science in the Spanish and Portuguese Empires*. Stanford: Stanford University Press.

Bleichmar, Daniela. 2005. "Books, Bodies, and Fields: Sixteenth-Century Transatlantic Encounters with New World *Materia Medica*." In Londa Schiebinger and Claudia Swan (eds.), *Colonial Botany: Science, Commerce, and Politics in the Early Modern* World, 83–99. Philadelphia: University of Pennsylvania Press.

Blunt, Wilfrid, and Sandra Raphael. 1979. *The Illustrated Herbal*. London: Frances Lincoln.

Bock, Hieronymus. 1539. *New Kreütterbuch*. Strasbourg: Rihel.

Bock, Hieronymus. 1546. *New Kreuterbuch*. 2nd edn. Strasbourg: Rihel.

Bonicalzi, F., M. Guidetti, M. Marcochi, P. P. Poggio, and P. Repetto. 1978. *Il Rinascimento e le riforme: Storia d'Italia e d'Europa*, vol. 3. Milan: Jaca Book.

Borgean, A. P., and William J. Angus (eds). 2001. *The World Wheat Book*. Paris: Lavoisier.

Botsford, Keith. 1989. "Staples: Some Considerations on the Nature of Staples Especially in Regard to Italy." In Harlan Walker (ed.), *Oxford Symposium on Food and Cookery, Proceedings*, 1–4. London: Prospect Books.

Bowen, Lloyd (ed.). 2006. *Family and Society in Early Stuart Glamorgan: The Household Accounts of Sir Thomas Aubrey of Llantrithyd, c. 1565–1641*. Cardiff: South Wales Record Society.

Boxer, C. R. 1963. *Two Pioneers of Tropical Medicine: Garcia d'Orta and Nicolás Monardes*. London: Wellcome Historical Medical Library. Available online: archive.org/details/b20457248 (accessed April 6, 2021).

Briant, Pierre. 2002. *From Cyrus to Alexander: A History of the Persian Empire*. Translated by Peter T. Daniels. Winona Lake, IN: Eisenbrauns.

Britten, James (ed.). 1881. *The Names of Herbes, by William Turner*. London: English Dialect Society.

Brown, A. G., I. Meadows, S. D. Turner, and D. J. Mattingly. 2001. "Roman Vineyards in Britain: Stratigraphic and Palynological Data from Wollaston in the Nene Valley, England." *Antiquity*, 75: 745–57.

Brown, David Alan, and Sylvia Ferino-Pagden. 2006. *Bellini, Giorgione, Titian and the Renaissance of Venetian Painting*. Washington, DC: National Gallery of Art.

Brunfels, Otto. 1530–6. *Herbarum vivae eicones*. 3 parts. Strasbourg: Johann Schott.

Brunfels, Otto. 1532–7. *Contrafayt Kräuterbuch*. 2 parts. Strasbourg: Johann Schott.

Burke, Peter. 1992. *The Fabrication of Louis XIV*. New Haven, CT: Yale University Press.

Burrows, Donald, and Rosemary Dunhill (eds.). 2002. *Music and Theatre in Handel's World: The Family Papers of James Harris*. Oxford: Oxford University Press.

Byrne, R., and J. H. McAndrews. 1985. "Pre-Columbian Purslane (*Portulaca oleracea* L.) in the New World." *Nature*, 253: 726–7.

Calzolari, Francesco. 1566. *Il viaggio di Monte Baldo della magnifica città di Verona*. Venice: Vincenzo Valgrisio.

Camerarius, Joachim. 1586. *De plantis epitome utilissima Petri Andreae Matthioli Senensis*. Frankfurt am Main: Feyerabend.

Camerarius, Joachim. 1590. *Symbolorum et emblematum ex re herbaria desumtorum centuria una*. Nuremberg.

Campbell, B. M. S., and M. Overton. 1991. *Land, Labour and Capital*. Manchester: Manchester University Press.

Canby, Sheila R. 2009. *Shah 'Abbas: The Remaking of Iran*. London: British Museum Press.

Caneva, Giulia. 1992. *Il Mondo di Cerere nella Loggia di Psiche*. Rome: Fratelli Palombi.

Caneva, Giulia, and G .M. Carpaneto (eds.). 2011. *Raffaello e l'immagine della natura*. Cinello Balsamo, Milan: Silvana Editoriale.

Cañizares Esguerra, Jorge. 2006. *Nature, Empire, and Nation: Explorations of the History of Science in the Iberian World*. Stanford: Stanford University Press.

Carasco, David (trans.). 2008. *The History of the Conquest of New Spain by Bernal Diaz del Castillo*. Albuquerque: University of New Mexico Press.

Cardanus, Hieronymus [Gerolamo Cardano]. 1550. *De subtilitate libri xxi*. Nuremberg: Ioh. Petreius.

Cartier, Jacques. 1545. *Brief recit de la navigation faicte es ysles de Canada*. Paris: Ponce Roffet dict Faucheur, & Anthoine le Clerc freres.

Cartwright, James J. (ed.). 1875. *The Memoirs of Sir John Reresby, of Thrybergh, bart., M. P. for York, &c., 1634–1689*. London: Longmans Green.

Casale, Gerardo. 1991. *Giovanna Garzoni, "insigne miniatrice" 1600–1670*. Rome: Jandi Sapi.

Casale, Gerardo (ed.). 1996. *Gli Incanti del Iride: Giovanna Garzoni pittrice nel Seicento*. San Severino Marche: Silvana.

Cassell, Anthony K., and Victoria Kirkham. 1991. *Diana's Hunt (Caccia di Diana): Boccaccio's First Fiction*. Philadelphia, PA: The University of Pennsylvania Press.

Celani, Enrico. 1902. "Sopra un Erbario di Gherardo Cibo conservato nella R. Biblioteca Angelica di Roma." *Malpighia*, 16: 181–226.

Cellauro, Louis. 2003. "Iconographical Aspects of the Renaissance Villa and Garden: Mount Parnassus, Pegasus and the Muses." *Studies in the History of Gardens and Designed Landscapes*, 23(1): 42–57.

Čermáková, Lucie, and Jana Černá. 2017. "Naked in the Old and the New World: Differences and Analogies in Descriptions of European and American herbae nudae in the Sixteenth Century." *Journal of the History of Biology*, 50: 1–38.

Cesalpino, Andrea. 1583. *De plantis libri XVI*. Florence: G. Marescottus.

Champlain, Samuel. 1613. *Voyages et descouvertures faites en la Nouvelle France*. Paris: Jean Berjon.

Ciancio, Luca. 2016. "The Many Gardens—Real, Symbolic, Visual—of Pietro Andrea Mattioli." In Juliette Ferdinand (ed.), *From Art to Science. Experiencing Nature in the European Garden, 1500–1700*, 34–45. Treviso: ZeL Edizioni.

Cieza de León, Pedro de. 1553. *Primera parte dela chrónica del Perú*. Seville.

Ciliberto, Michele, and Pier Davide Accendere (eds.). 2018. *Niccolò Machiavelli: Tutte le opere*. Florence: Bompiani.

Circa Instans. 1497. In *Practica Jo. Serapionis dicta Breviarium; Liber Serapionis de simplici medicina; Liber de simplici medicina dicta Circa instans; Practica Platearii*. Venice: mandato Octaviani Scoti per Bonetum Locatellum. Available online: daten.digitale-sammlungen.de/0006/bsb00061068/images/index.html (accessed April 6, 2021).

Clark, Peter (ed.). 1985. *The European Crisis of the 1590s*. London: Allen and Unwin.

Clemens, Raymond (ed.). [With an introduction by Deborah Harkness]. 2016. *The Voynich Manuscript*. New Haven: Yale University Press.

Clunas, Craig. 1996. *Fruitful Sites: Garden Culture in Ming Dynasty China*. London: Reaktion Books.

Clusius, Carolus (trans.) [Charles de l'Escluse]. 1557. *Histoire des plantes ... de Rembert Dodoens*. Antwerp: Jean Lo.

Clusius, Carolus (trans.). 1567. *Aromatum et simplicium aliquot medicamentorum apud Indos nascentium historia: ante biennium quidem Lusitanica linqua per dialogos conscripta, D. Garcia ab Horto auctore*. Antwerp: Plantin.

Clusius, Carolus (trans.). 1574. *Aromatum et simplicium aliquot medicamentorum apud Indos nascentium historia, D. Garcia ab Horto auctore*. 2nd edn. Antwerp: Plantin.

Clusius, Carolus. 1576. *Rariorum aliquot stirpium per Hispanis observatorum historia*. Antwerp: Plantin.

Clusius, Carolus (trans.). 1582. *Christophori a Costa ... Aromatum et medicamentorum in Orientali India nascentium liber*. Antwerp: Plantin.

Clusius, Carolus. 1583a. *Rariorum aliquot stirpium per Pannoniam, Austriam, and vicinas quasdam provincias observatarum historia*. Antwerp: Plantin.

Clusius, Carolus (trans.). 1583b. *Remberti Dodonaei Stirpium historiae pemptades sex*. Antwerp: Plantin.

Clusius, Carolus (trans.). 1589. *Petri Bellonii De neglecta stirpium cultura atque earum cognitione libellus*, Antwerp: Plantin.

Clusius, Carolus. 1601. *Rariorum plantarum historia*. Antwerp: Plantin-Moretus. [Roman page numbers refer to the appendix at the the end of the volume.]

Clusius, Carolus. 1605. *Exoticorum libri decem*. Leiden: Raphelengius.

Clusius, Carolus. 1611. *Curae posteriores*. Leiden: Raphelengius.

Collingham, Lizzie. 2005. *Curry: A Tale of Cooks and Conquerors*. Oxford: Oxford University Press

Collins, Minta. 2000. *Medieval Herbals. The Illustrative Tradition*. London: British Library.

Colonna, Francesco. 1499. *Hypnerotomachia Poliphili*. Venice: Aldus Manutius.

Cook, Harold. 2007. *Matters of Exchange. Commerce, Medicine, and Science in the Dutch Golden Age*. New Haven: Yale University Press.

Cook, Harold, and Timothy Walker. 2013. "Circulation of Medicine in the Early Modern Atlantic World." *Social History of Medicine*, 26(3): 337–51.

Cox-Rearick, Janet. 1984. *Dynasty and Destiny in Medici Art: Pontormo, Leo X, and the Two Cosimos*. Princeton: Princeton University Press.

Crawford, Matthew James. 2016. *The Andean Wonder Drug: Cinchona Bark and Imperial Science in the Spanish Atlantic, 1630–1800*. Pittsburgh: University of Pittsburgh Press.

Crollius, Oswaldus [Oswald Croll]. 1609. *Basilica Chymica ... additus est tractatus novus De signaturis rerum internis*. Frankfurt am Main: apud Claudium Marnium.

Crosby, Alfred W. 1972. *The Columbian Exchange: Biological and Cultural Consequences of 1492*. Westport, CT: Greenwood Press.

Crosby, Alfred W. 1987. *The Columbian Voyages, the Columbian Exchange, and their Historians*. Washington, DC: American Historical Association.

Cugnoni, Giuseppe. 1879. "Agostino Chigi il magnifico." *Archivio della Società Romana di Storia Patria*, 2: 37–83.

Cummins, John. 1992. *The Voyage of Christopher Columbus: Columbus' Own Journal of Discovery Newly Restored and Translated*. New York: St Martin's Press.

Cunningham, A., and Ole Peter Grell. 2000. *The Four Horsemen of the Apocalypse: Religion, War, Famine and Death in Renaissance Europe*. Cambridge: Cambridge University Press.

Cuttler, C. D. 1989. "Errata in Netherlandish Art: Jan Mostaert's 'New World' Landscape." *Simiolus*, 19: 191–7.

Dalby, Andrew. 2001. "Christopher Columbus, Gonzalo Pizarro, and the Search for Cinnamon." *Gastronomica*, 1(2): 40–9.

Dalby, Andrew. 2003. *Food in the Ancient World from A to Z*. London: Routledge.

Dalby, Andrew, and Maureen Dalby. 2012. *The Shakespeare Cookbook*. London: British Museum Press.

D'Ancona, Mirella Levi. 1977. *The Garden of the Renaissance: Botanical Symbolism in Italian Painting*. Florence: Leo S. Olschki.

Davies, Arthur. 1956. "The English Coast on the Map of Juan de la Cosa." *Imago Mundi*, 13.

Davies, Norman. 2005. *God's Playground: A History of Poland*. Oxford: OUP.

De Beer, E. S. (ed.). 1955. *The Diary of John Evelyn*. 6 vols. Oxford: Clarendon Press.

De Divitiis, Bianca. 2015. "Giuliano da Sangallo in the Kingdom of Naples: Architecture and Cultural Exchange." *Journal of the Society of Architectural Historians*, 74(2): 152–78.

De Toni, Ettore (ed.). 1940. *Pietro Antonio Michiel: I cinque libri di piante*. Venice: Carlo Ferrari.

De Vos, Paula. 2006. "The Science of Spices: Empiricism and Economic Botany in the Early Spanish Empire." *Journal of World History*, 17(4): 399–427.

Dembinska, Maria. 1999. *Food and Drink in Medieval Poland*. Philadelphia: University of Pennsylvania.

Denzer, Jörg. 2005. *Die Konquista der Augsburger Welser-Gesellschaft in Südamerika (1528–1556). Historische Rekonstruktion, Historiografie und lokale Erinnerungskultur*. Munich: C. H. Beck.

Dewar, Michael J. 2013. *Leisured Resistance: Villas, Literature and Politics in the Roman World*. London: Bloomsbury.

Di sana pianta. 1988. *Di sana pianta: Erbari e taccuini di sanità. Le radici storiche della nuova farmacologia*. Modena: Edizioni Panini.

Dix, Brian. 2011. "Experiencing the Past: The Archaeology of Some Renaissance Gardens." *Renaissance Studies*, 25(1): 151–82.

Dodart, Denis. 1676. *Mémoires pour servir à l'histoire des plantes*. Paris: Imprimerie Royale.

Dodoens, Rembert. 1554. *Cruijdeboeck*. Antwerp: Jan vander Loe.

Dodoens, Rembert [Rembertus Dodonaeus]. 1568. *Florum et coronariarum odoratarumque nonnullarum herbarum historia*. Antwerp: Plantin.

Dodoens, Rembert [Rembertus Dodonaeus]. 1569. *Florum et coronariarum odoratarumque nonnullarum herbarum historia*. 2nd edn. Antwerp: Plantin.

Dreiss, Meredith L., and Sharon Greenhill. 2008. *Chocolate: Pathway to the Gods*. University of Arizona Press.

Dryden, John (trans.). 1697. *The Works of Virgil, containing his Pastorals, Georgics, and Æneis, translated into English verse*. London: Jacob Tonson.

Du Bartas, Guillaume de Saluste. 1589. *La Seconde Sepmaine, reveüe*. Geneva: Jaques Chouet.

Duret, Claude. 1605. *Histoire admirable des plantes et herbes esmerveillables et miraculeuses en nature*. Paris: Nicolas Buon.

Duroselle-Melish, Caroline. 2016. "Centre and Periphery? Relations between Frankfurt and Bologna in the Transnational Book Trade of the 1600s." In Matthew Maclean and Sara Barker (eds.), *International Exchange in the Early Modern Book World*, 31–58. Leiden: Brill.

Duthie, Ruth. 1990. "The Planting of Some Seventeenth-Century Flower Gardens." *Garden History*, 18(2): 77–102.

Earle, Peter, 1989. *The Making of The Middle Class*. London: Methuen.

Eburne, Andrew. 2008. "The Passion of Sir Thomas Tresham: New Light on the Gardens and Lodge at Lyveden." *Garden History*, 36(1): 114–34.

Edelen, Georges (ed.). 1994. *William Harrison: The Description of England, 1587*. Washington: Folger Shakespeare Library.

Egerton, Frank. 2012. *Roots of Ecology: Antiquity to Haeckel*. Berkeley and Los Angeles: University of California Press.

Egmond, Florike. 2010. *The World of Carolus Clusius. Natural History in the Making, 1550–1610*. London: Pickering and Chatto.

Egmond, Florike. 2015. "Figuring Exotic Nature in Sixteenth-Century Europe: Garcia de Orta and Carolus Clusius." In Palmira Fontes Da Costa (ed.), *Medicine, Trade and Empire: Garcia de Orta's* Colloquies on the Simples and Drugs of India *(1563) in Context*, 167–93. Farnham: Ashgate.

Egmond, Florike. 2017. *Eye for Detail: Images of Plants and Animals in Art and Science, 1500–1630*. London: Reaktion Books.

Egmond, Florike. 2018. "Into the Wild: Botanical Fieldwork in the Sixteenth Century." In Arthur MacGregor (ed.), *Naturalists in the Field: Collecting, Recording and Preserving the Natural World from the Fifteenth to the Twenty-First Century* N, 116–211. Leiden and Boston: Brill.

Egmond, Florike, Paul Hoftijzer, and Rob Visser (eds.). 2007. *Carolus Clusius: Towards a Cultural History of a Renaissance Naturalist*. Amsterdam: Edita.

Elstob, Ivy (ed.). 1933. *The Garden Book of Sir Thomas Hanmer: Printed from the MS Volume of 1659*. London: Gerald Howe.

Epstein, E. P. 1998. "The Peasantries of Italy, 1350–1750." In Tom Scott (ed.), *The Peasantries of Europe*, 75–110. London: Longman.

Estienne, Charles. 1564. *L'Agriculture et maison rustique*. Paris: Jaques Du Puis.

Estienne, Charles. 1570. *L'Agriculture et maison rustique*. [Enlarged by Jean Liebaut and others.] Paris: Jaques Du Puys.

Everaert, J. G., J. E. Mendes Ferrão, and M. Cândida Liberato (eds.). 2001. *Manuel Godinho de Erédia: Summa de árvores e plantas da Índia intra Ganges (1612)*. Lisbon: CNCDP.

Everitt, Alan. 1967. "The Marketing of Agricultural Produce." In Joan Thirsk (ed.), *Agrarian History of England and Wales*, vol. IV, 466–592. Cambridge: Cambridge University Press.

Fanshawe, Lady Anne. 1907. *The Memoirs of Ann Lady Fanshawe, wife of the Right Honble. Sir Richard Fanshawe, Bart., 1600–72*. [Edited by Herbert Charles Fanshawe.] London: John Lane.

Fantuzzi, Giovanni. 1774. *Memorie della vita di Ulisse Aldrovandi*. Bologna.

Fausti, Daniela (ed.). 2004. *La complessa scienza dei semplici. Atti delle celebrazioni per il V centenario della nascita di Pietro Andrea Mattioli* [Atti dell'Accademia delle Scienze di Siena detta de' Fisiocritici, Supplemento al tomo XX, serie XV], 133–47. Siena: Accademia dei Fisiocritici Onlus.

Fausti, D. 2010. "Su alcune traduzioni cinquecentesche di Dioscoride: da Ermolao Barbaro a Pietro Andrea Mattioli." In Ivan Garofalo (ed.), *Sulla Tradizione Indiretta Dei Testi Medici Greci: Le Traduzioni*, 181–205. Pisa: Fabrizio Serra.

Ferguson, Dean T. 2004. "Nightsoil and the 'Great Divergence': Human Waste, the Urban Economy, and Economic Productivity, 1500–1900." *Journal of Global History*, 9(3): 379–402.

Ferino-Pagden, Sylvia. 2007. *Arcimboldo 1526–1593*. Milan: Skira.

Fernandez de Ybarra, A. M. 1907. "The Letter of Dr. Diego Alvarez Chanca, Dated 1494, Relating to the Second Voyage of Columbus to America." *Smithsonian Miscellaneous Collections*, 48: 428–57. Available online: repository.si.edu/handle/10088/26153 (accessed April 6, 2021).

Ferrarius, Ioannes Baptista [Giovanni Battista Ferrari]. 1633. *De florum cultura*. Rome: Stephanus Paulinus.

Ferrarius, Ioannes Baptista. 1646. *Hesperides, sive De malorum aureorum cultura et usu libri quatuor*. Rome: Hermannus Scheus.

Finan, John. 1948. "Maize in the Great Herbals." *Annals of the Missouri Botanical Garden*, 35(2): 149–91.

Findlen, Paula. 1994. *Possessing Nature: Museums, Collecting and Scientific Culture in Early Modern Italy*. Berkeley: University of California Press.

Findlen, Paula. 1999. "The Formation of a Scientific Community: Natural History in Sixteenth-Century Italy." In Anthony Grafton and Nancy Siraisi (eds.), *Natural Particulars. Nature and the Disciplines in Renaissance Europe*, 369–400. Cambridge, MA: MIT Press.

Finlay, Robert. 2010. *The Pilgrim Art: Cultures of Porcelain in World History*. Berkeley: University of California Press.

Fontes da Costa, Palmira (ed.). 2015. *Medicine, Trade and Empire. Garcia de Orta's* Colloquies on the Simples and Drugs of India *(1563) in Context*. Farnham: Ashgate.

Fontes da Costa, Palmira, and Henrique Leitão. 2008. "Portuguese Imperial Science: A Historiographical Review." In Daniela Bleichmar, Paula De Vos, Kristin Huffine, and Kevin Sheehan (eds.), *Science in the Spanish and Portuguese Empires*, 35–53. Stanford: Stanford University Press.

Foucault, Michel. 1966. *Les Mots et les choses*. Paris: Gallimard.

Foucault, Michel. 1970. *The Order of Things: An Archaeology of the Human Sciences*. London and New York: Routledge.

Fragoso, Juan. 1572. *Discursos de las cosas aromáticas, árboles y frutales, y de otras muchas medicinas simples que se traen de la India Oriental, y siruen al uso de medicina*. Madrid: Sebastian Yvañez.

Frame, Donald M. (trans.). 1943. *The Complete Works of Montaigne: Essays, Travel Journal, Letters*. Stanford: Stanford University Press.

Frampton, John. 1577. *Joyfull newes out of the newe founde worlde*. London.

Frampton, John. 1580. *Joyfull newes out of the newfound world*. London.

Francis, Jill. 2008. "Order and Disorder in the Modern Garden." *Garden History*, 36(1): 22–35.

Francis, Jill. 2018. *Gardens and Gardening in Early Modern England and Wales*. New Haven, CT: Yale University Press.

Fraser, Mike, and Liz Fraser. 2011. *The Smallest Kingdom: Plants and Plant Collectors at the Cape of Good Hope*. London: Kew Publishing.

Free, Melissa, and Alan Robock. 1999. "Global Warming in the Context of the Little Ice Age." *Journal of Geophysical Research*, 104(D16): 19057–70.

Fuchs, Leonhart. 1542. *De historia stirpium commentarii insignes*. Basel: Isingrin.

Funk, Holger. 2016. "The First Printed Latin Editions of Dioscorides's *De materia medica* (1478, 1512): An Inventory-Based Re-evaluation." *Archives of Natural History*, 43(2): 237–54.

Fussell, George, 1966. *Farming Technique from Prehistoric to Modern Times*. London: Pergamon.

Galinat, Walton C. 1992. "Maize: Gift from America's First Peoples." In Nelson Foster and Linda S. Cordell (eds.), *Chilies to Chocolate: Food the Americas Gave the World*, 47–60. Tucson: University of Arizona Press.

Gardiner, Richard. 1599. *Profitable Instructions for the Manuring, Sowing and Planting of Kitchin Gardens*. London.

Gart der Gesundheit. 1485. Mainz: Schöffer.

Gentilcore, David. 2010. *Pomodoro! A History of the Tomato in Italy*. New York: Columbia University Press.

Gentilcore, David. 2017. "The Impact of New World Plants, 1500–1800: The Americas in Italy." In Elizabeth Horodowich and Lia Markey (eds.), *The New World in Early Modern* Italy, *1492–1750*, 190–205. Cambridge: Cambridge University Press.

Gerard, John. 1597. *The Herball, or Generall Historie of Plants*. London: John Norton.

Gesnerus, Conradus (ed.). 1561. *Valerii Cordii Simesusii annotationes*. Strasbourg.

Gharipour, Mohamma. 2013. *Persian Gardens and Pavilions: Reflections on History, Poetry, and the Arts*. London and New York: I.B. Tauris.

Giesecke, Annette (ed.). 2022. *A Cultural History of Plants in Antiquity*. London: Bloomsbury.

Gimmel, Millie. 2008. "Reading Medicine in the Codex de la Cruz Badiano." *Journal of the History of Ideas*, 69(2): 169–92.

Givens, Jean A. 2006. "Reading and Writing the Illustrated *Tractatus de herbis*, 1280–1526." In Jean A. Givens, Karen Reeds, and Alain Touwaide (eds.), *Visualizing Medieval Medicine and Natural History, 1200–1550*, 115–45. Aldershot: Ashgate.

Givens, Jean A., Karen Reeds, and Alain Touwaide (eds.). 2006. *Visualizing Medieval Medicine and Natural History, 1200–1550*. Aldershot: Ashgate.

Godwin, Joscelyn (trans.). 1999. Francesco Colonna, *Hypnerotomachia Poliphili*. London: Thames and Hudson.

Goldgar, Anne. 2007. *Tulipmania: Money, Honor, and Knowledge in the Dutch Golden Age*. DG Chicago: University of Chicago Press.

Gonzáles Bueno, Antonio. 2007. "El Descubrimiento de la Naturaleza del Nuevo Mundo: Las Plantas Americanas en la Europa del siglo XVI." *Circumscribere: International Journal for the History of Science*, 2: 10–25.

Goody, Jack. 1993. *The Culture of Flowers*. Cambridge: Cambridge University Press.

Googe, Barnabe (trans.). 1577. *Foure Bookes of Husbandry, collected by M. Conradus Heresbachius*. London: printed by Richard Watkins. Available online: quod.lib.umich.edu/e/eebo/A03069.0001.001/ (accessed April 6, 2021).

Gray, Todd (ed.). 1996. *Devon Household Accounts, 1627–59, Part 2: Henry, Fifth Earl of Bath, and Rachel, Countess of Bath, of Tawstoc*. Exeter: Devon and Cornwall Record Society.

Greene, Edward Lee. 1983. *Landmarks of Botanical History*, vol. 2. Edited by Frank N. Egerton et al. Stanford: Stanford University Press.

Grewe, Rudolf. 1987. "The Arrival of the Tomato in Spain and Italy: Early Recipes." *Journal of Gastronomy*, 3: 67–81.

Grigson, Jane (ed.). 1989. *Giacomo Castelvetro: The Fruit, Herbs and Vegetables of Italy*. Totnes: Prospect Books.

G[rimeston], E. (trans.). 1604. *José de Acosta: The Naturall and Morall Historie of the East and West Indies*. London.

Gronovius, Johannes Fredericus. 1755. *Flora orientalis, sive recensio plantarum, quas ... Leonhardus Rauwolffus ... annis 1573–1574 et 1575 in Syria, Arabia, Mesopotamia, Babylonia, Assyria, Armenia and Iudaea crescentes observavit*. Leiden: De Groot.

Grove, Richard H. 1991. "The Transfer of Botanical Knowledge between Asia and Europe 1498–1800." *Journal of the Japan-Netherlands Institute*, 3: 160–76.

Grove, Richard H. 1995. *Green Imperialism. Colonial Expansion, Tropical Island Edens and the Origins of Environmentalism 1600–1860*. Cambridge: Cambridge University Press.

Grove, Richard H. 1996. "Indigenous Knowledge and the Significance of South-West India for Portuguese and Dutch Constructions of Tropical Nature." *Modern Asian Studies*, 30: 121–43.

Gruzinski, Serge. 1992. *Painting the Conquest. The Mexican Indians and the European Renaissance*. Paris: Flammarion.

Guerra, Francisco. 1966. "Drugs from the Indies and the Political Economy of the Sixteenth Century." *Analecta Medico-Historica*, 1: 29–54.

Guerrini, Luigi. 2006. *I Trattati Naturalistici di Federico Cesi*. Rome: Accademia dei Lincei.

Gunther, R. T. 1922. *Early British Botanists and their Gardens*. Oxford: Oxford University Press. Available online: archive.org/details/earlybritishbota1922gunt (accessed April 6, 2021).

Hainsworth, Peter (trans.). 2010. *The Essential Petrarch*. Indianapolis: Hackett.

Hakluyt, Richard (ed.). 1598–1600. *The Principal Navigations, Voyages, Traffiques and Discoveries of the English Nation*. 2nd edn. London: George Bishop.

Hanauer, James Edward. 1907. "About Coffee." In *Folk-lore of the Holy Land*, 291–6. London: Duckworth.

Hardie, Alison. 2004. "The Awareness Garden of Wang Shiheng in Yizhen." *Studies in the History of Gardens & Designed Landscapes*, 24(4): 272–9.

Hardie, Alison, and Duncan M. Campbell (eds.). 2020. *The Dumbarton Oaks Anthology of Chinese Garden Literature*. Washington, DC: Dumbarton Oaks.

Hardy, Gavin, and Laurence Totelin. 2016. *Ancient Botany*. London: Routledge.

Hargett, James M. (ed.). 2010. *Fan Chengda: Treatises of the Supervisor and Guardian of the Cinnamon Sea*. Seattle: University of Washington Press.

Hargett, James M. (trans.). 2010. "An Account of the Tree Peonies of Luoyang." In Alison Hardie and Duncan M. Campbell (eds.), *The Dumbarton Oaks Anthology of Chinese Garden Literature*, 133–44. Washington, DC: Dumbarton Oaks.

Hariot, Thomas. 1588. *A briefe and true report of the new found land of Virginia, of the commodities there found and to be raysed*. London: Theodore De Bry.

Hariot, Thomas. 1590. *A briefe and true report of the new found land of Virginia, of the commodities and of the nature and manners of the naturall inhabitants*, 2nd edn. Frankfurt: Theodore de Bry. Available online: www.gutenberg.org/files/4247/4247-h/4247-h.htm (accessed April 6, 2021).

Harkness, Deborah. 2007. *The Jewel House: Elizabethan London and the Scientific Revolution*. New Haven: Yale University Press.

Harris, Steven J. 1996. "Confession-Building, Long-Distance Networks, and the Organization of Jesuit Science." *Early Science and Medicine*, 1(3): 287–318.

Harris, Steven J. 2005. "Jesuit Scientific Activity in the Overseas Missions, 1540–1773." *Isis*, 96: 71–9.

Harrison, William. 1577. "Of the Foode and Diet of the Englishe." In Raphael Holinshed, *Chronicles*, vol. 1, 94–6. London: John Harrison. Available online: english.nsms.ox.ac.uk/holinshed/texts.php?text1=1577_0078 (accessed April 6, 2021).

Harrison, William. 1587. "Of the Aire, Soile, and Commodities of this Iland; Of the Food and Diet of the English; Of Gardens and Orchards." In Raphael Holinshed et al., *Chronicles*. 2nd edn, vol. 1, 108–12, 165–71, 208–10. London: John Harison. Available online: english.nsms.ox.ac.uk/holinshed/texts.php?text1=1587_0072, english.nsms.ox.ac.uk/holinshed/texts.php?text1=1587_0085, and english.nsms.ox.ac.uk/holinshed/texts.php?text1=1587_0099 (accessed April 6, 2021).

Hartlib, Samuel. 1650. *A Discours of Husbandrie used in Brabant and Flanders*. London: Samuel Hartlib.

Hartlib, Samuel. 1655. *Samuel Hartlib his Legacy of Husbandry*. London: Richard Wodnothe.

Hartt, Frederick, and David G. Wilkins. 2013. *History of Italian Renaissance Art: Painting, Sculpture and Architecture*. 5th edn. London: Thames and Hudson.

Harvey, John. 1972. *Early Gardening Catalogues*. London: Phillimore.

Harvey, John. 1974. *Early Nurserymen*. London: Phillimore.

Hawkes, J. G., and J. Francisco-Ortega. 1993. "The Early History of the Potato in Europe." *Euphytica*, 70(1–2): 1–7

Heckscher, William S. 1992. "Andrea Alciati's Laurel Tree and its Symbolic Traditions." *Emblematica*, 6: 209–10.

Heiser, Charles B. 1951. "The Sunflower among the North American Indians." *Proceedings of the American Philosophical Society*, 95: 432–48.

Hendrix, Lee, and Thea Vignau-Wilberg (eds.). 1992. *Mira calligraphiae monumenta: A Sixteenth-Century Calligraphic Manuscript Inscribed by Georg Bocskay and Illuminated by Joris Hoefnagel*. Los Angeles: Getty Publications.

Hendry, George W. 1934. "The Source Literature of Early Plant Introduction into Spanish America." *Agricultural History*, 8: 64–71.

Henrey, Blanche. 1975. *British Botanical and Horticultural Literature before 1800*. Oxford: Oxford University Press.

Hentznerus, Paulus [Paul Hentzner]. 1612. *Itinerarium Germaniae, Galliae, Angliae, Italiae*. Nuremberg.

Herbarium Apulei Platonici. *c*. 1481. Rome. Available online: daten.digitale-sammlungen.de/~db/0006/bsb00064270/images/ (accessed April 6, 2021).

Herbert, Thomas. 1638. *Some Yeares Travels into Divers Parts of Asia and Afrique*. London.

Herde, Simone, and Tilmann Walter. 2010. "Neues zur Biographie des Augsburger Artztes und Orientreisende Leonhard Rauwolf (1535?–1596)." *Sudhoffs Archiv*, 94(2): 129–56.

Heresbachius, Conradus [Conrad Heresbach]. 1570. *Rei rusticae libri quattuor*. Cologne: Ioannes Birckmannus.

Heresbachius, Conradus. 1573. *Rei rusticae libri quatuor*. Cologne: apud viduam Ioannis Birckmanni.

Herlihy, Patricia. 2012. *Vodka: A Global History*. London: Reaktion Books.

Hernández, Franciscus et al. 1651. *Rerum medicarum Novae Hispaniae thesaurus … ex Francisci Hernandez Novi Orbis medici primarii relationibus*. Rome: Mascardi.

Heurtel, Pascale, and Michèle Lenoir (ed.). 2016. *Les Vélins du Muséum national d'Histoire naturelle*. Paris: Muséum national d'histoire naturelle.

Hinds, A. B. (trans.). 1927. *The Lives of the Painters, Sculptors & Architects by Giorgio Vasari.* 4 vols. London: Dent.

Historical Manuscripts Commission. 1942. *Report on the MSS of Lord De L'Isle and Dudley Preserved at Penshurst Place*, vol. 4. London: HMSO.

Hobhouse, Penelope. 2004. *Gardens of Persia.* Carlsbad, CA: Kale Press.

Hoeniger, Cathleen. 2006. "The Illuminated *Tacuinum Sanitatis* Manuscripts from Northern Italy circa 1380–1400. Sources, Patrons, and the Creation of a New Pictorial Genre." In Jean A. Givens, Karen Reeds, and Alain Touwaide (eds.), *Visualizing Medieval Medicine and Natural History, 1200–1550*, 51–81. Aldershot: Ashgate.

Hornsey, I. S. 2003. *A History of Beer and Brewing.* Cambridge: Royal Society of Chemistry.

Hosking, Geoffrey. 2001. *Russia and the Russians.* Cambridge, MA: Harvard University Press.

How, Walter Wybergh, and Joseph Wells. 1912. *A Commentary on Herodotus.* Oxford: Clarendon Press.

Hulton, Paul, and David Beers Quinn. 1964. *The American Drawings of John White, 1577–1590, with Drawings of European and Oriental Subjects.* London: British Museum.

Hunt, John Dixon. 2000. *Greater Perfections: The Practice of Garden Theory.* Philadelphia: University of Pennsylvania Press.

Hunt, John Dixon, and Peter Ellis (eds.). 1975. *The Genius of the Place: The English Landscape Garden.* London: Elek.

Hurka, Herbert, and Barbara Neuffer. 2011. "Geschichte und Bedeutung von Herbarien." *Osnabrücker Naturwissenschaftliche Mitteilungen*, 37: 115–34.

Husselby, Jill, and Paula Henderson. 2002. "Location, Location, Location, Location! Cecil House in the Strand." *Architectural History*, 45: 159–93.

Hutton, Charles (trans.). 1803. *Recreations in Mathematics and Natural* Philosophy ... *first composed by M.* Ozanam ... *greatly enlarged, in a new edition, by the celebrated M.* Montucla, *and* now translated into English, and improved with many Additions and Improvements. London: G. Kearsley.

Hyll, Thomas. 1563. *A Most Briefe and Pleasaunt Treatyse, teaching howe to dress, sowe, and set a garden.* London. [There had been an earlier edition. The 1574 edition, titled *The Profitable Arte of Gardening*, is available online: babel.hathitrust.org/cgi/pt?id=uc1.31175035217820 (accessed April 6, 2021).

Hyll, Thomas. 1577. Dydymus Mountaine, *The Gardeners Labyrinth.* London: Henry Bynneman. Available online: archive.org/details/CAT10889940 (accessed April 6, 2021).

Innes, Mary M. (trans.). 1955. *Ovid: Metamorphoses.* Harmondsworth: Penguin.

Jacobus de Vitriaco [Jacques de Vitry]. 1597. *Libri duo quorum prior orientalis ... alter occidentalis historiae nomine inscribitur.* Douai: Balthazar Bellerus.

Jacques, David. 1999. "The *Compartment* System in Tudor England." *Garden History*, 27(1): 32–53.

Janick, Jules, and Giulia Caneva. 2005. "The First Image of Maize in Europe." *Maydica*, 50: 71–80.

Janick, Jules, and Harry S. Paris. 2006. "The Cucurbit Images (1515–1518) of the Villa Farnesina, Rome." *Annals of Botany*, 97(2): 165–76.

Jeanneret, Michel. 2001. *Perpetual Motion: Transforming Shapes in the Renaissance from da Vinci to Montaigne.* Translated by Nidra Poller. Baltimore: Johns Hopkins University Press.

Johnson, Thomas (ed.). 1633. *The Herball, or Generall Historie of Plants, gathered by John Gerarde.* London: Adam Islip Ioice Norton and Richard Whitakers.

Jones, John Winter, and George Percy Badger (eds). 1863. *The Travels of Ludovico di Varthema.* London: Hakluyt Society.

Justiniani, Augustinus [Agostino Giustiniani]. 1516. *Octaplus psalterii*. Genoa. Available online: archive.org/details/bub_gb_N7Gk1tKBXn8C (accessed April 6, 2021).

Kaufmann, Thomas DaCosta. 2010. *Arcimboldo, Visual Jokes, Natural History and Still Life Painting*. Chicago: University of Chicago Press.

Kayser, Wolfgang. 1963. *The Grotesque in Art and Literature*. Translated by Ulrich Weisstein. Bloomington: Indiana University Press.

Keeley, Edmund, and Philip Sherrard (trans.). 1967. *George Seferis: Collected Poems 1924–1955*. Princeton, NJ: Princeton University Press.

Kemp, Martin J. 1985. "Leonardo da Vinci: Science and the Poetic Impulse." *Journal of the Royal Society of Arts*, 133(5343): 196–214.

Kiple, K. F., and K. C. Ornelas (eds.). 2000. *The Cambridge World History of Food*. Cambridge: Cambridge University Press.

Kircherus, Athanasius. 1665. *Mundus subterraneus*. Amsterdam: Jansson.

Kircherus, Athanasius. 1667. *China … illustrata*. Amsterdam: Jansson.

Kusukawa, Sachiko. 2012. *Picturing the Book of Nature. Image, Text and Argument in Sixteenth-Century Human Anatomy and Medical Botany*. Chicago: University of Chicago Press.

Kuykendall, Erin D. 2019. "Plants, Pools, and Pavilions: 16th Century Safavid Garden Paintings and Expressions of Authority." PhD thesis, University of Texas at Austin, USA.

Kyle, Sarah. 2017. *Medicine and Humanism in Late Medieval Italy. The Carrara Herbal in Padua*. London: Routledge.

La Bruyère, Jean de. 1691. *Les Caractères de Theophraste traduite du Grec: avec les caractères ou les moeurs de ce siècle*. 6th edn. Paris: Estienne Michallet.

Ladner, Gerhardt B. 1961. "Vegetation Symbolism and the Concept of the Renaissance." In Millard Meiss (ed.), *De Artibus Opuscula XL: Essays in Honor of Erwin Panofsky*, 727–63, vol. 1. New York: New York University Press.

Lamb, H. H. 1982. *Climate, History, and the Modern World*. London: Methuen.

Lane, Carolina. 1980. "The Development of Pastures and Meadows During the Sixteenth and Seventeenth Centuries." *Agricultural History Review*, 28: 18–30.

Langgut, Dafna. 2017. "The Citrus Route Revealed: From Southeast Asia into the Mediterranean." *HortScience*, 52(6): 814–22.

Lanza, Fabrizia. 2011. *Olive: A Global History*. London: Reaktion Books.

Lawson, William. 1618. *A New Orchard and Garden*. London. [For the facsimile edition see Thick 2003.]

Lazzaro, Claudia. 1990. *The Italian Renaissance Garden: From the Conventions of Planting, Design, and Ornament to the Grand Gardens of Sixteenth-Century Central Italy*. New Haven, CT: Yale University Press.

Lazzaro-Bruno, Claudia. 1977. "The Villa Lante at Bagnaia: An Allegory of Art and Nature." *The Art Bulletin*, 59(4): 553–60.

Leith-Ross, Prudence. 1984. *The John Tradescants: Gardeners to the Rose and Lily Queen*. London: Peter Owen.

Le Moyne de Morgues, Jacques. 1591. *Brevis narratio eorum quae in Florida Americae provincia Gallis acciderunt. Americae pars secunda*. Frankfurt: Theodore de Bry.

Leonicenus(o), Nicolaus. 1492. *De Plinii et aliorum in medicina erroribus*. Ferrara: Impr. per magistrum Laurentium de valentia et Andream de castronovo.

Le Roy Ladurie, E. 1976. *The Peasants of Languedoc*. Champaign, IL: University of Illinois Press.

Léry, Jean de. 1578. *Histoire d'un voyage fait en la terre du Bresil*. La Rochelle: Chuppin.

Libellus. 1964. *Martín de la Cruz, Libellus de medicinalibus Indorum herbis: manuscrito azteca de 1552*. Mexico City: Instituto Mexicano del Seguro Social.

Liber Serapionis. 1473. *Liber Serapionis agregatus in medicinis simplicibus*. Milan. Available online: daten.digitale-sammlungen.de/0007/bsb00070094/images/index.html (accessed April 6, 2021).

Liber Serapionis. 1497. In *Practica Jo. Serapionis dicta Breviarium; Liber Serapionis de simplici medicina; Liber de simplici medicina dicta Circa instans; Practica Platearii*. Venice: mandato Octaviani Scoti per Bonetum Locatellum. Available online: daten.digitale-sammlungen.de/0006/bsb00061068/images/index.html (accessed April 6, 2021).

Li, Hui-Lin (trans.). 2020. "Picking Branches," from *A Treatise of Vase Flowers*. In Alison Hardie and Duncan M. Campbell (eds.). *The Dumbarton Oaks Anthology of Chinese Garden Literature*, 154–5. Washington, DC: Dumbarton Oaks.

Linschoten, Johannes Huygen van. 1596. *Itinerario, voyage ofte schipvaert van Jan Huygen van Linschoten, naer Oost ofte Portugaels Indien*. Amsterdam: Cornelis Claesz.

Linschoten, Johannes Huygen van. 1598. *Iohn Hvighen van Linschoten, his discours of voyages into ye Easte and West Indies*. London: Wolfe.

Livermore, Harold V. (trans.). 1966. *Royal Commentaries of the Incas and General History of Peru, by Garcilaso de la Vega, El Inca*. 2 vols. Austin: University of Texas Press.

Lobel, Matthias de. 1576. *Plantarum seu stirpium historia*. Antwerp: Plantin. [*Stirpium observationes* is a separately titled section of this work.]

Lobel, Matthias de. 1591. *Icones stirpium*. Antwerp: Plantin.

Lockhart, G. Wallace. 1997. *The Scots and Their Oats*. London: Birlinn.

López Piñero, José Maria, José Luis Frequet Febrer, María Luz López Terrada, and José Pardo Tomás. 1992. *Medicinas, drogas y alimentos vegetales des nuevo mundo. Textos e imágenes españolas que los introdujeron en Europa*. Madrid: Ministerio de Sanidad y Consumo.

Lubbock, Tom. 2008. "Dürer, Albrecht: *The Large Turf* (1503)." *The Independent*, Friday January 18. Available online: https://www.independent.co.uk/arts-entertainment/art/great-works/durer-albrecht-the-large-turf-1503-770976.html (accessed April 6, 2021).

Lucas, Anthony T. 1960. "Irish Food Before the Potato." *Gwerin* 3. Cardiff: Gee and Son.

Lyte, Henry (trans.). 1578. *A Niewe Herball or Historie of Plantes ... by ... Rembert Dodoens*. Antwerp: H. Loe; London: G. Dewes.

Mabberley, David J. 2017. *Mabberley's Plant-Book: A Portable Dictionary of Plants, their Classificiation and Uses*. 4th edn. Cambridge: Cambridge University Press.

Macchioni, Nicola, Benedetto Pizzo, and Chiara Capretti. 2016. "An Investigation into Preservation of Wood from Venice Foundations." *Construction and Building Materials*, 111: 652–61.

MacDougall, Elisabeth B. 1991. "A Paradise of Plants: Exotica, Rarities, and Botanical Fantasies." In Joy Kenseth (ed.), *The Age of the Marvelous*, 145–57. Hanover: Hood Museum of Art, Dartmouth College.

Maclean, Ian. 1998. "Foucault's Renaissance Episteme: An Aristotelian Counterblast." *Journal of the History of Ideas*, 59(1): 149–66.

MacNutt, Francis Augustus (trans.). 1912. *De orbe novo: the eight decades of Peter Martyr d'Anghera*. 2 vols. New York: Putnam.

Majori, Giancarlo. 2012. "Short History of Malaria and its Eradication in Italy with Short Notes on the Fight Against Infection in the Mediterranean Basin." *Mediterranean Journal of Hematology and Infectious Diseases*, 4(1). Available online: www.ncbi.nlm.nih.gov/pmc/articles/PMC3340992/ (accessed April 6, 2021).

Mangani, Giorgio, and Lucia Tongiorgi Tomasi. 2013. *Gherardo Cibo, dilettante di botanica e pittore di "paesi", arte scienza e illustrazione botanica nel XVI secolo*. Ancona: Il Lavoro.

Mariani Canova, Giordana, Elsa Mariella Cappelletti, Giancarlo Cassina, Ernesto Riva, and Roberta Sarzetto. 2006. *Codex Bellunensis. Erbario Bellunese del XV secolo*. 2 vols. Feltre: Parco Nazionale Dolomiti Bellunesi.

Markey, Lia. 2016. *Imagining the Americas in Medici Florence*. University Park: Penn State University Press.

Markham, Clements (trans.). 1913. *Colloquies on the Simples and Drugs of India by Garcia da Orta*. London: Hakluyt Society.

Markham, Gervase. 1613. *The English Husbandman*. London: John Browne. Available online: www.gutenberg.org/ebooks/22973 (accessed April 6, 2021).

Markham, Gervase. 1656. *The English Housewife*. London: E. Brewster and George Sawbridge. [The 1st edition is entitled *The English Hus-wife*, published as vol. 2 of *Country Contentments*, 1615. Available online: archive.org/details/b30337628/page/n127 (accessed April 6, 2021). The 1623 edition is entitled *Countrey Contentments, or The English huswife*. Available online: quod.lib.umich.edu/e/eebo/A06913.0001.001 (accessed April 6, 2021).

Marlowe, Christopher. 1604. *The Tragicall History of D. Faustus ... written by Ch. Marl.* London: Thomas Law.

Mason, Peter. 2001. *The Lives of Images*. London: Reaktion Books.

Mason, Peter. 2009. *Before Disenchantment. Images of Exotic Animals and Plants in the Early Modern World*. London: Reaktion Books.

Mason, Peter. 2015. *The Ways of the World. European Representations of Other Cultures from Homer to Sade*. Canon Pyon: Sean Kingston Publishing.

Mathew, K. S. 1997. "The Portuguese and the Study of Medicinal Plants in India in the Sixteenth Century." *Indian Journal of the History of Science*, 32(4): 369–76.

Mattioli, Pietro Andrea. 1544. *Di Pedacio Dioscoride Anazarbeo libri cinque della historia et materia medicinale tradotti in lingua volgare italiana*. Venice: Nicolò de Bascarini.

Mattioli, Pietro Andrea. 1554. *Commentarii in libros sex Pedacii Dioscoridis* Anazarbei, *De medica materia: adiectis quàm plurimis plantarum and animalium imaginibus*. Venice: Valgrisi.

Mattioli, Pietro Andrea. 1565. *Commentarii in sex libros Pedacii Dioscoridis Anazarbei De medica materia*. Venice: Valgrisi.

Mattioli, Pietro Andrea. 1577. *I discorsi ... nelli sei libri di Pedacio Dioscoride Anazarbeo della materia medicinale*. Venice: appresso gli heredi di Vincenzo Valgrisi.

Mattioli, Pietro Andrea. 1744. *Discorsi di m. Pietro Andrea Mattioli sanese, ne' sei libri di Pedacio Dioscoride Anazarbeo, della materia medicinale: colle figure delle piante, ed animali cavate dal naturale*. Venice: Niccolò Pezzana.

Maudslay, A. P. (trans.). 1908. *The True History of the Conquest of New Spain, by Bernal Díaz del Castillo, one of its conquerors*. 4 vols. London: Hakluyt Society.

Mayhew, Alan. 1973. *Rural Settlement and Farming in Germany*. London: Batsford.

McClure, Norman (ed.). 1939. *The Letters of John Chamberlain*. Philadelphia: American Philosophical Society.

Meusnier de Querlon, Anne-Gabriel (ed.). 1774. *Journal du voyage de Michel de Montaigne en Italie*, Rome.

Miller, Edward (ed.). 1971. *The Agrarian History of England and Wales, vol. III*. Cambridge: Cambridge University Press.

Mills, J. V. G. 1970. *Ying-yai sheng-lan: "The Overall Survey of the Ocean's Shores" [1433]*. Cambridge: Cambridge University Press.

Misson de Valbourg, Henri. 1698. *Mémoires et observations faites par un voyageur en Angleterre sur ce qu'il y a de plus remarquable*. The Hague. [Probably by François Maximilien Misson.]

Modlen, William (ed.). 1912. *The Tragical History of Dr Faustus, text of 1604, by Christopher Marlowe*. London: Macmillan.
Monardes, Nicolás. 1565. *Dos libros. El uno trata de todas las cosas que traen de nuestras Indias Occidentales, que sirven al uso de medicina …* Seville: Hernando Diaz. [2nd edn, 1569.]
Monardes, Nicolás. 1571. *Segunda parte del libro des las cosas que se traen de nuestras Indias Occidentales ….* Seville: Alonso Escrivano.
Monardes, Nicolás. 1574. *Primera, segunda y tercera partes de la historia medicinal de las cosas que se traen de nuestras Indias Occidentales*. Seville: Alonso Escrivano.
Montalboddo, Fracanzano. 1507. *Paesi novamente retrovati et novo mondo da Alberico Vesputio Florentino*. Vicenza: Henrico Vicentino.
Montanari, Massimo. 1984. "Rural Food in Late Medieval Italy." In Harry Kühnel (ed.), *Bauerlihe Sachkultur Des Spatmittelalters [International Congress], Krems an der Donau, 21–24 Sept. 1982*. Vienna: Verlag Der Österreichischen Akademie Der Wissenschaften.
Montanari, Massimo. 1994. *The Culture of Food*. New York: Columbia University Press.
Moody, Joanne (ed.). 1998. *The Diary of Lady Margaret Hoby, 1599–1605*. Stroud: Sutton.
Moretus, Balthasar II (ed.). 1644. *Cruydt-boeck Remberti Dodonaei volghens sijne laetste verbeteringhe*. Antwerp: Plantin.
Morgan, Luke. 2007. *Nature as Model: Salomon de Caus and Early Seventeenth-Century Landscape Design*. Penn Studies in Landscape Architecture. Philadelphia: University of Pennsylvania Press.
Morin, Pierre. 1651. *Catalogues de quelques plantes à fleurs qui sont de présent au jardin de Pierre Morin le jeune, dit Troisième, scitué au faux-bourg Saint Germain, proche la Charité*. Paris: François le Cointe.
Morison, S. E. 1978. *The Great Explorers: The European Discovery of America*. Oxford.
Moryson, Fynes. 1617. *An Itinerary: Containing His Ten Years Travel Through the Twelve Dominions of Germany, Bohemia, Switzerland, Netherland, Denmark, Poland, Italy, Turkey, France, England, Scotland and Ireland*. London: J. Beale.
Muriel, Josefina. 1990. *Hospitales de la Nueva España. Tomo I. Fundaciones del Siglo XVI*. Mexico City: UNAM/Cruz Roja Mexicana.
Muylle, Jan. 2016. "De huisraad en de bibliotheek van Antoon van Arenberg, graaf van Seninghem. Sporen van een hoogadellijke verzameling (Brussel, 1617)." *De Gulden Passer*, 94(1): 7–86.
Myers, Kathleen Ann, and Nina M. Scott (trans.). 2007. *Fernández de Oviedo's Chronicle of America: A New History for a New World*. Austin: University of Texas Press.
Nabhan, Gary Paul. 1993. *Songbirds, Truffles, and Wolves: An American Naturalist in Italy*. New York: Pantheon Books.
Navarrete, Martin Fernandez de (ed.). 1858. *Colección de los viages y descubrimientos*, 2nd edn., vol. 1. Madrid. [First edition, 1825, less complete: *Colección de los viages y descubrimientos que hicieron por mar los españoles desde fines del siglo XV*, vol. 1.]
Néraudau, Jean Pierre. 1986. *L'Olympe du Roi-Soleil: Mythologie et idéologie royal au Granad Siècle*. Paris: Société d'Edition Les Belles Lettres.
Nonaka, Natsumi. 2017. *Renaissance Porticoes and Painted Pergolas in Early Modern Italy*. London: Routledge.
Nonni, Giorgio (ed.). 1982. *Costanzo Felici, Lettere a Ulisse Aldrovandi*. Urbino: QuattroVenti.
Norden, John. 1618. *The Surveiors Dialogue*. London: Thomas Snodham.
Nuovo receptario composto dal famossisimo chollegio degli eximii doctori della arte et medicina della inclita cipta di Firenze. 1498. Florence: Per la Compagnia del Dragho.

Nutton, Vivian. 1993. "Humoralism." In William Bynum and Roy Porter (eds.), *Companion Encyclopedia of the History of Medicine*, vol. 1, 281–91. London: Routledge.

Ogilvie, Brian. 2006. *The Science of Describing. Natural History in Renaissance Europe*. Chicago: Chicago University Press.

Ó Gráda, Cormac. 2009. *Famine: A Short History*. Princeton, NJ: Princeton University Press.

Olariu, Dominic. 2015. *The Misfortune of Philippus de Lignamine's Herbal or New Research Perspectives in Herbal Illustrations From an Iconological Point of View*. [Max Planck Institute for the History of Science, Preprint 469.] Berlin: MPIHS.

Olmi, Giuseppe. 1978. *Ulisse Aldrovandi, Scienza e Natura nel secondo cinquecento*. Trento: Liber Università dagli Studi di Trento.

Olmi, Giuseppe. 1992. *L'Inventario del Mondo, Catalogazione della natura e luogi del sapere nella prima età moderna*. Bologna: Il Mulino.

Olmi, Giuseppe, and Fulvio Simoni. 2018. *Ulisse Aldrovandi, libri e immagini di storia naturale nella prima età moderna*. Bologna: Bononia University Press.

Orta, Garcia de. 1563. *Colóquios dos simples e drogas he cousas medicinaes da India*. Goa: Joannes de Enden.

Ortus [Hortus] Sanitatis. 1491. Mainz: J. Mayenbach.

Oviedo, Gonzalo Fernández. 1526. *Oviedo Dela natural hystoria delas Indias*. Toledo.

Pächt, Otto. 1950. "Early Italian Nature Studies and the Early Calendar Landscape." *Journal of the Warburg and Courtauld Institutes*, 13: 13–47.

*Palladius, Blosius [*Biagio Pallai*]. 1512. Suburbanum Augustini Chisii*. Rome. Available online: books.google.it/books?id=ipNWAAAAcAAJ (accessed April 5, 2021).

Palma, Nuno, and Jaime Reis. 2014. *Portuguese Demography and Economic Growth, 1500–1850*. Prepared for the "Accounting for the Great Divergence" conference, University of Warwick in Venice, May 22–4.

Pardo Tomás, José. 2012. "Opening Bodies in a New World: Anatomical Practices in Sixteenth-Century New Spain." In Giuseppe Olmi and Claudio Pancino (eds.), *ANATOME. Sezione, scomposizione, raffigurazione del corpo fra Medioevo e Età Moderna*, 185–202. Bologna: Bononia University Press.

Pardo Tomás, José. 2013a. "Conversion Medicine: Communication and Circulation of Knowledge in the Franciscan Convent and College of Tlatelolco, 1527–1577." *Quaderni Storici*, 48(1): 21–42.

Pardo Tomás, José. 2013b. "Representación e imaginación de la nueva Espana—las pinturas de las relaciones geográficas de Indias." In Elke Köppen and Mauricio Sánchez Menchero (eds.), *Los trazos de las ciencias. Circulación de conocimiento en imágines*, 13–60. Mexico City: UNAM.

Pardo Tomás, José. 2014. "Antiguamente vivían más sanos que ahora. Explanations of the native mortality in the Relaciones Geográficas de Indias." In John Slater, Maríaluz López Terrada, and José Pardo Tomás (eds.), *Medical Cultures of the Early Modern Spanish Empire*, 41–65. Farnham: Ashgate.

Pardo Tomás, José, and María Luz López Terrada. 1993. *Las primeras noticias sobre plantas americanas en las relaciones de viajes y crónicas de Indias (1493–1553)*. Valencia: Instituto de Estudios Documentales e Históricos sobre la Ciencia.

Pardo Tomás, José, and José María López Piñero. 1996. *La influencia de Francisco Hernández (1515–1587) en la constitucion de la botanica y de la materia medica modernas*. Valencia: Instituto de Estudios Documentales e Históricos sobre la Ciencia.

Paris, Harry S., et al. 2006. "First Known Image of Cucurbita in Europe, 1503–1508." *Annals of Botany*, 98(1): 41–7.

Paris, Harry S., et al. 2012. "Occidental Diffusion of Cucumber (*Cucumis sativus*) 500–1300 CE: Two Routes to Europe." *Annals of Botany*, 109(1): 117–26.

Parkinson, John. 1629. *Paradisi in sole Paradisus terrestris: A garden of all sorts of pleasant flowers*. London: Humphrey Lownes, Robert Young.

Parry, Glyn. 2004. "William Harrison." In *Oxford Dictionary of National Biography*. Oxford: Oxford University Press. Available online: https://www-oxforddnb-com.udel.idm.oclc.org/view/10.1093/ref:odnb/9780198614128.001.0001/odnb-9780198614128-e-12453 (accessed April 6, 2021)

Pavord, Anna. 1999. *The Tulip*. London: Bloomsbury.

Pavord, Anna. 2005. *The Naming of Names*. London: Bloomsbury.

Payne, W., and Sidney J. Herrtage (eds.). 1878. *Fiue Hundred Pointes of Good Husbandrie, by Thomas Tusser: the edition of 1580 collated with those of 1573 and 1577*. London: Trubner.

Peacock, John. 2006. *The Look of Van Dyck: The Self-Portrait with a Sunflower and the Vision of the Painter*. Aldershot: Ashgate.

Peck, Linda. 2005. *Consuming Splendor*. Cambridge: Cambridge University Press.

Pelliot, Paul. 1933. "Les grands voyages maritimes chinois au début du XVème siècle." *T'oung Pao*, 30: 237–452.

Pena, Petrus, and Matthias de Lobel. 1571. *Stirpium adversaria nova*. Antwerp: Christoph Palantin.

Peterson, Laura Favrot. 1993. *The Paradise Garden Murals of Malinalco*. Austin: University of Texas Press.

Petrus Martyr ab Angleria. 1511. *Legatio Babylonica, Oceani decas, Poemata epigrammatica*. Seville: Corumberger.

Petrus Martyr ab Angleria. 1530. *De orbe novo … decades*. Compluto: Michael d'Eguia.

Phillips, Rod. 2014. *Alcohol: A History*. Chapel Hill: University of North Carolina Press.

Piergiovanni, Angela R., and Lucia Lioi. 2010. "Italian Common Bean Landraces: History, Genetic Diversity and Seed Quality." *Diversity*, 2: 837–62.

Pigafetta, Filippo. 1591. *Relatione del reame di Congo et delle circonvicine contrade, tratta dalli scritti and ragionamenti di Odoardo Lopez*. Rome: Bartolomeo Grassi.

Poggius Florentinus [Poggio Bracciolini]. 1492. *India recognita*. Milan: Udalricus Scinzenzeler.

Porta, Ioannes Baptista [Giambattista della Porta]. 1588. *Phytognomonica*. Naples: Horatius Salvianus.

Prance, Sir Ghillean, and Mark Nesbitt (eds.). 2005. *The Cultural History of Plants*. London: Routledge.

Prest, John. 1981. *The Garden of Eden: The Botanic Garden and the Re-Creation of Paradise*. New Haven and London: Yale University Press.

Ptak, Roderich. 1995. "Asian Trade in Cloves, circa 1500: Quantities and Trade Routes—ua Synopsis of Portuguese and Other Sources." In A. Dutra and Joao Camilo dos Santos (eds.), *The Portuguese and the Pacific*, 149–69. Santa Barbara: Center for Portuguese Studies, University of California at Santa Barbara.

Puerto, Javier. 2003. *La Leyende Verde. Naturaleza, sanidad e ciencia en la corte de Felipe II (1527–1598)*. Castilla y León: Junta de Castilla y León.

Pyle, Cynthia. 1996. "The Art and Science of Renaissance Natural History: Thomas of Cantimpré, Pier Candido Decembrio, Conrad Gessner, and Teodoro Ghisi in Vatican Library MS Urb. lat. 276." *Viator*, 27: 265–321.

Pyle, Cynthia. 2010. "Bridging the Gap: A Different View of Renaissance Humanism and Science." In Rens Bod, Jaap Maat, and Thijs Weststeijn (eds.), *The Making of the Humanities 1*, 39–57. Amsterdam: University of Amsterdam Press.

Quinn, D. B. 1974. *England and the Discovery of America, 1481–1620*. New York: Knopf.
Rabel, Daniel. 1627. *Theatrum Florae in quo ex toto orbe selecti mirabiles, venustiores ac praecipui flores, tanquam ab ipsius deae sinu, proferuntur*. Paris: Pierre Mariette.
Rabelais, François. 1546. *Tiers livre des faicts et dicts heroïques du bon Pantagruel*. Paris.
Ramusio, Giovanni Battista (ed.). 1550–9. *Navigationi et viaggi*. Venice: Giunti.
Rauwolf, Leonhart. 1582–3. *Aigentliche Beschreibung der Raiß ... so er vor diser Zeit gegen Auffgang inn die Morgenländer, fürnemlich Syriam, Iudaeam, Arabiam, Mesopotamiam, Babyloniam, Assyriam, Armeniam &c. nicht ohne geringe mühe unnd große gefahr selbs volbracht*. Laugingen: Rainmichel, Willers.
Rawson, Jessica. 1984. *Chinese Ornament: The Lotus and the Dragon*. London: British Museum Publications.
Rea, John. 1665. *Flora, seu De florum cultura, or A complete florilege*. London: Thomas Clarke.
Reeds, Karen. 1991. *Botany in Medieval and Renaissance Universities*. New York: Garland.
Renaux, Alain. 2008. *Louis XIV's Botanical Engravings*. Lund Humphries.
Richter, Irma A. (trans.). 1977. *Selections from the Notebooks of Leonardo da Vinci*. Oxford: Oxford University Press.
Riddell, John R. D. 1986. "John Parkinson's Long Acre Garden 1600–1650." *Journal of Garden History*, 6(2): 112–24.
Ridolfi, Angelo. 1810. *Pensieri intorno allo scopo di Nicolò Machiavelli nel libro Il Principe*. Milan: Destefanis.
Riley, Gillian. 2007. *The Oxford Companion to Italian Food*. Oxford and New York: Oxford University Press.
Roberts, R. 1965. "The Early History of the Import of Drugs into England." In F. N. L. Poynter (ed.), *The Evolution of Pharmacy in Britain*, 165–86. London: Pitman Medical Pub.
Robertson, James Alexander. 1906. *Magellan's voyage around the world, by Antonio Pigafetta: the original text of the Ambrosian MS., with English translation*. 3 vols. Cleveland: Arthur H. Clark.
Robinsheaux, Thomas. 1998. "The Peasantries of Western Germany, 1300–1750." In Tom Scott (ed.), *The Peasantries of Europe*, 111–42. London: Longman.
Robinson, Simon. 2007. "Chili Peppers: Global Warming." *Time*, June 14. Available online: http://content.time.com/time/specials/2007/article/0,28804,1628191_1626317_1632291,00.html (accessed April 7, 2021).
Rossetti, Giovanni Battista. 1584. *Dello scalco*. Ferrara: Domenico Mammarolo. Available online: www.mori.bz.it/gastronomia/Rossetti%20-%20Dello%20scalco.pdf (accessed April 7, 2021).
Rowland, Ingrid D. 1997. "The Intellectual Background of the School of Athens: Tracking Divine Wisdom in the Rome of Julius II." In Marcia Hall (ed.), *Raphael's School of Athens*, 131–70. Cambridge: Cambridge University Press.
Rowland, Ingrid D. (trans.). 1999. *Vitruvius: Ten Books on Architecture*. Cambridge: Cambridge University Press.
Rowland, Ingrid D. 2005a. "Il giardino *trans Tiberim* di Agostino Chigi." In Carla Benocci (ed.), *I giardini Chigi tra Siena e Roma: dal Cinquecento agli inizi dell'Ottocento*, 57–72, 421. Siena: Protagon Editrice per la Fondazione Monte dei Paschi.
Rowland, Ingrid D. 2005b. *The Roman Garden of Agostino Chigi*. Groningen: Gerson Lectures Foundation.
Rowland, Ingrid D. 2009. "The Flowering Genius of Maria Sibylla Merian." *The New York Review of Books*, April 9.

Rozzo, Ugo. 2011. "La fortuna della Naturalis Historia di Plinio nell' editoria del XV secolo." *Archives internationales d'histoire des sciences*, 61(166–7): 73–113.

Ruggles, D. Fairchild. 2008. *Islamic Gardens and Landscapes*. Philadelphia: University of Pennsylvania Press.

Ruiz, Teofilo F. 1998. "The Peasantries of Iberia, 1400–1800." In Tom Scott (ed.), *The Peasantries of Europe*. London: Longman.

Rumeu de Armas, Antonio (ed.). 1989. *Libro copiador de Cristóbal Colón*. Madrid: Testimonio Compañia Editorial.

Russo, Lucio. 2004. *The Forgotten Revolution: How Science was Born in 300 BC and Why it had to be Reborn*. Translated by Silvio Levy. Berlin: Springer.

Sakellariou, Eleni. 2012. *Southern Italy in the Late Middle Ages*. Leiden: Brill.

Sandrini, Giuseppe (ed.). 2007. *Francesco Calzolari: Il viaggio di Monte Baldo*. Verona: Alba Pratalia.

Scaliger, Julius Caesar. 1557. *Exotericarum exercitationum liber quintus decimus, de subtilitate, ad Hieronymum Cardanum*. Paris: Michael Vascosanus.

Schaffrath, Ulrich. 2012. "Läuse, Muscheln und Tabak: Das Herbar Ratzenberger." *Philippia. Abhandlungen und Berichte aus dem Naturkundemuseum im Ottoneum zu Kassel*, 15(3): 191–214.

Schiebinger, Londa. 2007. "Prospecting for Drugs: European Naturalists in the West Indies." In Londa Schiebinger and Claudia Swan (eds.), *Colonial Botany: Science, Commerce, and Politics in the Early Modern World*, 119–33. Philadelphia: University of Pennsylvania Press.

Schmölz-Häberlein, Michaela. 2013. "Transfer und Aneignung außereuropäischer Pflanzen im Europa des 16. und frühen 17. Jahrhunderts: Akteure, Netzwerke, Wissensorte." *Zeitschrift für Agrargeschichte und Agrarsoziologie*, 61(2): 11–26.

Scot, Reginald. 1578. *A Perfite Platforme of a Hoppe Garden*. London: Henrie Denham. Available online: picryl.com/topics/perfite+platforme (accessed April 7, 2021).

Scott, Hamish (ed.). 2015. *The Oxford Handbook of Early Modern European History*, vol. 1, 1350–1750. Oxford: Oxford University Press.

Scott, Tom. 2015. "The Agrarian West." In Hamish Scott (ed.). *The Oxford Handbook of Early Modern European History*, vol. 1, 1350–1750. Oxford: Oxford University Press.

Screech, M. A. (trans.). 2006. *François Rabelais: Gargantua and Pantagruel*. London: Penguin Books.

Seferis, George. 1950. *Ποιήματα 1924–1946*. Athens: Ikaros.

Serés, Guillermo. n.d. *Bernal Díaz del Castillo, Historia verdadera de la conquista de la Nueva España. Aparato de variantes*. Available online: www.rae.es/sites/default/files/Aparato_de_variantes_Historia_verdadera_de_la_conquista_de_la_Nueva_Espana.pdf (accessed April 7, 2021).

Serres, Olivier de. 1600. *Theatre d'agriculture et mesnage des champs*. Paris.

Sgammellotti, Antonio, and Giulia Caneva. 2017. *I colori della prosperità: frutti del vecchio e nuovo mondo*. Rome: Bardi Edizioni.

Shulman, Chaim C. 2018. *A Tale of Three Thirsty Cities: The Innovative Water Supply Systems of Toledo, London and Paris in the Second Half of the Sixteenth Century*. Leiden: Koninklijke Brill.

Sider, Sandra. 2005. *Handbook to Life in Renaissance Europe*. Oxford: Oxford University Press.

Skelton, R. A. 1962. "Cartography of the Voyages." In James Williamson (ed.), *The Cabot Voyages and Bristol Discovery Under Henry VII*, 297–307. Cambridge: Cambridge University Press.

Slater, John, Maríaluz López Terrada, and José Pardo Tomás (eds.). 2014. *Medical Cultures of the Early Modern Spanish Empire*. Farnham: Ashgate.

Smith, Bruce D., and Richard A. Yarnell. 2009. "Initial Formation of an Indigenous Crop Complex in Eastern North America at 3800 B.P." *PNAS*, 106: 6561–6.

Smith, R. E. F., and David Christian. 1984. *Bread and Salt*. Cambridge: Cambridge University Press.

Snowden, Frank M. 1989. *The Fascist Revolution in Tuscany, 1919–1922*. Cambridge: Cambridge University Press.

Snowden, Frank M. 2006. *The Conquest of Malaria: Italy 1900–1962*. New Haven, CT: Yale University Press.

Snyder, James. 1976. "Jan Mostaert's West Indies Landscape." In F. Chiappelli (ed.), *First Images of America: The Impact of the New World on the Old*, 2 vols. Vol. 1, 495–502. Berkeley: University of California Press.

Sokolov, Ray. 1991. *Why We Eat What We Eat*. New York: Simon and Shuster.

Spishak, S. C. 1998. "Review of Richard H. Britnell, *The Commercialisation of English Society 1000–1500* (Manchester: Manchester University Press, 1996)." *Medieval Review*, June.

Stacy, Jeanne. 1991. *Goose Fat and Garlic*. London: Kyle Cathie.

Stafleu, Frans A. 1987. "Die Geschichte der Herbarien." *Botanische Jahrbücher für Systematik, Pflanzengeschichte und Pflanzengeographie*, 108: 155–66.

Stanley of Alderley, Lord. 1874. *The first voyage round the world, by Magellan, translated from the accounts of Pigafetta and other contemporary writers*. London: Hakluyt Society.

Stanley, Thomas (trans.). 1665. *Claudius Ælianus his Various History*. London: Thomas Dring. Available online: quod.lib.umich.edu/e/eebo/A26482.0001.001 (accessed April 7, 2021).

Stephanus, Carolus [Charles Estienne]. 1554. *Praedium rusticum*. Paris: Stephanus.

Stow, John. 1592. *The Annales of England*. London: Ralfe Newbery. [Later editions 1600, 1602, 1605.]

Strong, Roy. 1979. *The Renaissance Garden in England*. London: Thames and Hudson.

Strong, Roy. 2000. *The Artist and the Garden*. London: Yale University Press.

Stroup, Alice. 1990. *A Company of Scientists: Botany, Patronage, and Community at the Seventeenth-Century Parisian Royal Academy of Sciences*. Berkeley: University of California Press.

Strubel, Armand. 2011. "Symbolique et Poétique de la Chasse." In *Gaston Fébus, Prince Soleil*, 68–113. Paris: Editions de la Rmn, Bibliothèque Nationale de France.

Stubbings, Frank H. 1946. "Xerxes and the Plane-Tree." *Greece and Rome*, 50: 63–7.

Subtelny, Maria E. 2007. "Visionary Rose: Metaphorical Application of Horticultural Practice in Persian Culture." In Michel Conan and W. John Kress (eds.), *Botanical Progress, Horticultural Innovation and Cultural Changes*, 13–25. Washington, DC: Dumbarton Oaks Research Library and Collection.

Surflet, Richard (trans.). 1600. *Maison rustique, or The countrie farme ... by Charles Stevens and John Liebault*. London: Bonham Norton.

Surflet, Richard (trans.). 1606. *Maison rustique, or The countrey farme ... by Charles Stevens and John Liebault*. London: John Norton, John Bill.

Surflet, Richard (trans.) and Gervase Markham (ed.). 1616. *Maison rustique, or The countrey farme ... by Charles Stevens and John Liebault*. London: John Bill.

Sylvester, Josuah (trans.). 1621. *Du Bartas his Divine Weekes and Workes ...* London: Humfrey Lounes.

Tabernaemontanus, Jacobus Theodorus. 1588–91. *Neuw Kreuterbuch*. 2 vols. Frankfurt am Main.

Taegio, Bartolomeo. 1559. *La villa: dialogo*. Milan: Francesco Moscheni.

Teutenberg, Hans J. (ed.). 1992. *European Food History*. Leicester: Leicester University Press.

Thick, Malcolm. 1985. "Market Gardening in England and Wales." In Joan Thirsk (ed.), *The Agrarian History of England and Wales,* vol. V, part ii. Cambridge: Cambridge University Press.

Thick, Malcolm (ed.). 2003. *William Lawson: A New Orchard and Garden (1618).* Totnes: Prospect Books.

Thick, Malcolm. 2010. *Sir Hugh Plat: The Search for Useful Knowledge in Early Modern London.* Totnes: Prospect Books.

Thiem, Jon (trans.). 1991. *Lorenzo de' Medici: Selected Poems and Prose.* University Park, PA: Pennsylvania State University Press.

Thijsse, Gerard. 2016. "'Tusschen pampier geleyt'. Ontstaan, verspreiding en gebruik van de vroegste herbaria." In Linda IJpelaar and Claudine A. Chavannes-Mazel (eds.), *De Groene Middeleeuwen. Duizend jaar gebruik van planten (600–1600),* 64–93 [with notes 279–81]. Eindhoven: Lecturis.

Thirsk, Joan. 1967a. "Farming Techniques." In Joan Thirsk (ed.), *The Agrarian History of England and Wales,* vol. IV. Cambridge: Cambridge University Press.

Thirsk, Joan (ed.). 1967b. *The Agrarian History of England and Wales,* vol. IV: 1500–1640. Cambridge: Cambridge University Press.

Thirsk, Joan (ed.). 1985. *The Agrarian History of England and Wales,* vol. V part ii: 1640–1750*: Regional Farming Systems.* Cambridge: Cambridge University Press.

Thirsk, Joan. 2007. *Food in Early Modern England.* London: Hambledon Continuum.

Thomas, Hugh. 1993. *The Conquest of Mexico.* London: Hutchinson.

Thomas, Keith. 2009. *The Ends of Life: Roads to Fulfilment in Early Modern England.* Oxford: Oxford University Press.

Thuilius, Joannes (ed.). 1621. *V. cl. Andreae Alciati Emblemata cum commentariis amplissimis.* Padua: Petro Paulo Tozzi.

Tongiorgi Tomasi, Lucia. 1983. "Projects for Botanical and Other Gardens: A 16th-Century Manual." *The Journal of Garden History,* 3(1): 1–34.

Tongiorgi Tomasi, Lucia. 1989. "Gherardo Cibo: Visions of Landscape and the Botanical Sciences in a Sixteenth-Century Artist." *Journal of Garden History,* 9(4): 199–216.

Tongiorgi Tomasi, Lucia, and Gretchen A. Hirschauer. 2002. *The Flowering of Florence: Botanical Art for the Medici.* Washington: National Gallery of Art. Available online: www.nga.gov/research/publications/pdf-library/flowering-of-florence.html (accessed April 7, 2021).

Tournefort, Pitton [de]. 1694. *Elemens de botanique, ou Méthode pour connoître les plantes.* Paris: Imprimerie Royale.

Touwaide, Alain. 2015. "Il manoscritto più misterioso—l'erbario Voynich." In Marina Formica (ed.), *Villa Mondragone seconda Romai,* 139–58. Rome: Palomba Editori.

Transylvanus, Maximilianus. 1523. *De Moluccis insulis.* Cologne.

Turner, William. 1548. *The Names of Herbes in Greke, Latin, Englishe, Duch and Frenche.* London: John Day.

Turner, William. 1551. *A New Herball* [part 1]. London: Steven Mierdman.

Turner, William. 1562. *The Seconde Parte of William Turners Herball.* Cologne: Arnold Birckman.

Turner, William. 1568. *The First and Seconde Partes of the Herbal ... with the Third Parte.* Cologne. Available online: quod.lib.umich.edu/e/eebo2/A14059.0001.001 (accessed April 7, 2021).

Tussenbroek, Gabri von. 2017. "Timber-framed Town Houses in the Northern Netherlands Before 1600: Construction and Geographical Distribution." *Vernacular Architecture,* 48(1): 44–72.

Tusser, Thomas. 1557. *A Hundreth Good Pointes of Husbandrie*. London. Available online: www.luminarium.org/renascence-editions/tusser1.html (accessed April 7, 2021).

Tusser, Thomas. 1573. *Five Hundreth Pointes Of Good Husbandrie*. London: Richard Tottell.

Ulloa, Alfonso (trans.). 1571. *Historie del S. D. Fernando Colombo*. Venice.

Unger, Richard. 2004. *Beer in the Middle Ages and the Renaissance*. Philadelphia: University of Pennsylvania Press.

Unterkircher, Franz (ed.). 2004. *Tacuinum sanitatis in medicina: Codex Vindobonensis series nova 2644 der Österreichischen Nationalbibliothek*. Graz: Akademische Druck- und Verlagsanstalt.

Unwin, Tim. 1991. *Wine and the Vine*. London: Routledge.

Vande Walle, W. F. (ed.). 2001. *Dodonaeus in Japan: Translation and the Scientific Mind in the Tongukawa Period*. Leuven: Leuven University Press.

Van Riemsdijk, Willem. 2015. "Christine Bertolf en Dodonaeus: Netwerken in de zestiende eeuw." Available online: www.stinze-stiens.nl/blog/ (accessed April 7, 2021).

Varela, Consuelo (ed.). 1984. *Cristóbal Colón. Textos y documentos completos. Relaciones de viajes, cartas y memoriales*. 2nd edn. Madrid: Alianza Editorial.

Varey, Simon, and Rafael Chabrán. 1995. "Mexican Medicine Comes to England." *Viator*, 26: 333–55.

Varey, Simon, Rafael Chabrán, and Dora B. Weiner (eds.). 2000. *Searching for the Secrets of Nature: The Life and Works of Dr. Francisco Hernández*. Stanford: Stanford University Press.

Varthema, Ludovico di. 1510. *Itinerario*. Rome. [Cited by folios of the Milan 1523 edn.]

Vasari, Giorgio. 1568. *Le vite de' piu eccellenti pittori, scultori, et architettori*. 2nd edn. Florence: Giunti.

Vassberg, David E. 1984. *Land and Society in Golden Age Castile*. Cambridge: Cambridge University Press.

Vaughan, Rowland. 1610. *Most Approved and Long Experienced Water-Workes, containing the manner of winter and summer drowning of medow and pasture*. London: George Eld.

Vega, Garcilaso de la. 1609. *Primera parte de los commentarios reales que tratan del origen de los Yncas*. Lisbon: Pedro Crasbeeck.

Veracini, Cecilia, and Dante Martins Teixeira. 2017. "Perception and Description of New World Non-human Primates in the Travel Literature of the Fifteenth and Sixteenth Centuries: A Critical Review." *Annals of Science*, 74(1): 25–63.

Verhulst, Adriaan. 1990. *Precis d'histoire rurale de la Belgique*. Bruxelles: Editions de l'Universite de Bruxelles.

Visone, Massimo. 2016. "Poggio Reale rivisitato: preesistenze, genesi e trasformazioni in età vicereale." In Encarnación Sánchez García (ed.), *Rinascimento meridionale: Napoli e il viceré Pedro de* Toledo, 771–98. Naples: Tullio Pironti.

Waite, Arthur Edward (trans.). 1894. *The Hermetic and Alchemical Writings of … Paracelsus the Great*, vol. 1. London: James Elliott. Available online: archive.org/details/hermeticalchemic00para (accessed April 7, 2021).

Walcher, Heidi A. 1997. "Between Paradise and Political Capital: The Semiotics of Safavid Isfahan." *Middle Eastern Natural Environments Journal*, 103: 330–48.

Walker, Timothy. 2008. "Acquisition and Circulation of Medical Knowledge Within the Early Modern Portuguese Empire." In Daniela Bleichmar, Paula De Vos, Kristin Huffine, and Kevin Sheehan (eds.), *Science in the Spanish and Portuguese Empires*, 247–70. Stanford: Stanford University Press.

Walker, Timothy. 2013. "The Medicines Trade in the Portuguese Atlantic World: Acquisition and Dissemination of Healing Knowledge from Brazil (*c.* 1580–1800)." *Social History of Medicine*, 26(3): 403–31.

Walker, Timothy. 2015. "'Enduring Echoes of Garcia da Orta', the Royal Hospital Gardens in Goa and Evolving Hybridization in Portuguese Colonial Medical Culture." In Palmira Fontes Da Costa (ed.), *Medicine, Trade and Empire: Garcia de Orta's* Colloquies on the Simples and Drugs of India *(1563) in Context*, 213–36. Farnham: Ashgate.

Walker, Timothy. 2016. "Global Cross-Cultural Dissemination of Indigenous Medical Practices through the Portuguese Colonial System: Evidence from Sixteenth to Eighteenth-Century Ethno-Botanical Manuscripts." In Helge Wendt (ed.), *The Globalization of Knowledge in the Iberian Colonial World* [Max Planck Research Library for the History and Development of Knowledge Proceedings 10], 162–92. Berlin: MPIHS.

Wallis, Patrick. n.d. "Exotic Drugs and English Medicine: England's Drug Trade, c.1550–c.1800." Available online: eprints.lse.ac.uk/28577/1/WP143.pdf (accessed April 7, 2021).

Wallis, Patrick. 2006. "Apothecaries and the Consumption and Retailing of Medicines in Early Modern London." In Louise Hill Curth (ed.), *From Physick to Pharmacology: Five Hundred Years of British Drug Retailing*, 13–27. Aldershot: Ashgate.

Walpole, Horace (ed.). 1757. *A Journey into England by Paul Hentzner in the year MDXCVIII*. Translated by Richard Bentley. Strawberry Hill: Horace Walpole.

Walravens, Hartmut. 1973. "Konrad Gessner in chinesischem Gewand. Darstellungen fremder Tiere im K'un-yü t'u-shuo des P. Verbiest (1623–1688)." *Gesnerus*, 30: 87–98.

Weiner, Dora. 2002. "The World of Dr Francisco Hernández." In Simon Varey, Rafael Chabran, and Dora Weiner (eds.), *Searching for the Secrets of Nature: The Life and Works of Dr. Francisco Hernández*, 3-10. Stanford: Stanford University Press.

Wettengl, Kurt (ed.). 1993. *Georg Flegel 1566–1638 Stilleben*. Stuttgart: Verlag Gerd Hatje.

Wheelock, Arthur K. 1999. *From Botany to Bouquets: Flowers in Northern Art*. Washington, National Gallery of Art. Available online: www.nga.gov/research/publications/pdf-library/from-botany-to-bouquets.html (accessed April 7, 2021).

Wilhelm, Stephen. 1974. "The Garden Strawberry." *American Scientist*, 62: 264–71.

Williams, Clare (trans.). 1937. *Thomas Platter's Travels in England, 1599*. London: Jonathan Cape. Available online: https://www.e-manuscripta.ch/bau/doi/10.7891/e-manuscripta-18678 (accessed April 7, 2021).

Williamson, James A. 1962. *The Cabot Voyages and Bristol Discovery Under Henry VII*. Cambridge: Cambridge University Press.

Wills, Gary. 1994. *Cincinnatus: George Washington and the Enlightenment*. New York: Doubleday.

Wilson, N. G. (trans.). 1997. *Aelian: Historical Miscellany*. Cambridge, MA: Harvard University Press.

Worlidge, John. 1700. *Systema horti-culturae*. London: Thomas Dring.

Worth Estes, J. 2000. "The Reception of American Drugs in Europe, 1500–1650." In Simon Varey, Rafael Chabrán, and Dora B. Weiner (eds.), *Searching for the Secrets of Nature: The Life and Works of Dr. Francisco Hernández*, 111–21. Stanford: Stanford University Press.

Wotton, Sir Henry. 1624. *The Elements of Architecture*. London: John Bill.

Woudstra, Jan, and Sally O'Hallaran. 2008. "Sir John Reresby's Garden Notebook and Garden (1633–44) at Thrybergh, Yorkshire." *Garden History*, 36(1): 135–93.

Zahn, G. 1902. "Das Herbar des Dr. Caspar Ratzenberger (1598) in der Herzoglichen Bibliothek zu Gotha." *Mitteilungen des Thüringischen Botanischen Vereins. N. F.*, 16: 50–121.

Zalum, Margherita. 2013. "Plantings." In Elizabeth Hyde (ed.), *A Cultural History of Gardens in the Renaissance*, 73–96. London: Bloomsbury.

Zecchino, Francesco. 2002. "La Villa di Poggioreale, residenza degli Aragonesi a Napoli." *Delpinoa*, n.s. 44: 3–16.

Zemanek, Alicia. 1998. "Renaissance Botany and Modern Science." In Z. Mirek and A. Zemanek (eds.), *Studies in Renaissance Botany*, 9–47. Kraków: W. Szafer Institute.

Županov, Ines. 2002. "Drugs, Health, Bodies and Souls in the Tropics; Medical Experiments in Sixteenth-Century Portuguese India." *Indian Economic and Social History Review*, 39(1): 1–43.

Županov, Ines, and Angela Barreto Xavier. 2014. "Quest for Permanence in the Tropics: Portuguese Bioprospecting in Asia (16th–18th Centuries)." *Journal of the Economic and Social History of the Orient*, 57(4): 511–48.

CONTRIBUTORS

Andrew Dalby, once a librarian at Cambridge University Library, UK, lives in France, writes on food history (*Siren Feasts*, 1996; *Empire of Pleasures*, 2000; *Food in the Ancient World from A to Z*, 2003; *The Breakfast Book*, 2015), and translates historical sources on farming and food (*Cato on Farming*, 1998; *Tastes of Byzantium*, 2010; *Geoponika*, 2011; *The Treatise of Walter of Bibbesworth*, 2012). His latest book, on which he collaborated with his daughter Rachel, is *Gifts of the Gods: A History of Food in Greece* (2017).

Florike Egmond is a Dutch historian who lives in Rome. She specializes in the cultural and visual aspects of early modern natural history, and has worked for various research projects based at Leiden University, The Netherlands. Her most recent book is *Eye for Detail: Images of Plants and Animals in Art and Science, 1500–1630* (2017).

Jill Francis is an independent researcher whose first monograph, *Gardens and Gardening in Early Modern England and Wales* was published in 2018. She has devised and runs garden history courses in the Midlands area and also lectures in early modern history at a variety of institutions, including the universities of Birmingham and Worcester.

Annette Giesecke is a specialist in the history, meaning, and representation in literature and the arts of ancient Greek and Roman gardens and designed landscapes. Her work extends to Near Eastern garden traditions and cultural uses of plants in antiquity. She is Professor of Classics at the University of Delaware, USA and is an Archaeological Institute of America National Lecturer. Her books include *A Cultural History of Plants in Antiquity* (ed. and contrib., 2022), *The Mythology of Plants: Botanical Lore from Ancient Greece and Rome* (2014), and *The Good Gardener? Nature, Humanity and the Garden* (ed. and contrib., 2015).

Elizabeth Hyde is Professor and Chair of the Department of History at Kean University, New Jersey, USA. Her first book, *Cultivated Power: Flowers, Culture, and Politics in the Reign of Louis XIV* (2005) explores the collection, cultivation, and political importance of flowers in early modern France. She edited *A Cultural History of Gardens in the Renaissance, 1400–1650* (2013) and is currently writing *Of Monarchical Climates and Republican Soil: Nature, Nation, and Botanical Diplomacy in the Franco-American Atlantic World.*

David Marsh was awarded a PhD for a thesis on “The Gardens and Gardeners of Later Stuart London” in 2005. A freelance lecturer and researcher, he is co-convenor of the Garden History seminar at London University’s Institute of Historical Research and a trustee of the Gardens Trust, for whom he writes a weekly garden history blog.

Luke Morgan is Director of Art History at Monash University, Australia and an elected Fellow of the Australian Academy of the Humanities. His most recent book is *The Monster in the Garden: The Grotesque and the Gigantic in Renaissance Landscape Design* (2016).

Gillian Riley is a freelance food historian. Her publications include *The Oxford Companion to Italian Food* (2007), *Food in Art: From Prehistory to the Renaissance* (2015), *A Feast for the Eyes* (1995), and translations of Giacomo Castelvetro, *The Fruit, Herbs and Vegetables of Italy* (1989, 2012) and Maestro Martino, *Libro de Arte Coquinaria* (2005).

Ingrid D. Rowland is based in Rome as a professor in the Department of History and the School of Architecture of the University of Notre Dame, London, UK. Her books include *Giordano Bruno: Philosopher/Heretic* (2008), *Giordano Bruno*: *On the Heroic Frenzies* (2014), *From Pompeii: The Afterlife of a Roman Town* (2014), *The Collector of Lives: Giorgio Vasari and the Invention of Art* (with Noah Charney, 2017), and *The Divine Spark of Syracuse* (2019).

Malcolm Thick is an historian of food and agriculture, and a regular attendee of the Oxford Symposium on Food and Cookery and the Leeds Symposium on Food History and Traditions. His most recent publication, "The Sale of Produce from Non-Commercial Gardens in Late Medieval and Early Modern England" (2018, *Agricultural History Review*, 66 (1): 1–17), won the 2019 Sophie Coe prize for food history. Malcolm is a Fellow of the Royal Historical Society.

INDEX

Note: Page locators in *italic* refer to figures.